Practical Electrical Equipment and Installations in Hazardous Areas

Other titles in the series

Practical Cleanrooms: Technologies and Facilities (David Conway)

Practical Data Acquisition for Instrumentation and Control Systems (John Park, Steve Mackay)

Practical Data Communications for Instrumentation and Control (John Park, Steve Mackay, Edwin Wright)

Practical Digital Signal Processing for Engineers and Technicians (Edmund Lai)

Practical Electrical Network Automation and Communication Systems (Cobus Strauss)

Practical Embedded Controllers (John Park)

Practical Fiber Optics (David Bailey, Edwin Wright)

Practical Industrial Data Networks: Design, Installation and Troubleshooting (Steve Mackay, Edwin Wright, John Park, Deon Reynders)

Practical Industrial Safety, Risk Assessment and Shutdown Systems for Instrumentation and Control (Dave Macdonald)

Practical Modern SCADA Protocols: DNP3, 60870.5 and Related Systems (Gordon Clarke, Deon Reynders)

Practical Radio Engineering and Telemetry for Industry (David Bailey)

Practical SCADA for Industry (David Bailey, Edwin Wright)

Practical TCP/IP and Ethernet Networking (Deon Reynders, Edwin Wright)

Practical Variable Speed Drives and Power Electronics (Malcolm Barnes)

Practical Centrifugal Pumps (Paresh Girdhar and Octo Moniz)

Practical E-Manufacturing and Supply Chain Management (Gerhard Greef and Ranjan Ghoshal)

Practical Grounding, Bonding, Shielding and Surge Protection (G. Vijayaraghavan, Mark Brown and Malcolm Barnes)

Practical Hazops, Trips and Alarms (David Macdonald)

Practical Industrial Data Communications: Best Practice Techniques (Deon Reynders, Steve Mackay and Edwin Wright)

Practical Machinery Safety (David Macdonald)

Practical Machinery Vibration Analysis and Predictive Maintenance (Cornelius Scheffer and Paresh Girdhar)

Practical Power Distribution for Industry (Jan de Kock and Cobus Strauss)

Practical Process Control for Engineers and Technicians (Wolfgang Altmann)

Practical Power Systems Protection (Les Hewitson, Mark Brown and Ben. Ramesh)

Practical Telecommunications and Wireless Communications (Edwin Wright and Deon Reynders)

Practical Troubleshooting of Electrical Equipment and Control Circuits (Mark Brown, Jawahar Rawtani and Dinesh Patil)

Practical Batch Process Management (Mike Barker and Jawahar Rawtani)

Practical Hydraulics (Ravi Doddannavar, Andries Barnard)

Practical Electrical Equipment and Installations in Hazardous Areas

Geoffrey Bottrill Bottrill and Associates, London, United Kingdom

Derek Cheyne PrEng, BSc (ElecEng), Chief Electrical and Instrument Engineer, South Africa

G. Vijayaraghavan Vijayaraghavan and Associates, Mumbai, India

Series editor: Steve Mackay FIE (Aust), CPEng, BSc (ElecEng), BSc (Hons), MBA, Gov.Cert.Comp. Technical Director – IDC Technologies

AMSTERDAM • BOSTON • HEIDELBERG • LONDON
NEW YORK • OXFORD • PARIS • SAN DIEGO
SAN FRANCISCO • SINGAPORE • SYDNEY • TOKYO

Newnes is an imprint of Elsevier

ELSEVIER

Newnes

Newnes
An imprint of Elsevier
Linacre House, Jordan Hill, Oxford OX2 8DP
30 Corporate Drive, Burlington, MA 01803

First published 2005

British Library Cataloguing in Publication Data
Botrill, G.
  Practical electrical equipment and installations in hazardous areas
  (Practical professional)
  1. Electric apparatus and appliances – installation – safety measures
  2. I. Title   II. Cheyne, D.   III. Vijayaraghavan, G.
  3. 621. 3'1'042'0289

Library of Congress Cataloguing in Publication Data
A catalogue record for this book is available from the Library of Congress

ISBN 0 7506 6398 7

For information on all Newnes Publications
visit our website at www.newnespress.com

Typeset by Integra Software Services Pvt. Ltd, Pondicherry, India
www.integra-india.com
Transferred to Digital Printing in 2010

Contents

Preface

This book provides delegates with an understanding of the hazards involved in using electrical equipment in potentially explosive atmospheres. It is based on the newly adopted international IEC 79 Series of Standards that are now replacing the older national standards and are directly applicable to most countries in the world (including North America). Explosion-proof installations can be expensive to design, install and operate. The wider approaches described in these standards can significantly reduce costs whilst maintaining plant safety. The book explains the associated terminology and its correct use. It covers from area classification through to the selection of explosion-protected electrical apparatus, describing how protection is achieved and maintained in line with these international requirements. Standards require that engineering staff and their management are trained effectively and safely in hazardous areas and this book is designed to help fulfil that need.

This book is aimed at anyone involved in design, specification, installation, commissioning, maintenance or documentation of industrial instrumentation, control and electrical systems. This includes:

- Tradespersons working in potentially explosive areas
- Electrical and instrument tradespersons
- Instrumentation and control engineers
- Electrical engineers
- Instrumentation technicians
- Design engineers.

We would hope that you will gain the following from this book:

- A good understanding of terminology used with hazardous areas
- An understanding of the hazards of using electrical equipment in the presence of flammable gases, vapors and dusts
- A basic knowledge of explosion protection to IEC Standards
- The ability to do a simple hazardous area classification
- Details of the types of apparatus that can be used in a given hazardous area
- How to design and install safe working systems in hazardous areas
- An understanding of the safety and operational aspects of hazardous areas
- A knowledge of the system limitations in using hazardous areas protection
- A brief review of the key areas of the national codes of practice.

You will need a basic understanding of instrumentation and electrical theory for the book to be of greatest benefit. No previous knowledge of hazardous area installation is required.

1

Introduction

1.1 Introduction

Our Industry free from injuries and accidents.

… is the vision statement of the Mines Inspectorate, Safety and Health, Queensland Government, Australia.

This is not only true of the mining industry as such; similar vision or mission statements abound in surface industries as well. The basic urge of the industry stakeholders to implement loss prevention and reduction measures form the backbone of workshops such as this.

Having stated this, it is also true that:

'*Organizations have very short memories when it comes to safety.*' This is the prime driver of loss prevention measures.

'*The future is merely a reflection of past.*' This adage is as true for organizations as it is for individuals. All incidents contain lessons, which may help to prevent accidents causing damage to others in future. All mishaps, incidents, or *near-miss situations*, when escalated to 'Loss producing events – of life or property' are termed '*ACCIDENTS*'.

Industrial accidents are as old as industry itself and so are *preventive measures*. The *standards* a person finds enshrined in various industrial fields are results of such painfully researched and documented activities. These principally lay down dos and don'ts in a systematic manner, acceptable to all stakeholders in the industry. Thus, the Standards for Explosive Areas or Atmospheres have also evolved diversely worldwide, based on the local needs of the industries for the overall safe operation of the plants.

Explosion and fire are two of the major constituents of these mishaps, major or minor. Depending upon the environment, these can be termed 'Accidents' or fade away as simply 'incidents' or '*Near Misses*' in the safety officers' statistics.

Fire, from the time first discovered by Stone-Age man, has been a great friend, when under control; and a foe, when uncontrolled. The chance rubbing of two firestones by our ancestors led to discovery of fire (Figure 1.1). This has been a lifeline for the survival of the human race.

Fire is at our service for cooking, heating, as a weapon against powerful enemies of nature, etc. … the list can be endless. However, when left to its own antics it has razed to the ground full communities, forests and buildings. Hence, it has been man's endeavor to understand its nature, causes, etc., so that appropriate preventive measures are taken, since –

Prevention is better than cure

Fire

Rubbing of stone

Figure 1.1
Generation of fire

This has led to extensive research, studies and surveys to find ways and means to categorize and assimilate knowledge across generations and continents to prevent explosions and fires in all types of '*hazardous environment*' found in various industries.

This course on '*Potential Hazardous Environment and Electrical Equipment*' will make an attempt to take you all through a journey, which at the end of it will leave you well equipped to handle situations arising in –

- Designing
- Selecting
- Manufacturing
- Installing
- Operating and maintaining
- Troubleshooting and repairing electrical equipment.

Furthermore, we will strive to give insight into how to avoid fires or explosions involving electrically powered equipment in explosive or hazardous atmosphere due to foreseeable or avoidable human errors like

- Improper installation, selection and design
- Lack of proper maintenance
- Improper use
- Carelessness or oversight.

The first step logically is to start defining and understanding some of the terms used in the whole gambit of the loss prevention in accidents due to explosion and fire.

FIRE is nothing but a rapid oxidation–reduction reaction (combustion) which results in the production of heat and generally visible light.

EXPLOSION is a violent and sudden expansion of gases produced by rapid combustion; that very strong force when shut in a small space and generally associated with a loud, sharp noise and a supersonic shock wave.

For example – bursting of balloons, boiler explosion, combustion of a gas mixture, detonation of high explosives.

Hazards as colloquially understood are of two types – natural and manmade.

The natural ones like blizzards, flash floods, earthquakes, heat waves, hurricanes, tornadoes, volcanic eruption, etc., cannot be prevented and only countermeasures toward mankind's preparedness to mitigate and reduce the loss consequent to their happening are possible.

It is the manmade hazards like explosion and fire, which are subjects of continuous scientific studies and are the subjects of our discussion during the next few days.

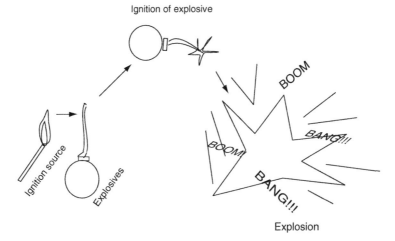

Figure 1.2
Explosion

> *HAZARDOUS area/environment – as defined in a dictionary, is the one which might be potentially prone to . . .*

- an accident
- create a dangerous situation
- be fraught with/beset with dangerous situations
- be a flashpoint
- be explosive
- be inflammable or flammable
- be radioactive, etc.

Despite the best efforts in designing, installing and operating, due to human error the breakdown of instruments and equipment could lead to one of the conditions described above. This can happen wherever the combustible, flammable and radioactive materials are processed or stored, on account of their being prone to leakage or their ability to create an explosive atmosphere in conjunction with oxygen from air or some oxidizing agent. Invariably, electricity is required to run the wheels of production processes in any industry. The fundamental nature of electricity to create a *spark or generate energy* leads to a situation where, in the ambivalent 'explosive atmosphere', ignition is triggered leading to explosion or fire (Figure 1.2).

Picture of a pit explosion

(Source: R. Stahl)

1.2   Approach

The chemistry of electricity associated with explosion has been known to mankind since the turn of the twentieth century. Preventive measures have been deployed based on the geographical locations. Different continents have developed different methods, termed 'Standards', such as NFPA 70 – NEC (North America); CENELEC – EN 50 (Europe), etc. These evolved based on the specific needs of the local industry, in conjunction with manufacturers to reduce the risk associated with usage of electricity in hazardous area. However, as globalization is taking place and manufacturers are selling their wares across the continents, a need has been felt to integrate these, and efforts are ongoing in this direction.

Explosions cannot be totally prevented wherever the explosive environment, necessitated by the production processes, is present. All human endeavors are fallible.

Nevertheless, we should develop these preventive measures to an extent where explosions become such a rare commodity that the gains far outweigh the losses in a big way. The preventative measures should be developed, where it in effect reduces the risk of an explosion to an acceptable level.

An explosion occurs when there is a confluence of the three basic elements – namely (Figure 1.3)

1. Fuel, any combustible material
2. Oxygen, generally available for free in nature as 21% in air. It can also be supplied by oxidizers such as acids
3. Ignition source – chemical, mechanical, nuclear or electrical.

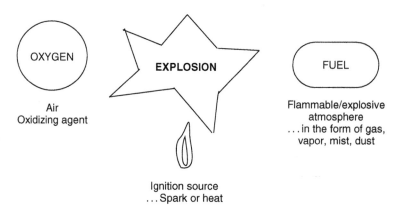

Figure 1.3
The explosion triangle

As stated above, with the abundance of air in nature we can only apply control over the other two elements to reduce risk of explosion.

The electrical ignition energy can be imparted by spark or heat generated by any of the following means,

- Resistance heating
- Dielectric heating
- Induction heating
- Leakage current heating
- Energy from arcing
- Static electricity heating
- Lightning.

The technology is currently based on the identification of the risk of an explosive atmosphere being present in a particular place. This is coupled with the identification of the likelihood of electrical equipment within the explosive atmosphere, malfunctioning in a way that would cause it to become a source of ignition coincident with the presence of that explosive atmosphere. The objectives are not just to identify these coincidences but to utilize the information so obtained to influence the design of particular process plants and similar operational situations. This will help to minimize the risk of an explosion due to electrical installations. In this approach, the areas normally prone to have an explosive atmosphere, due to the requirement of varies processes involved, are identified. Similarly, the areas where its likelihood is low but identifiable are marked up. It is needless to say that this is not an end in itself but should be deployed as a part of 'overall safety strategy' for the plant.

As a part of this technology, once the potential hazardous areas are identified and marked-up, only the minimum number of justifiable, necessary and essentially required electrical equipment/appliances/installations are housed in these identified areas. These are then so protected as to render the overall risk of explosion sufficiently low.

1.3   Historical development

The use of electricity in the explosive environment was initially done in the mining industry and it is there where the first safeguards were employed. Contrary to the earlier approach of burning the fire damp (methane) gases which eliminated the risk of explosion, the present approach is to minimize the risk of ignition and thereby explosion. This was done by 'specialists' among the miners who lighted the methane which, being lighter than air, generally used to gather at highest point near the roof of the working area. The warning of the presence of methane was given when the flame of the lantern changed color. The 'fire men' were the younger miners, who, covered with wet sacking, would go ahead of the others with a long lighted taper. This made the mine safe for miners to work. Although this was an effective method, it was somewhat barbaric in nature and fell into disrepute.

This technique was further refined and the lighter methane gas or its mixture was removed by the method of venting. The safety lamps were introduced to minimize the risk of explosion. The Davy safety lamp (Figure 1.4) is one of the examples. This device must have saved innumerable lives over the years.

Around the late nineteenth century and in the early part of the twentieth century the use of electricity in mining began. This led to the identification of other risks. To begin with, electricity was used in mining for lighting purposes and as a motive power. The control equipment required to run these had potential risks of explosion due to hot surfaces and sparks, in an atmosphere of methane and coal dust. Initially, before World War II, extensive research work was done in Germany and Britain and two techniques were generally employed – flameproof enclosures, which housed the equipment having electrical energy capable of igniting the gases and vapors, and techniques for signaling. These released energy levels which were insufficient to ignite the gas and vapors present in the mines. It is the second method that later became known as 'intrinsically safe circuits'.

Originally these were developed for the mining industry for prevention of explosions from methane and coal dust, but later the same was slowly applied to the surface industry.

It slowly became apparent that whereas, in mines, only coal dust and methane gas presented the hazardous conditions, on the surface a myriad of situations depending upon the type of industry and processes presented themselves; hence the need was felt to classify the 'hazardous areas'. This led to the technique of area classification to identify the risks associated with each type of explosive atmosphere.

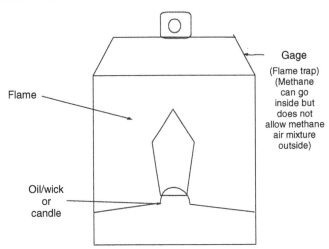

Figure 1.4
Principle of Davy safety lamp

1.4   Bird's eye view of legislation and research in other countries

United States of America

In the United States of America the National Fire Protection Association (NFPA) was formed in 1896 with the aim to reduce the burden of fire on quality of life by advocating scientifically based consensus codes and standards. It also carries out research and education for fire and related safety issues. The association was incorporated in 1930 under laws of the Commonwealth of Massachusetts.

The electrical section was added in 1948. The National Electricity Code (NEC) under NFPA 70 defines rules and regulations regarding use of electrical equipment. The sections 500 read through to 517 deal with installation, testing, operation and maintenance of electrical equipment in hazardous area.

In addition, various government laboratories, university laboratories, private and industrial laboratories do research and education in USA. One of the most prominent is the Underwriters Laboratories Inc. This was founded in 1894. It is a not-for-profit corporation having as its sole objective the promotion of public safety through the conduct of –

> *scientific investigation, study, experiments, and tests, to determine the revelation of various materials, devices, products, equipment, constructions, methods, and systems to hazards appurtenant thereto or to the use thereof affecting life and property and to ascertain, define, and publish standards, classifications, and specifications for materials, devices, products, equipment, constructions, methods, and systems affecting such hazards, and other information tending to reduce or prevent bodily injury, loss of life, and property damage from such hazards.*

It is pertinent to note that it has no stock, nor shareholders, and exists solely for the service it renders in the fields of fire, crime, explosions and casualty prevention.

The role of the Federal Government was minimal in fire protection prior to 1974. However, in 1974 Congress passed the Federal Fire Prevention and Control Act. Under this act 12 Executive Branch departments and 10 independent agencies are supposed to administer the various provisions of act. The NFPA enjoys a cooperative relationship with these agencies. A number of agencies rely upon NFPA standards and participate in the NFPA standards-making process.

The Congress, in 1970, established the Occupational Safety and Health Administration (OSHA) within Department Of Labour to oversee development and implementation of mandatory occupational safety and health standards – rules and regulations applicable at the workplace. The Mine Safety and Health Administration was established in 1977 with a functional scope similar to that of OSHA, but with a focus on mining industry.

United Kingdom

In the United Kingdom the first legislation covering the use of electrical equipment in explosive atmospheres came into being through The Electricity (Factories Act) Special Regulations 1908 and 1944, Regulation 27, which states that –

All conductors and apparatus exposed to the weather, wet, corrosion, inflammable surroundings or explosive atmosphere, or used in any process or for any special purpose other than for lighting or power, shall be so constructed or protected, and such special precautions shall be taken as may be necessary adequately to prevent danger in view of such exposure or use.

Even Regulation 6 of 'Electricity at Work Regulation 1989' is in the same spirit of placing the responsibility of achieving the objective on the owner of industry without specifying the methods to be adopted. In the UK, it is not illegal to use equipment in hazardous area although it is not certified for such use. Thus, as long as the owner maintains adequate records of plant safety this clause gets satisfied.

This is in variance to the one being followed in the USA and Germany and other parts of Europe, where specifics are also formulated. Both approaches have withstood the test of time and there is not much evidence of putting one method over the other.

In the UK much work has been done in the area of electrical installation safety in hazardous atmosphere by the Safety in Mines Research Establishment, the Electrical Research Association (now ERA Technology Ltd), the Fire Protection Association, Institution of Fire Engineers, Loss Prevention Council and The Institute of Petroleum.

Electrical code evolution

In order to understand how the electrical code is evolving and what guides this evolution we need to look back in history and their development till date. In the early 1900s, when contractors were busy electrifying industrial buildings, electrical wires were run through existing gas pipes, resulting in today's conduit system of wiring. This formed the basis of wiring in North America, and the codes and standards were made to suit the safety requirement pertinent to these practices.

While this was being done on the American continent, the International Electro technical Commission (IEC) was founded in Switzerland. The IEC is supposed to be the 'United Nations' of the electrical industry. Its ultimate goal is to unify worldwide electrical codes and standards. Few IEC practices were incorporated into the NEC or Commission of the European Communities (CEC) mainly because North America operated on different voltages and frequencies than most of the rest of the world.

Advent of hazardous area in surface industries

With the advent of automobiles and airplanes in the early 1920s the need was felt to create facilities to refine fuels. Because volatile vapors from gasoline and electrical sparks did not safely mix, the first hazardous area classification was invented. In the USA Division 1 described areas being normally hazardous.

Thus, a new industry with the goal of protecting electrical equipment in hazardous areas was born. Explosion-proof enclosures, intrinsic safety (IS), oil immersion, and wire-gauze (mesh) enclosures for mining lanterns were the first types of protection developed.

World War II brought many changes in Europe and North America. Metal shortages in Europe prompted more plastic use in electrical equipment, and the first construction standards for explosion-protected electrical equipment appeared in Germany.

At about the same time, North American industries determined that hazardous area classifications needed to be expanded. A Division 2 was needed to describe locations that were not normally hazardous to allow use of less-expensive equipment and less-restrictive wiring methods.

In the 1960s, the European Community was founded to establish a free-trade zone in Europe. To reach this goal, technical standards needed to be harmonized. As a result the European Community for Electro technical Standardization (CENELEC) was established.

By this time the German chemical industry had departed from the traditional conduit or pipe wiring system and migrated towards cable as a less-expensive alternative. This change in the wiring method led to the zone classification system later adopted in 1972 by most European countries in a publication known as IEC 79-10. This action led to the different methods of classifying hazardous areas as well as protective, wiring and installation techniques, which form the basis of the present IEC classification.

1.5   The certification

It is a very costly affair in terms of both cost and resources for testing each electrical installation for design validation and conformance to standards, hence making it an impractical preposition. Thus, a system of certification has been evolved for validation of use of an electrical installation in the hazardous atmosphere of the plant.

This technology primarily consists of –

- The classification of the area of the plant where electrical installation is to be done.
- The classification of the electrical equipment used in that area of the plant.

The international and national codes and standards have been developed keeping in mind the practices followed across generations and across continents. Adherence to them gives the user a level of confidence for safe operation of electrical equipment under the specified conditions. This, however, makes it obligatory on the part of the inspecting and classifying agency that the same set of standards and codes is used for area classification or categorization and for establishing the suitability/applicability of the selection of equipment to be operated in it.

These classifications will be considered in depth in this workshop. This will enable you to apply the how, why, where and when for a given situation.

The certification process merely states conformance with the standard to which the equipment has been assessed. It does not imply that the equipment is safe. A few examples of the logos/stamping are illustrated in Figure 1.5.

Currently, internationally acceptable markings are used to identify explosion-protection (Ex) equipment. This leads to uniformity in the industry and gives a confidence level to the user, vis-à-vis, the suitability and integrity of quality and design and lessens the work of manufacturer in getting each piece approved all over the globe.

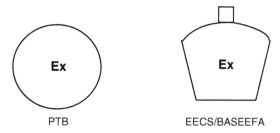

Typical certification logos

Figure 1.5
Typical certification logos

As described above, it is at present not mandatory in the UK to carry out any specific certification for electrical equipment used in hazardous areas. However, this will change with the European 'ATEX' Directive coming into force from 1 July 2003. Thereafter, all equipment, of whatever origin, sold in and installed throughout Europe, must have been assessed for safe use and marked accordingly.

The safety standards expect the plant management to keep all documentation. This classification system helps in collating documentation to meet the plant safety requirements.

As already stated, a worldwide attempt is in process to converge local and international standards for the design and installation of explosion-protection systems into the one of the International Electro-technical Committee Standards series IEC 79.

As a sequel, many have adopted this standard. Even the USA has incorporated the relevant classification requirements into NEC to plug the gap. Over a period, all are expected to fall in line.

Thus, for plants that are going to install globally accepted technology or the manufacturers who are going to supply electrical equipment worldwide will be perforce required to adhere to IEC 79.

1.6   Conclusion

Before concluding this chapter, it is important to ponder on the following few points: During the course

- Information will be presented
- Education will be given
- But when you will implement change will happen and then we can say learning has taken place.

2

Electrical energy, ignition and flammability

2.1   Electrical energy and ignition

Electricity ...

...What is it?

One of the more obvious answers could be – *Electricity is one of the basic physical forces of the universe – like gravity*. It is one of the most important forms of energy, consisting of oppositely charged electrons and protons that produce light, heat, magnetic force and chemical changes.

We harness electricity and use it in our everyday life. If it weren't for electricity, there wouldn't be any television, radio ... or computers. Even this presentation would not be possible.

Electrical systems, if properly designed, installed, maintained and carefully used, are both convenient and safe, otherwise they have the potential to be a serious workplace hazard, exposing employees to such dangers as fires, explosions and personal injury, i.e. electric shock, electrocution, etc.

Generally electrical hazards are limited to 'electrical shock' and many employee health and safety codes do not go further. Nevertheless, this book (and associated books) focus on electrical flashover and electrical blast hazards. These can result in serious and extensive loss of life and property.

The working with electricity itself is a hazard and its usage in hazardous areas requires that you are fully alert about the dangers when designing, installing and maintaining these systems. Often the victim is not someone whose job is to install or repair electrical equipment but the employee who didn't realize that worn electrical equipment needed replacement or that exposed live parts were in the work vicinity.

In order to understand the chemistry between 'electricity and explosion/fire' it is essential to first refresh our memory regarding often-used electrical terms.

2.2   The basics of electricity

Some of the basic principles, concepts and often-repeated technical terms associated with electricity will be discussed. For more details please refer Appendix G. Electrical terms like amps, voltage, watts, kilowatts, KVA, single-phase, etc. are explained hereunder.

2.2.1   Voltage

The electricity in your home, unless you live on a farm or on a large mansion, is most likely to come in as a 240 V 50 cycle alternating current supply. The voltage is a measure of pressure (force) and as with pressure in a water pipe, voltage ensures that electricity continues to flow through a cable.

Voltage measurements may be expressed in the following units:

- volts (V)
- kilovolts (kV)
- millivolts (mV)
- microvolts (μV).

For example:

1 kV = 1000 V
1 mV = 0.001 V
1 μV = 0.000001 V.

Methods of producing a voltage include:

Friction

The friction between two given surfaces (like plastics, etc.) gives rise to electrostatic electricity, which while discharging the charges (electrons) may give rise to a small spark. This is to be kept in mind by designers, and suitable precautions to be taken for hazardous area application, as in highly explosive atmospheres it may be catastrophic.

The best way to avoid this is to securely ground all apparatus and their enclosures.

Pressure (piezoelectricity)

Some type of crystals, ionic crystals such as quartz, Rochelle salts and tourmaline, have the properties of generating emf when mechanical force is applied to them. These crystals have the remarkable ability to generate a voltage whenever stresses are applied to their surfaces. Thus, if a crystal of quartz is compressed, charges of opposite polarity will appear on two opposite surfaces of the crystal. If the force is reversed and the crystal is stretched, charges will again appear, but will be of the opposite polarity from those produced by squeezing. If a crystal of this type is given a vibratory motion, it will produce a voltage of reversing polarity between two of its sides.

This phenomenon, called the Piezoelectric effect, is shown in Figure 2.1.

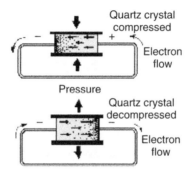

Figure 2.1
Illustration showing generation of electron due to compression and de-compression of quartz crystal

Some of the common devices that make use of piezoelectric crystals are microphones, phonograph cartridges and oscillators used in radio transmitters, radio receivers, and sonar equipment.

This principle is likely to find use by designers of instrumentation to conform to IS protection concept due to low levels of power generated and capability of crystal to convert mechanical energy to electrical energy also (reverse).

Heat (thermoelectricity)

When a length of metal, such as copper, is heated at one end, electrons tend to move away from the hot end toward the cooler end. This is true of most metals. However, in some metals, such as iron, the opposite takes place and electrons tend to move toward the hot end. These characteristics are illustrated in Figure 2.2. The negative charges (electrons) are moving through the copper away from the heat and through the iron toward the heat. They cross from the iron to the copper through the current meter to the iron at the cold junction. This device is generally referred to as a Hot Junction.

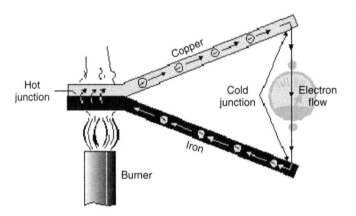

Figure 2.2
Voltage produced by heat

The designer of hazardous area needs to keep in view this phenomenon so that no inadvertent voltage gets generated in explosive atmosphere due to rise in temperature between such dissimilar materials, causing large currents to flow in enclosures, etc.

Light (photo electricity)

When light strikes the surface of a substance, it may dislodge electrons from their orbits around the surface atoms of the substance. This occurs because light has energy, the same as any moving force.

Some substances, mostly metallic ones, are far more sensitive to light than others. That is, more electrons will be dislodged and emitted from the surface of a highly sensitive metal, with a given amount of light, than will be emitted from a less-sensitive substance. Upon losing electrons, the photosensitive (light-sensitive) metal becomes positively charged, and an electric force is created. Voltage produced in this manner is referred to as a photoelectric voltage (Figure 2.3).

The power capacity of a photocell is very small. However, it reacts to light-intensity variations in an extremely short time. This characteristic makes the photocell very useful in detecting or accurately controlling a great number of operations. For instance, the photoelectric cell, or some form of the photoelectric principle, is used in television cameras, automatic manufacturing process controls, door openers, burglar alarms, etc.

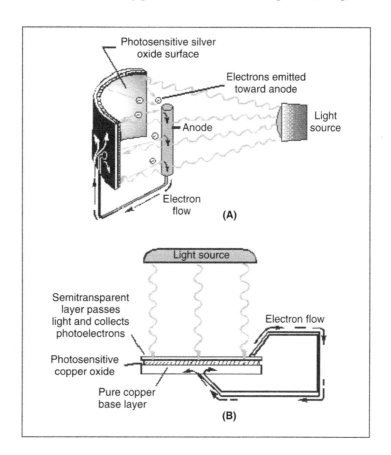

Figure 2.3
Voltage produced by light

This principle finds extensive use by designers of instrumentation to conform to IS protection concept.

Chemical action (battery)

Voltage may be produced chemically when certain substances are exposed to chemical action.

If two dissimilar substances (usually metals or metallic materials) are immersed in a solution that produces a greater chemical action on one substance than on the other, a difference of potential will exist between the two. If a conductor is then connected between them, electrons will flow through the conductor to equalize the charge. This arrangement is called a primary cell. The two metallic pieces are called electrodes and the solution is called the electrolyte. The voltaic cell illustrated in Figure 2.4 is a simple example of a primary cell. The difference of potential results from the fact that material from one or both of the electrodes goes into solution in the electrolyte, and in the process, ions form in the vicinity of the electrodes. Due to the electric field associated with the charged ions, the electrodes acquire charges.

The amount of difference in potential between the electrodes depends principally on the metals used.

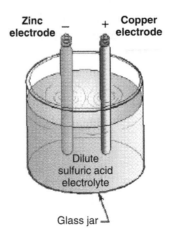

Figure 2.4
A typical voltaic cell

The type of electrolyte and the size of the cell have little or no effect on the potential difference (PD) produced.

There are two types of primary cells,

1. The wet cell
2. The dry cell.

In a wet cell the electrolyte is a liquid. A cell with a liquid electrolyte must remain in an upright position and is not readily transportable. An automotive battery is an example of this type of cell.

The dry cell, much more commonly used than the wet cell, is not actually dry, but contains an electrolyte mixed with other materials to form a paste. Flashlights and portable radios are commonly powered by dry cells. Batteries are formed when several cells are connected together to increase electrical output.

Designers of hazardous area today have access to totally sealed and maintenance free cells and batteries. This helps in reducing the hazards in explosive atmosphere and their selection need to be adequately addressed.

Magnetism (electromagnetic induction generator)

Magnets or magnetic devices are used for thousands of different jobs. One of the most useful and widely employed applications of magnets is in the production of vast quantities of electric power from mechanical sources. The mechanical power may be provided by a number of different sources, such as gasoline or diesel engines, and water or steam turbines. However, generators employing the principle of electromagnetic induction do the final conversion of these source energies to electricity. These generators, of many types and sizes, are discussed in other modules in this series. The important subject to be discussed here is the fundamental operating principle of all such electromagnetic-induction generators (Figure 2.5).

The principles of the industrial motors are also based upon the above theory of magnetism but in the reverse direction. The above goes a long way in explaining the mystery behind principles of operation of motors. The grasp of above will help designers in understanding the concept of explosion-proof (the concept will be explained in Chapter 3) motor enclosures.

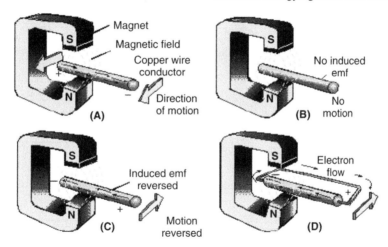

Figure 2.5
Principle of voltage produced by magnetism

2.2.2   Potential difference

In most electrical circuits only the difference of potential between two points is of importance and the absolute potentials of the points are of little concern. Very often it is convenient to use one standard reference for all of the various potentials throughout a piece of equipment. For this reason, the potentials at various points in a circuit are generally measured with respect to the metal chassis on which all parts of the circuit are mounted. The chassis is considered to be at zero potential and all other potentials are either positive or negative with respect to the chassis. When used as the reference point, the chassis is said to be at *ground potential*.

Occasionally, rather large values of voltage may be encountered, in which case the volt becomes too small a unit for convenience. In a situation of this nature, the kilovolt (kV) is frequently used. In other cases, the volt may be too large a unit, as when dealing with very small voltages. For this purpose the millivolt (mV) and the microvolt (μV) are used.

2.2.3   Electric current (amperage)

The *electric current* can be defined as a directed movement of electrons in a conductor or circuit. Let us try to answer the following question,

An electric current flows from what potential to what potential?

When a difference in potential exists between two charged bodies that are connected by a conductor, electrons will flow along the conductor. This flow is from the negatively charged body to the positively charged body, until the two charges are equalized and the PD no longer exists.

An analogy of this action is shown in the two water tanks connected by a pipe and valve in Figure 2.6. At first the valve is closed and all the water is in tank A. Thus, the water pressure across the valve is at maximum. When the valve is opened, the water flows through the pipe from A to B until the water level becomes the same in both tanks. The water then stops flowing in the pipe, because there is no longer a difference in water pressure between the two tanks.

The electrons (negative charges) move through a conductor in response to an electric field. Electron current is defined as the directed flow of electrons. The direction of electron movement is from a region of negative potential to a region of positive potential.

But electric current is said to flow from positive to negative when dealing with the circuit outside the source producing it. The direction of current flow in a material is determined by the polarity of the applied voltage.

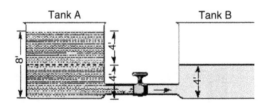

Figure 2.6
Water analogy of electric differences of potential

Electron movement through an electric circuit is directly proportional to the difference in potential or electromotive force (emf), across the circuit, just as the flow of water through the pipe in Figure 2.6 is directly proportional to the difference in water level in the two tanks.

A fundamental law of electricity is that the *electron flow is directly proportional to the applied voltage*. If the voltage is increased, the flow is increased. If the voltage is decreased, the flow is decreased.

The *ampere* is the basic unit used to indicate an electric current. A current of one ampere is said to flow when one coulomb of charge (6.28×10^{18} electrons) passes a given point in 1 s of time.

Ampere is a measure of the rate of flow of electric current. The unit of electric current is Ampere (amp).

Current measurements may be expressed in the following units:

- Ampere (A)
- Milliampere (mA)
- Microampere (μA).

2.2.4   Electrical power

Power, whether electrical or mechanical, pertains to the rate at which work is being done. Work is done whenever a force causes motion. When a mechanical force is used to lift or move a weight, work is done. However, force exerted without causing motion, such as the force of a compressed spring acting between two fixed objects, does not constitute work.

Previously, it was shown that voltage is an electrical force, and that voltage forces current to flow in a closed circuit. However, when voltage exists but current does not flow because the circuit is open, no work is done. This is similar to the spring under tension that produces no motion.

If we compare again electrical and water supplies, pressure is present in the water pipe even when the tap is turned off. Opening the tap will cause a certain amount of water to flow. Switching on a light or an electrical device will cause a current to flow and we measure the quantity of the current in terms of the amps.

When voltage causes electrons to move, work is done. The instantaneous rate at which this work is done is called the electric power rate, and is measured in watts. Multiply the pressure or force by the rate of current flow and we can calculate the total amount of electricity consumed:

$$(P) \text{ volts} \times (I) \text{ amps} = (W) \text{ watts}$$

A total amount of work may be done in different lengths of time. For example, a given number of electrons may be moved from one point to another in 1 s or in 1 h, depending on the rate at which they are moved. In both cases, total work done is the same. However, when the work is done in a short time, the wattage, or instantaneous power rate, is greater than when the same amount of work is done over a longer period of time.

As stated, the basic unit of power is the watt. Power in watts is equal to the voltage across a circuit multiplied by current through the circuit. This represents the rate at any given instant at which work is being done. Because a watt is too small a unit for most practical purposes, we use the kilowatt (=1000 W). A kilowatt used for 1h, for example – an hour's use of a one bar electric room heater will consume 1 KW-hour, which is the unit for the charging of electrical power in our electricity bills.

The amount of power changes when, either voltage or current, or both voltage and current, are caused to change. In practice, the only factors that can be changed are voltage and resistance. In explaining the different forms that formulas may take, current is sometimes presented as a quantity that is changed.

Remember, if current is changed, it is because either voltage or resistance has been changed.

2.2.5 Electrical resistance

It is the opposition of a conductor to the flow of electrons. It is known that the directed movement of electrons constitutes a current flow. It is also known that the electrons do not move freely through a conductor's crystalline structure. Some materials offer little opposition to current flow, while others greatly oppose current flow. This opposition to current flow is known as resistance (R), and the unit of measure is the ohm.

It is the ratio between the PD across a conductor and the resulting current that flows. The unit for electrical resistance is the ohm.

The standard of measure for $1\,\Omega$ is the resistance provided at $0\,^{\circ}\text{C}$ by a column of mercury having a cross-sectional area of $1\,\text{mm}^2$ and a length of 106.3 cm. A conductor has $1\,\Omega$ of resistance when an applied potential of 1 V produces a current of one ampere. The symbol used to represent the ohm is Ω.

Resistance, although an electrical property, is determined by the physical structure of a material. The resistance of a material is governed by many of the same factors that control current flow.

The resistance of a material is determined by the type, the physical dimensions and the temperature of the material that is,

- A good conductor contains an abundance of free electrons.
- As the cross-sectional area of a given conductor is increased, the resistance will decrease.
- As the length of a conductor is increased, the resistance will increase.
- In a material having a positive temperature coefficient, the resistance will increase as the temperature is increased.

2.2.6 Power factor

Unfortunately, electrical supplies are not as simple as depicted above. Many machines, which run on high electrical pressure and current, produce their own 'backwash' that reduces efficiency. This means that more electricity has to be used to overcome this backwash. The measurement of efficiency over total power consumed is called the power factor. The ratio of the power (kW) divided by the power factor is expressed as KVA.

This number will always be larger than the kilowatt number because it reflects the power required for running the machinery and overcoming the inefficiencies.

Electrical circuit

Figure 2.7 is a very simplified version of electrical circuit. Here the operation of everyday-used flashlight is taken as an example. The normal dry cell, the switch and bulb are shown in two states –

1. One with switch in 'Off' position. That is no current flowing and hence no power being consumed, even though the voltage is present.
2. The second when with switch in 'On' position with current flowing and lighting the bulb. The chemical power being converted to electrical power and thence to light energy.

In real life much more complicated circuits are likely to be used. But the principle will remain same that a current will flow when the circuit gets completed. Any breakage in circuit will lead to sparking, which can give ignition energy (this is further explained elsewhere in the chapter) in an explosive atmosphere.

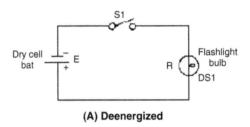

(A) Deenergized

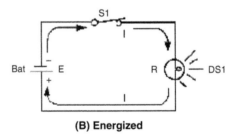

(B) Energized

Figure 2.7
A typical flashlight circuit

2.2.7 Summary

Summing up the brief description on electricity, we can say that an alternating current is a result of the technique of electromagnetic electrical generation, by which electricity is produced. It means that electricity flows like a continual incoming tide, wave upon wave, but in this case, the waves come at a frequency of 50 Hertz (Hz). This wavelike motion peaking in power, declining and resurging, could be seen in the instability and flicker of some older fluorescent lights.

2.2.8 Three-phase and single-phase supply

Each of your next-door neighbors also receive a 240 V similar supply, but not identical to your own. The probability is that they are on a different phase. *Three phases* are the basis on which electricity is produced by an alternator. It is not the object of this explanation to discuss why it is produced in this way, but merely to warn that there are three phases of electricity, normally referred to by color – *red*, *yellow* and *blue*. The importance of this process is that we now have three sets of waves producing our incoming tide. Therefore, the pressure is higher. The voltage between two phases is not, as one might expect, 2×240 V, but is actually 415 V because the phases peak in turn and not simultaneously.

Certain machines, like lift motors, air conditioning plant and other equipment found in buildings, need a three-phase supply. Others, like personal computers, electric kettles, are single-phase machines.

Under normal conditions, touching a single point of a 240 V electricity supply should only shock a reasonably healthy person. However, being in contact with a three-phase 415 V supply would kill him.

Practically all non-domestic buildings and industrial establishments are supplied with a three-phase supply at the intake point, normally at ground floor level or in the basement in case of buildings and at a designated substation in case of industries.

From here,

- Three-phase supply is taken to those machines, which require it.
- Single-phase supplies are taken to office floors to operate lighting and desk-type machinery.

Difficulties arise because the total loads, which are connected to each phase, need to be balanced in order to achieve optimization of the generation system and make best use of the distribution infrastructure.

Therefore, on large office floors there might be two or three differently phased 240 V supplies installed. Apart from helping to balance the load, it gives the additional advantage of ensuring that some part of the building is operational, should a fuse of one of the phases trip. It does enforce a discipline, however, to ensure that no devices accidentally receive supply from two phases, so as to ensure the safety of the occupants.

2.3   Electrical heat energy

We have briefly seen that how electricity is generated and utilized. It is in the process of its transmission and utilization that it has to traverse a long path before it is expended in process as to mechanical or chemical energy or heat energy or in any other form.

While dealing with devices in hazardous area the designers should have adequate knowledge of the sources of heat generation. The electrical energy being such a common feature in any process plant, it is necessary to study how it can be converted into heat energy.

The following describe various means in which electrical energy can be converted to heat energy:

- *Resistance heating*:   It is the electrical resistance, which is proportional to the energy required to move a unit quantity of electrons through the substance against the forces of electrons capture and collision. It is this energy expenditure, which appears in the form of heat. Heat generated is proportional to resistance and to square of current.
- *Dielectric heating*:   The free electrons move toward positive anode and free protons toward cathode. The distribution of normal atomic or molecular arrangement represents energy expenditure and as long as it is 'uni-directional', it poses no problem. However, once it becomes alternating or pulsating we have problem on our hand. If the frequency of alternation becomes too high, the heating of the di-electric could be substantial.

- *Induction heating*:   The alternating magnetic field induces current in adjacent conductor, thereby causing current (due to mutual induction), which generates heat.
- *Leakage current heating*:   Practically leakage current flows through all insulators and if these currents exceed the safety limit due to any reason with high voltages being applied, it may result in localized heating and consequent deterioration in material properties and breakdown.
- *Heat from arcing*:   Whenever an electric current that is carrying current is broken, an arc appears. These arcs have high temperature, and heat released is sufficient to cause ignition. This is one of the most common causes of 'electrical fires'.
- *Static electricity heating*:   Static charges, which accumulate over surfaces of two materials, are capable of generating sparks, which can ignite gases, vapors, mist, etc.
- *Heat generated by lightning*:   Lightning discharges are known to release tremendous amount of heat energy, whenever they discharge to ground.

2.4   Sources of ignition

The ignition sources can be any of the following:

- Hot surfaces
- Electrical arcs and sparks
- Electrostatic discharge
- Electromagnetic radiation
- Atmospheric discharge (lightning)
- Mechanical friction or impact sparks
- Ultrasonic
- Adiabatic compression (shock waves)
- Ionizing radiation
- Optical radiation
- Chemical reactions
- Open flames.

Out of the above the first three causes could be the source of ignition due to use of electrical equipment in hazardous area.

Generally, a potential source of ignition from an *electrical system* based on the above can be broadly classified into:

- *Arc/spark*:   An electric discharge, usually accompanied by a popping sound, that occurs when air or gas between two charged conductors becomes highly ionized, enough to breakdown, conducting current through a distinct, luminous channel.
- *Hot component*:   Due to any of the reasons as enumerated above, that releases energy sufficient to ignite a combustible mixture surrounding it.

The ignition source due to electrical system may occur in any of four mechanisms:

1. Discharge of capacitive circuits
2. Interrupting (opening) of inductive circuits
3. Opening or closing of resistive circuits with slow intermittent interruption increasing the ignition capability (hazard)
4. High temperature sources.

The ignition mechanisms may occur in relay contacts, switch contacts, fuses, short circuits (from damage or component failure) and arc-over between components or conductors.

The components or circuits that present a potential ignition source may be designed in a variety of ways in order to prevent ignition of a hazardous atmosphere.

2.4.1   Ignition by heat

As the temperature of a flammable mixture is increased, there comes a point at which it will ignite due to heat level and the energy associated with that level. This point is referred to as the ignition temperature. It is also known as 'auto-ignition' and 'spontaneous ignition' temperatures. The former comes from older standards and the latter is used in circles that are more scientific. The ignition temperature of a flammable material is normally given at the point of the 'most easily ignitable mixture'. As the concentration changes from this point, the temperature at which it occurs will rise further but the Minimum Easily Ignitable Mixture (MEIM) point is always considered.

Ensuring that surface temperatures of apparatus that are exposed to a vapor or gas do not exceed the gas ignition temperature can prevent ignition. The definition is stated in the following table:

Property	Definition
Ignition temperature	The lowest temperature of a flammable gas or vapor at which ignition occurs when tested as described in International Electro technical Commission (IEC) publication IEC 60079-4

There is an important difference between flashpoint and ignition temperature.

Property	Definition
Flashpoint (Alternative):	The minimum temperature at which a material gives off sufficient vapor to form an explosive atmosphere . . . (under test conditions) The minimum temperature at which sufficient vapor is given off from the surface of a liquid in still air such that it may be ignited by a small flame

Where a flammable material is in its 'liquid' or 'solid' state then it cannot be ignited until a vapor forms above the surface of the liquid, which can then mix with the air. If the temperature of the liquid remains reliably below the flashpoint then it does not form any hazard. This condition may be difficult to maintain if spillage is likely onto a heated pipe or vessel, for example.

2.4.2   Ignition by energy

Sufficient energy occurring in the presence of a flammable mixture will cause ignition to take place. Energy can occur in two forms, 'spark energy' or 'flame energy'. Flame energy will be discussed in the section covering other methods of Ex Protection.

Past research into the combustibility of vapors has been formalized into various explosion-protection standards over the years. The testing methods used are described in the standards and are still used by authorities (undertaking equipment testing) in some circumstances. The methods always determine and use the most easily ignitable mixture.

Minimum ignition energy (*MIE*) – *for a gas* is the minimum energy required to ignite the most easily ignitable mixture of that gas.

It is needless to point out that different gas/air mixtures require different amounts of energy to ignite the gas/air mixture. If a source of ignition, such as a spark, has energy below this, it cannot cause an explosion. The minimum ignition energies of gases are typically the range of 0.019 mJ (for hydrogen) to 0.29 mJ (for methane). To illustrate the small amounts of energy required for ignition the human body can store 40–60 mJ electrical energy.

The original work by the two scientists, Wheeler and Thornton, developed the technique of assessing the ease of electrical ignition of gases and vapors. This same technique was used to quantify the minimum igniting voltages and currents. The more modern 'spark ignition test' method described in the IEC 79 (and other national) standards is now used and will be summarized.

A flammable mixture surrounds a revolving contact arrangement. This uses a tungsten wire brushing on a slotted cadmium disk to make and break a circuit 1600 times during the course of a test. The voltage across the contact is derived from a constant voltage supply. A variable resistor placed in series with the supply limits the short-circuit current, through the contact, when it is closed. A range of voltages and currents are then applied to the switching contact to determine at what level ignition occurs. A graph is plotted, taking the curve below the lowest points. Different flammable mixtures reveal different curves. The curves are replaced by tabulated figures for the more precise interpretation by apparatus design engineers and certifying authorities (see later).

The material of the disk was found to influence the combustibility of the mixture. If cadmium, zinc, aluminum or magnesium were used then the spark produced for the same current/voltage was larger. These materials are often found in association with electrical circuits and therefore the standards publish curves and figures, which include these materials. These are referred to on the curves.

2.5 Flammability

The basic reason for explosions to take place is presence of the ignition source in an atmosphere where the fuel and oxidant have been already allowed to pre-mix. In case this premixed mixture is confined in an enclosed space, a rapid pressure rises, and explosions are the result.

Thus in order for combustion to take place, in the fuel and air mixture, the concentration of fuel should be present within the limits of *flammability*. Once fuel is ignited, the flame propagates until all combustible/unburnt mixture is exhausted.

The limits of flammability are the extremes of concentration within which the ignitable vapor must fall for combustion to start. Once the flame is ignited then it will continue to propagate at the specified pressure and temperature. It is to be noted that above or below these limits there is insufficient oxygen or insufficient fuel to sustain combustion.

Upper flammability limit (*UFL*) The highest (rich) concentration of a given vapor in air, above which ignition will not take place.

Lower flammability limit (*LFL*) The lowest (lean) concentration of a given vapor in air, below which ignition will not take place.

It is very difficult to theoretically calculate and predict the exact values of these limits. Hence, they are generally determined by experiment. The latest internationally accepted values are to be found in AS/NZS 60079.20. Previously, NFPA 325M was used as the reference document.

The range of gas/air mixtures between the LFL and the UFL is the explosive range (see Figure 2.8). Gas/air mixtures outside this range are non-explosive or non-flammable under normal atmospheric conditions.

Below the LFL the percentage of gas by volume, in the gas/air mixture, is too low to explode. Above the UFL the percentage of gas by volume, in the gas/air mixture, is too high to explode.

If the gas/air mixture is pre-heated and/or raised to a higher pressure the LFL will be reduced and the UFL increased thereby widening the explosive range.

Generally in oxygen-enriched atmospheres the LFL is relatively unaffected, but the UFL is increased thereby widening the explosive range as well as increasing the violence of the explosion.

The graph shows the relationship between the ignition energy and concentration level of air/gas mixtures of three types – viz propane–air, ethylene–air, hydrogen–air. The graph (Figure 2.8) illustrates that beyond the extremes of concentration the mixture does not burn.

It can also be noticed from this graph that within the limits of flammability there exists a '*minimum easily ignitable mixture*' (*MEIM*) for each type of mixture, for the given environmental conditions, where ignition takes place with minimum energy (most easily ignitable).

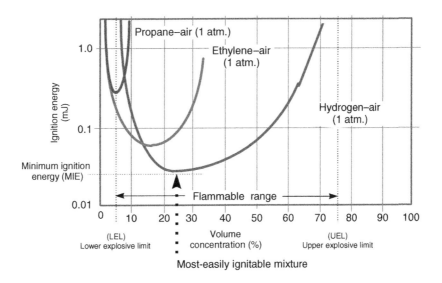

Figure 2.8
Flammability limits

For example, hydrogen/air mixture between 4 and 74% by volume of hydrogen at 21 °C and atmospheric pressure is highly combustible and explosive in nature. The following graph (Figure 2.9) will illustrate that when the temperature of the mixture is increased the flammability range widens and on decrease in temperature, the range narrows down. It can be seen that by lowering the temperature the ignitable mixture can be made non-flammable. In addition, it can be inferred that, either by placing it above or below the limits of flammability for a specific environment condition, the ignitable mixture can be made non-ignitable.

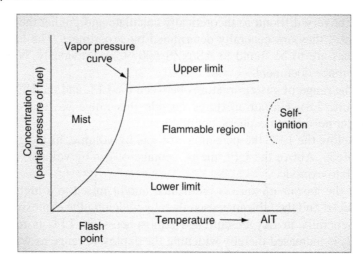

Figure 2.9
Flammability range

Further, it can also be noticed from the above graphs that for liquid fuels in equilibrium with their vapors in air, a maximum temperature exists for each fuel below which there is insufficient vapor for ignition to take place. This is termed as 'lower flashpoint'. The flashpoint temperatures for liquid vapor increase with the increase in pressure (Figure 2.10).

Flammability data of apparatus group reference gases

Material	Boiling Point (°C)	Flashpoint (°C)	Ignition Temperature (°C)	Explosive Limits		Relative Vapor Density (air = 1)	Apparatus Group
				LFL	UFL		
Acetylene	−83	Gas	305	2.3	100	0.9	IIC†
Ethylene	−104	Gas	425	2.3	36	0.97	IIB
Hydrogen	−252	Gas	560	4	77	0.07	IIC
Methane	−162	Gas	537	4.4*	17.0*	0.55	I
Propane	−42	−104/Gas	470	1.7	10.9	1.56	IIA

NOTE: The data given in this table are derived from IEC 60079.20 (previously derived from NFPA 325M).

* For coal mines the LFL and UFL for methane is taken to be 5 and 15%, as these are the values written into coal mine legislation, particularly in New South Wales.

† Additional requirements apply to acetylene (see As 2380.2).

Figure 2.10
Other relevant properties of flammable gases

2.5.1   Materials – relative vapor density

The density of the vapor will also influence the behavior of flammable vapors.

The relative vapor density of a gas can be defined as, *the mass of a given volume of pure vapor or gas compared with the mass of the same volume of dry air, at the same temperature and pressure.*

The density of air is considered as one and a relative vapor density less than one indicated that the vapor is lighter than air. Thus, such gas or vapor will rise in a comparatively still atmosphere.

On the other hand a relative vapor density greater than one indicates that the gas or vapor is heavier than air, and will tend to sink and fall toward the ground. Thus the gas or vapor may travel at low levels for a considerable distance.

The above properties and facts of propagation of flammable vapors are used to assess where gas will travel to and collect before considering the concentration level that may be formed at the point of collection.

The density of the vapor will also influence the behavior of flammable vapors. For example, if the vapor densities are in the range of about 0.75–1.25, proper care need to be taken in designing of the plant. Gases or vapors in this range, particularly if released slowly, may be rapidly diluted to a low concentration and their movement will be similar to that of the air in which they are effectively suspended.

2.5.2   Other properties affecting general safety

Materials or substances may also be toxic/asphyxiants and/or corrosive as well as flammable. Special handling precautions would be needed in addition to those for Ex protection.

Hydrogen sulfide, H_2S, is a good example of a noxious gas found in abundance during the oil refining processes. It is extremely poisonous. Levels of toxicity and flammability are shown in the table below for comparison.

H_2S Level	PPM	%
Smell detection level	0.1	0.000001
Safe working level	10	0.0001
Serious health danger level	50	0.0005
Lower flammable limit	40,000	4
Upper flammable limit	460,000	46

Where different gases and vapors can become mixed, methods for calculating how the properties of the mix will change in relation to the component characteristics are given in IEC 79-10. This can change the properties of flammability, toxicity and/or corrosiveness.

It is becoming law in many countries that manufacturers and suppliers make material safety data available to communicate safe handling information to users. This system has been formalized for many years in the materials transport industry but has only recently become more general involving all that come into any contact with the material.

2.6   Flammability principles

. . . are summarized in a nutshell hereunder,

An oxidizing agent, a combustible material and an ignition source are essential for combustion.

The combustible material needs to be heated to its piloted ignition temperature before it can be ignited or support flame spread (Figure 2.11).
The combustion will continue till,

The combustible material is consumed, or

The oxidizing agent concentration is lowered to below the concentration necessary to support combustion, or

The flames are chemically inhibited or sufficiently cooled to prevent further reaction, or

Sufficient heat is removed or prevented from reaching the combustible material.

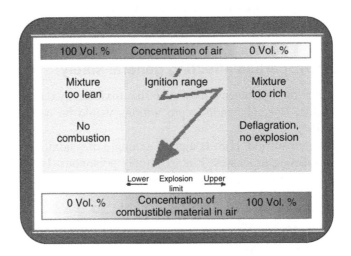

Figure 2.11
Flammable atmosphere and concentration of combustible material

A clear understanding of the nature of flammable materials and properties affecting their ability (or inability) to ignite is necessary before ways of ensuring that ignition cannot occur are devised and operated successfully.

There are some 750 flammable materials found in common industrial use. These have been tested over a period and compiled into reference lists that are included in IEC 79 and other standards.

The categorization of the properties of gases and vapors has led to an internationally recognized 'classification' system. This makes the job of defining a hazard and selecting equipment to operate in association with the hazard more effective and reliable.

3

Area classification

3.1   General

In September 1998, the following made the headlines in Australian papers:

ESSO blamed for Australian gas explosion

Former High Court Justice Dawson had passed the following remarks concerning the September 1998 gas explosion in Victoria:

The major cause of the accident was the failure of ESSO to equip its employees with appropriate knowledge to deal with the events which occurred.

The lack of 'adequate procedures for the identification of hazards in Gas Plant 1 contributed to the occurrence of the explosion and fire'.

The idea of bringing this up early in the chapter is to emphasize the importance of the identification of hazards and hazardous areas. If this is not done in a systematic manner with due care and diligence then the results can be catastrophic as it happened in ESSO's Gas Plant 1.

Thus, it is very important that all of us have the right answers for the following question:

… What is Area Classification, vis-à-vis, hazardous locations or areas?

And to answer this it is necessary first to understand the philosophy an objectives of 'area classification'.

The first step is to list and identify the areas in the plant where there is possibility of a conducive atmosphere for explosion or fire to occur. Based on the knowledge so acquired, the design, selection and operation of the equipment has to be influenced in such a way that the risk of fire or explosion taking place is minimized.

The area classification needs to be used as a tool to achieve the ultimate objective of ensuring overall plant safety and minimizing the risk of loss to life and property. This principle guides the management in arriving at compromises, checks and balances between ease and convenience of operation, vis-à-vis, security against risk of explosion or fire during normal and possible abnormal operations of the plant. The balance needs to be struck between presence of an explosive and ignitable environment in an unrestricted area and the limits prescribed for its classification, because a too-conservative approach may make the whole exercise so costly that the viability of the project might be questioned.

It is useful to understand what a non-hazardous (safe) area is.

An area classified as non-hazardous has a small probability of a flammable mixture being present. It is also called a 'safe area' and includes most control rooms.

3.2 Principles of safety

The 'HAZOP study' (hazard and operability studies) is carried out during initial design, renovation and modernization stages. These need to be followed up with 'periodic risk assessment studies' during operation of the plant.

While carrying out these studies 'the principle of safety' with reference to area classification needs to be adhered to. The 'safety first' principle dictates that the quantity of release of flammable material in and around electrical installations is to be kept to a minimum, in case it is not possible to eliminate it, so that the extent of area under influence of hazardous classification is also kept to a minimum. While estimating or assessing this, the normal operation of the plant is anyway to be kept in mind, but due attention is also to be paid to frequency, quantity and duration of 'mal-operation/ abnormal operation' of the plant, resulting in the release of hazardous material.

While assessing the situation, a conscious and distinct effort should be made not to confuse the maintenance-related operations with abnormal operations, as the permit-to-work system is supposed to ensure safe operating conditions during maintenance.

Even in emergency conditions, emphasis should be placed on weeding out and eliminating the root cause of the manifestation of ignitable or combustible atmosphere, i.e.

- Disconnection or isolation of defective electrical equipment
- Shutdown of processes
- Isolation of process vessels having ignitable material
- Containment of spillage.

Then a review needs to be done to augment with additional ventilation, wherever the process and location of the installation permit, to quickly reduce the intensity of hazardous atmosphere or environment.

Safety and minimization of risk to life and property demand,

- That ignitable, explosive or hazardous environments be isolated from the 'ignition source'.
- That the source of ignition be eliminated.

Whenever it is not possible to reliably and safely implement at least one of the above, then steps should be taken to put in place procedures whereby the operations are implemented in such a way so as to minimize the above conditions occurring simultaneously to an acceptable safe level.

In this chapter, we shall seek to identify the principles on which the classification rests and the roles to be played by those who are involved in design and operation of the plant and machinery.

3.3 Hazards and hazardous areas

Hazards can be

- *Urban structure fires*: Perhaps the most common human-caused hazard (often a disaster) is fire in large occupied buildings. Causes can be accidental or deliberate, but unless structures have been built to safe fire standards, and sound emergency procedures are used, heavy loss of life can result. Disastrous fires have affected most countries. Notable overseas cases include a high-rise building fire in Sao Paulo, Brazil; the Kings Cross Station inferno in London and Bradford Soccer Stadium, both in England; and hotel fires all around the world, such as at Pattaya, Thailand, in July 1997, when 100 died. These, and many like them, have cost thousands of lives, injuries and untold property. In August 1981, 19 people died in the Rembrandt Hotel fire, Sydney, NSW, Australia.

There is greater potential for disaster, due to the use or movement of hazardous materials, than from most other technological hazards. For example, a whole community is more likely to be affected by a toxic gas leak than by deaths and injuries caused in a major transport accident.

- *BLEVE*: An entire community was involved at Mississauga, Ontario, Canada when 250 000 had to be evacuated to avert disaster following a train accident which triggered a series of BLEVEs (boiling liquid expanding vapor explosions). Liquefied gas BLEVEs have occurred in Cairns (1987 – 1 dead, 24 injured) and in Sydney, where, fortunately, there were no casualties.

- *Other explosions*: Great loss of life occurred in Halifax, Nova Scotia, Canada in 1917 when a ship carrying explosives collided with another. The resulting explosion destroyed large sections of the town and killed 1963 people! Australia's most disastrous explosion was in the *Mt Kembla mine*, Wollongong, in 1902, when 95 miners died. One of the worst non-mining explosions occurred in 1974 at the Mt St Candice Convent in Hobart, when seven died in a boiler explosion.

- *Toxic emission*: Not all hazardous materials accidents involve transport, and some can result in worse disasters. During 1984, cyanide gas escaped from a fertilizer factory in Bhopal, India. The resulting deadly cloud caused the deaths of approximately 2000 people living close-by. In Australia in August 1991, the Coode Island fire burnt 8.6 million liters of chemicals in the heart of Melbourne and loomed as a potential disaster. Good luck rather than good management resulted in winds dispersing toxic fumes away from residential areas. Over 250 workers were evacuated from nearby ships and factories but only 2 injuries occurred (to fire fighters).

While hazards are associated with all of the above conditions, areas are only considered hazardous (classified) locations under conditions defined by the IEC 60079-10 or SABS 0108: 1996, as applicable.

An area may also be considered 'hazardous' for various other reasons. These may also include the use of electrical equipment in the vicinity of water, the risk of personal injury from moving or falling parts, or even the presence of biological hazards.

Hazardous areas, for the purpose of this course are those in which there exists a risk of explosion because flammable atmospheres are likely to be present. These conditions can be manmade as in petrochemical plants or refineries, or occur naturally, as in coal mining. It is therefore necessary to ensure that no electrical equipment and instrumentation installed in a hazardous area can form a spark or hot surface, which would ignite the flammable atmospheres.

The determination that areas can be classified as hazardous locations is based on the following:

- The possible presence of an explosive atmosphere such as flammable gases, vapors or liquids, combustible dusts or ignitable fibers and flyings
- The likelihood that the explosive atmosphere is present when equipment is operating
- The ignition-related properties of the explosive atmosphere that is present.

This approach while classifying hazardous locations is used in IEC 60079-10 worldwide and also in South Africa (South African Bureau of Standards), the United States (National Electrical Code), Canada (Canadian Electrical Code), Europe (CENELEC EN 60079-10) and Australia (Australian Code).

The hazardous locations information provided during this course is intended to answer questions associated with South African and IEC system of classification.

3.4 Basic properties of combustible and ignitable material

An adequate and functional knowledge of the following is required in order to understand and carry out the hazardous area classification:

- The electrical equipment involved in process
- The production process
- Safety procedures related to process
- Properties of ignitable and combustible (flammable) materials.

The electrical equipment … An attempt has been made to familiarize and update you with working knowledge of the electrical equipments commonly used in daily life and principles of electricity in Chapter 2 of this book.

The production processes … are so many and varied. They are domain-specific and hence out of the purview of this course.

The safety procedures … are again a full course by themselves and reference shall be made wherever they touch our domain of discussions.

Thus, it leaves us with the task of understanding the properties of hazardous (flammable, ignitable, combustible or explosive) materials. This chapter is devoted to this study.

A flammable material is defined in IEC 79-0 as follows:

A gas, vapour, liquid or solid that can react continuously with atmospheric oxygen and may therefore sustain fire or explosion when such reaction is initiated by a suitable spark, flame or hot surface.

Thus the hazardous materials of concern are,

- Gas
- Vapor/mist
- Dust/flakes or fibers.

Each of the above is having distinct properties and nature. These behave differently under similar circumstances. Thus, their nature needs to be understood so that the classification done is accurate.

3.4.1 Flammable gases

Gas is generally found in the form of vapors at ambient temperatures and pressures. To liquefy them at ambient conditions not only does pressure needs to be changed but also they need to be cooled. Hence they do not generally occur in the form of liquid and, if they are, they will rapidly vaporize once released. If sufficient ventilation is present then their effect will be minimal once their release is stopped.

The features as described hereunder influence the approach to be adopted in respect of electrical installation with which they come in contact:

- Relative density with respect to air. This determines how the gas will disperse when no other influence is present.
- MIE of the gas/air mixture.
- Minimum experimental safe gap defines the burning characteristics of the ideal gas/air mixture (which may differ from the most easily ignitable mixture) insofar as its ability to burn through small gaps is concerned.
- Ignition temperature is the minimum temperature at which the ideal gas/air mixture will ignite.

The release velocity of gas also plays a significant role in determining the area. As a low velocity release of gas in poorly ventilated building or space may give rise to inefficient mixing with air and thus gas/air mixture will vary from place to place giving rise to large and unpredictable hazardous areas; whereas gas released at high velocity will disperse quickly and form a more predictable zone of influence.

3.4.2   Flammable vapor

Generally, flammable vapors are similar in nature to gases. The difference being that they can be liquefied by pressure alone and thus are more easily liquefied. The flammable vapors are present over the flammable liquid (partial pressure in liquids) at temperatures well below the boiling point of the liquid. As and when this vapor pressure exceeds the lower explosive limit (LEL), an explosive atmosphere can exist although only the liquid is released. At times, this explosive atmosphere can exist for quite sometime even after the release of liquid stops.

3.4.3   Flammable mists

A mist is formed by release of flammable liquids at high pressure such that very fine particles (droplets) of liquids are produced. The mist has a significant existence time, and the particles will remain in suspension in air for a longer period. They are to be treated in similar manner as gas and vapors for the purpose of area classification.

3.4.4   Flammable liquids

It is well known that liquid does not burn and hence this word is misnomer. Generally, flammable liquid implies that a liquid, which at ambient temperature and pressure or temperature and pressure at which it is handled or stored, has a vapor pressure sufficient to liberate enough vapor to form an explosive atmosphere.

3.4.5   Combustible dusts

The dust which can be easily ignited when mixed with air is termed combustible dust. In layer form also it is ignited easily. This is distinct from explosives and should not be confused with it. The behavior of dust is distinctly different from gas or vapors, as when released it is in a cloud form and then gradually settles down on surfaces in layer form.

The parameters used to identify combustible dust are,

- *Cloud ignition energy*:   The minimum energy which is required in the form of an arc or spark to ignite a gas cloud.
- *Cloud ignition temperature*:   The minimum temperature at which an ideal mixture of dust in suspension with air will ignite. This can depend on particle size.
- *Layer ignition temperature:*   The minimum temperature at which a layer of the dust of specific thickness will ignite and burn.
- *Particle size*:   The size of particle from which a dust is formed and which has an effect on its ignition capability.

3.5   Basis of area classification

Area classification cannot be set rigidly into any standard. Each installation will be different in some respect and therefore each site must be examined on its individual merits. Having said that, different industries do use similar approaches so there are some accepted industry standard ways of looking at parts of a plant. Vessels, pipe-work, storage tanks and processing units are common in many industries and so some experience has been gained in the handling of these. Codes of practice and guideline documents have been released by various organizations to assist in this area. IEC 79-10: 1995 now lays down some formalized approaches to area classification but does not provide all the answers. The guidelines are always open to interpretation.

The Institute of Petroleum has published recommendations for petrol forecourts. In this situation, a uniform approach to the problem may be adopted due to the similarity of situations. The industry is expected to follow these rules closely. The tanker industry conveying LPG is another example of where more guidelines can be issued. The offshore industry also has some recommendations on standard layouts of rigs. Other larger organizations, mainly within the petrochemical industry, do have more rigid internal rules because of their standardization on processing methods.

There are three situations that can occur in an operating plant with reference to hazardous areas:

1. A situation where an explosive atmosphere is present always or for long periods because of operational requirement, i.e. *continuous*.
   Typical examples are interiors of process vessels or equipment processing gases, vapor, mists or dusts where air is also present, the interiors of stock tanks where liquids, gases or vapors are present with air, and silos and mills where dust and air are both present.
2. A situation where explosive atmosphere occurs frequently, or if infrequently may persist for a considerable time, i.e. *primary*.
   Typical examples are – areas surrounding rotating seals of machines which are subject of wear, manual loading points on such things as vessels where gases, vapors, mists, liquids and dusts are loaded, paint-spraying facilities, etc.
3. A situation in which explosive atmosphere occurs rarely and normally result from failure of equipment or procedures, i.e. *secondary*.
   Typical examples are failure of a gasketed joint in a process piping releasing gas, liquid or mist, overflowing of a vessel due to failure of process control functions releasing a liquid, powder-handling rooms where dust deposits can form and then be agitated into clouds, etc.

The above criteria are used in classifying the areas under 'Source of Release' methodology of classification.

3.5.1 Some typical examples of sources of release

Open surface of liquid

In most cases, the liquid temperature will be below the boiling point and the vapor release rate will depend principally on the following parameters:

- Liquid temperature
- Vapor pressure of the liquid at its surface temperature
- Dimensions of the evaporation surface.

Virtually instantaneous evaporation of a liquid (for example from a jet or spray)

Since the discharged liquid vaporizes virtually instantaneously, the vapor release rate is equal to the liquid flow rate and this depends on the following parameters:

- Liquid pressure
- Geometry of the source of release.

Where the liquid is not instantaneously vaporized, the situation is complex because droplets, liquid jets and pools may create separate sources of release.

Leakage of a gas mixture

The gas release rate is affected by the following parameters:

- Pressure within the equipment which contains the gas
- Geometry of the source of release
- Concentration of flammable gas in the released mixture.

For typical process plants, certain examples are given here to clarify the principles governing the sources of release

I. Sources giving a continuous grade of release

- The surface of a flammable liquid in a fixed roof tank, with a permanent vent to the atmosphere
- The surface of a flammable liquid which is open to the atmosphere continuously or for long periods (for example an oil/water separator).

II. Sources giving a primary grade of release

- Seals of pumps, compressors or valves if release of flammable material during normal operation is expected
- Water drainage points on vessels which contain flammable liquids, which may release flammable material into the atmosphere while draining off water during normal operation
- Sample points which are expected to release flammable material into the atmosphere during normal operation
- Relief valves, vents and other openings, which are expected to release flammable material into the atmosphere during normal operation.

III. Sources giving a secondary grade of release

- Seals of pumps, compressors and valves where release of flammable material during normal operation of the equipment is not expected.
- Flanges, connections and pipe fittings, where release of flammable material is not expected during normal operation.
- Sample points which are not expected to release flammable material during normal operation.
- Relief valves, vents and other openings, which are not expected to release flammable material into the atmosphere during normal operation.

3.6   Zonal classification

We have understood the type of sources or release of hazardous material (gas/liquid/mist/vapor/dust) and can now proceed further to understand the zonal classification.

3.6.1   Gases, vapor and mists

Areas where there is the likelihood of the presence of explosive gas/air mixtures are referred to as zones. Zones are classified as shown in the table below. The higher the number in this 'zonal classification' the smaller is the risk of an explosion. This is as per IEC 79:

Zone 0	An area in which an explosive gas/air mixture is continually present or present for long periods
Zone 1	An area in which a gas/air mixture is likely to occur in normal operation
Zone 2	An area in which a gas/air mixture is not likely to occur in normal operation, and if it occurs, it will exist only for a short time

These zones represent the varying probabilities of explosive gas/air mixtures being present. A plant will have areas in which gas/air mixtures are more likely to occur than in others. This is an important consideration because where there is a greater risk of the presence of a gas, the Ex protection afforded to apparatus operating in that area may need to be more stringent.

3.6.2   Dusts

In respect of dust, the situation had been much fluid. In recent times effort has been made to address this by classifying the zones in a way, which is, similar to that adopted for gas and vapor.

Zone 20	An area in which combustible dust, as a cloud, is present continuously or frequently, during normal operation, in sufficient quantity to be capable of producing an explosive concentration of combustible dust in mixture with air, and/or where layers of dust of uncontrollable and excessive thickness can be formed

Zone 21	Zone 21 is a zone not classified as Zone 20 in which combustible dust, as a cloud, is present continuously or frequently, during normal operation, in sufficient quantity to be capable of producing an explosive concentration of combustible dust in mixture with air
Zone 22	Zone 22 is a zone not classified as Zone 21 in which combustible dust, as a cloud, is present continuously or frequently, during normal operation, in sufficient quantity to be capable of producing an explosive concentration of combustible dust in mixture with air

The above classification is as given in IEC 1241. Generally it is considered that 1 mm or less thickness of dust is not likely to result in formation of explosive atmosphere.

3.7 Plant operations – normal and abnormal

The terms '*normal*' and its expressed or implied opposite '*abnormal*' require some explanation.

'Normal' is not intended to mean 'ideal' or 'perfect', or other similar connotations; it does mean 'actual' or 'real' applied to the conditions, as they exist in any given plant:

- The actual standard of design used
- The achieved state of maintenance
- The expected environmental limitations
- The usual operations and operating practices employed, etc.

In modern plants handling flammable materials, it is of course the main objective of design, maintenance and operating philosophy to ensure that there are few ways in which a flammable atmosphere can occur. This is to be achieved by:

- Proper choice of process equipment
- Safe disposal of vented products
- Provision of special ventilation arrangements
- Good maintenance
- Good production supervision and other similar precautions.

Where such precautions can be regarded as the normal state of affairs they will be reflected in the absence of, or the reduction in, the number of areas designated Zone 1 and in the extent of these areas (as elaborated in later paragraphs dealing with area classification). Where the normal state of affairs is less-rigorously controlled the expectation of Zone 1 classification and the extent thereof will be higher.

At the other end of the scale 'abnormal' does not refer to such catastrophic events as the bursting of a process vessel or large pipeline (unless there are special circumstances which make such possibilities likely and therefore a necessary consideration); the process of area classification is not concerned with such eventualities. Between these extremes, however, and 'normally' as already defined there will be circumstances recognizable as abnormal events, which though they may occur at some time will do so infrequently.

In the well-ordered modern plants, referred to above, examples would be the collapse of a pump gland, the failure of a pipe gasket, the loss of control of the manual draining operation of a tank, the fracture of a small branch pipe or the accidental spillage of small quantities of flammable liquid. They are usually the unintended, unpredictable, non-catastrophic events in a plant. They are in most instances the kinds of faults, which it is expected, will be avoided by good design and preventive maintenance; if despite this they do occur, matters will have been so arranged that they will be rapidly rectified. With conditions so well controlled these events will therefore be both infrequent and have short duration.

3.8 Area classification – gas and vapors

A three-dimensional region or space is defined as an 'area' for the purpose of our discussion during this course.

Area classification is a method of analyzing and classifying the environment where explosive gas atmospheres may occur so as to facilitate the proper selection and installation of apparatus to be used safely in that environment, taking into account gas groups and temperature classes.

It may not be possible to classify areas into Zone 0, 1 or 2 by a simple examination of a plant or plant design. Hence, a systematic analysis and detailed approach is required to determine the possibility of an explosive gas atmosphere occurring, which is elaborated hereunder.

3.8.1 Sources of release

The basic elements for establishing the hazardous zone types are the identification of the source of release and the determination of the grade of release.

Since an explosive gas atmosphere can exist only if a flammable gas or vapor is present with air, it is necessary to decide if any of these flammable materials can exist in the area concerned. Generally speaking, such gases and vapors (and flammable liquids and solids which may give rise to them) are contained within process equipment, which may or may not be totally enclosed. It is necessary to identify where a flammable atmosphere can exist inside a process plant, or where a release of flammable materials can create a flammable atmosphere outside a process plant.

Each item of process equipment (for example, tank, pump, pipeline, vessel, etc.) should be material, it will clearly not give rise to a hazardous area around it. The same will apply if the item contains a flammable material but cannot release it into the atmosphere (for example, an all-welded pipeline is not considered to be a source of release).

If it is established that the item may release flammable material into the atmosphere, it is necessary, first of all, to determine the grade of release in accordance with the definitions, by establishing the likely frequency and duration of the release. It should be recognized that the opening-up of parts of enclosed process systems (for example, during filter changing or batch filling) should also be considered as sources of release when developing the area classification.

By means of this procedure, each release will be graded either (refer basis of area classification),

- Continuous
- Primary or
- Secondary.

Having established the grade of the release, it is necessary to determine the release rate and other factors, which may influence the type and extent of the zone.

3.8.2   Type of zone

The likelihood of the presence of an explosive gas atmosphere and hence the type of zone depends mainly on the grade of release and the ventilation. A continuous grade of release normally leads to Zone 0, a primary grade to Zone 1 and a secondary grade to Zone 2.

3.8.3   Extent of zone

The extent of the zone is mainly affected by the following chemical and physical parameters, some of which are intrinsic properties of the flammable material; others are specific to the process. For simplicity, the effect of each parameter listed below assumes that the other parameters remain unchanged.

Consideration should always be given to the possibility that a gas which is heavier than air may flow into areas below ground level, for example, pits or depressions and that a gas which is lighter than air may be retained at high level, for example, in a roof space.

Where the source of release is situated outside an area or in an adjoining area, the penetration of a significant quantity of flammable gas or vapor into the area can be prevented by suitable means such as:

- Physical barriers
- Maintaining a static overpressure in the area relative to the adjacent hazardous areas, so preventing the ingress of the hazardous atmosphere
- Purging the area with a significant flow of air, so ensuring that the air escapes from all openings where the hazardous gas or vapor may enter.

Release rate of gas or vapor

The greater the release rates the larger the extent of the zone. The release rate itself depends on other parameters, namely:

- *Geometry of the source of release*: This is related to the physical characteristics of the source of release, for example, an open surface, leaking flange, etc.
- *Release velocity*:   For a given source of release, the release rate increases with the release velocity. In the case of a product contained within process equipment, the release velocity is related to the process pressure and the geometry of the source of release. The size of a cloud of flammable gas or vapor is determined by the rate of flammable vapor release and the rate of dispersion. Gas and vapor flowing from a leak at high velocity will develop a cone-shaped jet, which will entrain air and be self-diluting. The extent of the explosive atmosphere will be almost independent of wind velocity. If the release is at low velocity or if its velocity is destroyed by impingement on a solid object, it will be carried by the wind and its dilution and extent will depend on wind velocity.
- *Concentration*: The release rate increases with the concentration of flammable vapor or gas in the released mixture.
- *Volatility of a flammable liquid*:   This is related principally to the vapor pressure and the heat of vaporization. If the vapor pressure is not known, the boiling point and flashpoint can be used as a guide.

An explosive atmosphere cannot exist if the flashpoint is above the relevant maximum temperature of the flammable liquid. The lower the flashpoint, the greater may be the extent of the zone. If a flammable material is released in a way that forms a mist (for example, by spraying), an explosive atmosphere may be formed below 'the flashpoint of the material', for example.

Flashpoints of flammable liquids are not precise physical quantities, particularly where mixtures are involved.

Some liquids (for example, certain halogenated hydrocarbons) do not possess a flashpoint although they are capable of producing an explosive gas atmosphere. In these cases, the equilibrium liquid temperature, which corresponds to the saturated concentration at the LEL, should be compared with the relevant maximum liquid temperature.

- *Liquid temperature*: The vapor pressure increases with temperature, thus increasing the release rate due to evaporation.

Lower explosive limit (LEL)

For a given release volume, the lower the LEL the greater will be the extent of the zone.

Ventilation

With increased ventilation, the extent of the zone will be reduced. Obstacles, which impede the ventilation, may increase the extent of the zone. On the other hand, some obstacles, for example dykes, walls or ceilings, may limit the extent. We shall be discussing this in more detail.

Relative density of the gas or vapor when it is released

If the gas or vapor is significantly lighter than air, it will tend to move upward. If significantly heavier, it will tend to accumulate at ground level. The horizontal extent of the zone at ground level will increase with increasing relative density, and the vertical extent above the source will increase with decreasing relative density.

Other parameters to be considered

- Climatic conditions
- Topography.

3.8.4   Openings

Openings as possible sources of release

Openings between areas should be considered as possible sources of release. The grade of release will depend upon:

- The zone type of the adjoining area
- The frequency and duration of opening periods
- The effectiveness of seals or joints
- The difference in pressure between the areas involved.

Openings classification

Openings are classified as A, B, C and D with the following characteristics:

1. *Type A*: Openings not conforming to the characteristics specified for types B, C or D.

   Examples:

   - Open passages for access or utilities, for example, ducts, pipes through walls, ceilings and floors
   - Fixed ventilation outlets in rooms, buildings and similar openings of types B, C and D which are opened frequently or for long periods.

2. *Type B*: Openings that are normally closed (for example, automatic closing) and infrequently opened, and which are close fitting.

3. *Type C*: Openings normally closed and infrequently opened, conforming to Type B, which are also fitted with sealing devices (for example, a gasket) along the whole perimeter; or two openings Type B in series, having independent automatic closing devices.

4. *Type D*: Openings normally closed conforming to Type C, which can only be opened by special means or in an emergency.

   Type D openings are effectively sealed, such as in utility passages (for example, ducts, pipes) or can be a combination of one opening Type C adjacent to a hazardous area and one opening Type B in series.

The following table describes the effect of openings on grade of release.

Zone Upstream of Opening	Opening Type	Grade of Release of Openings Considered as Sources of Release
Zone 0	A	Continuous
	B	(Continuous)/primary
	C	Secondary
	D	No release
Zone 1	A	Primary
	B	(Primary)/secondary
	C	(Secondary)/no release
	D	No release
Zone 2	A	Secondary
	B	(Secondary)/no release
	C	No release
	D	No release

Source: IEC 79

Note: For grades of release shown in brackets, the frequency of operation of the openings should be considered in the design.

3.8.5　Ventilation

Gas or vapor released into the atmosphere can be diluted by dispersion or diffusion into the air until its concentration is below the lower explosion limit. Hence, the designs of artificial ventilation systems are of paramount importance in the control of the dispersion of releases of flammable gases and vapors. Ventilation, i.e. air movement leading to replacement of the atmosphere in a (hypothetical) volume around the source of release by fresh air will promote dispersion. Suitable ventilation rates can also avoid persistence of an explosive gas atmosphere, thus influencing the type of zone.

3.8.6　Main types of ventilation

Ventilation can be accomplished by the movement of air due to the wind and/or by temperature gradients or by artificial means such as fans. So two main types of ventilation are thus recognized:

1. Natural ventilation
2. Artificial ventilation, general or local.

3.8.7　Natural ventilation

This is a type of ventilation that is accomplished by the movement of air caused by the wind and/or by temperature gradients. In open-air situations, natural ventilation will often be sufficient to ensure dispersal of any explosive atmosphere, which arises in the area. Natural ventilation may also be effective in certain indoor situations (for example, where a building has openings in its walls and/or roof).

For outdoor areas, the evaluation of ventilation should normally be based on an assumed minimum wind speed of 0.5 m/s, which will be present virtually continuously. The wind speed will frequently be above 2 m/s.

Examples of natural ventilation:

- Open air situations typical of those in the chemical and petroleum industries, for example, open structures, pipe racks, pump bays and the like.
- An open building which, having regard to the relative density of the gases and/or vapors involved, has openings in the walls and/or roof so dimensioned and located that the ventilation inside the building, for the purpose of area classification, can be regarded as equivalent to that in an open air situation.
- A building which is not an open building but which has natural ventilation (generally less than that of an open building) provided by permanent openings made for ventilation purposes.

3.8.8　Artificial ventilation

The air movement required for ventilation is provided by artificial means, for example, fans or extractors. Although artificial ventilation is mainly applied inside a room or enclosed space, it can also be applied to situations in the open air to compensate for restricted or impeded natural ventilation due to obstacles.

The artificial ventilation of an area may be either general or local and, for both of these, differing degrees of air movement and replacement can be appropriate.

With the use of artificial ventilation, it is possible to achieve:

- Reduction in the extent of zones
- Shortening of the time of persistence of an explosive atmosphere
- Prevention of the generation of an explosive atmosphere.

Artificial ventilation makes it possible to provide an effective and reliable ventilation system in an indoor situation. An artificial ventilation system, which is designed for explosion protection, should meet the following requirements:

- Its effectiveness should be controlled and monitored.
- Consideration should be given to the classification immediately outside the extract system discharge point.
- For ventilation of a hazardous area the ventilation air should usually be drawn from a non-hazardous area.
- Before determining the dimensions and design of the ventilation system, the location, grade of release and release rate should be defined.

In addition, the following factors will influence the quality of an artificial ventilation system:

- Flammable gases and vapors usually have densities other than that of air, thus they will tend to accumulate near to either the floor or ceiling of an enclosed area, where air movement is likely to be reduced
- Changes in gas density with temperature
- Impediments and obstacles may cause reduced, or even no air movement, i.e. no ventilation in certain parts of the area.

Examples of general artificial ventilation:

- A building which is provided with fans in the walls and/or in the roof to improve the general ventilation in the building
- An open-air situation provided with suitably located fans to improve the general ventilation of the area.

Examples of local artificial ventilation:

- An air/vapor extraction system applied to an item of process equipment which continuously or periodically releases flammable vapor
- A forced or extract ventilation system applied to a small, ventilated local area where it is expected that an explosive atmosphere may otherwise occur.

3.8.9   Degree of ventilation

The effectiveness of the ventilation in controlling dispersion and persistence of the explosive atmosphere will depend upon the degree and availability of ventilation and the design of the system.

The most important factor is that the degree or amount of ventilation is directly related to the types of sources of release and their corresponding release rates. This is irrespective of the type of ventilation, whether it is wind speed or the number of air changes per time unit. Thus, optimal ventilation conditions in the hazardous area can be achieved and the higher the amount of ventilation in respect of the possible release rates, the smaller will be the extent of the zones (hazardous areas), in some cases reducing them to a negligible extent (non-hazardous area).

The methods developed allow the determination of the type of zone by:

- Estimating the minimum ventilation rate required to prevent significant build up of an explosive atmosphere and using this to calculate a hypothetical volume, which, with an estimated dispersion time, allows determination of the degree of ventilation.
- Determining the type of zone from the degree and availability of ventilation and the grade of release.

Although primarily of direct use in indoor situations, the concepts explained may assist in outdoor locations also.

3.8.10   Availability of ventilation

The availability of ventilation has an influence on the presence or formation of an explosive atmosphere. Thus the availability (as well as the degree) of ventilation needs to be taken into consideration when determining the type of zone.

Three levels of availability of the ventilation should be considered:

- *Good*:   Ventilation is present virtually continuously.
- *Fair*:   Ventilation is expected to be present during normal operation. Discontinuities are permitted, provided they occur infrequently and for short periods.
- *Poor*:   Ventilation, which does not meet the standard of fair or good, but discontinuities are not expected to occur for long periods.

Ventilation that does not even meet the requirement for poor availability must not be considered to contribute to the ventilation of the area.

Natural ventilation

For outdoor areas, the evaluation of ventilation should normally be based on an assumed minimum wind speed of 0.5 m/s, which will be present virtually continuously. In which case the availability of the ventilation can be considered as good.

Artificial ventilation

In assessing the availability of artificial ventilation, the reliability of the equipment and the availability of, for example, standby blowers should be considered. Good availability will normally require, on failure, automatic start-up of standby blower(s). However, if provision is made for preventing the release of flammable material when the ventilation has failed (for example, by automatically closing down the process), the classification determined with the ventilation operating need not be modified, i.e. the availability may be assumed good.

Combining the concepts of degree of ventilation and level of availability results in a quantitative method for the evaluation of zone type.

3.8.11   Calculation of release rates and extent of hazardous area

In industry there is no unanimous approach of mathematical calculation of such accidental releases due to inadequate information from experimentation on mixing of releases with air when such releases are accidental, rather than deliberate. Generally mathematical approach is built around calculation of fluid flow from orifices specially designed to ensure the maximum flow for minimum effort.

We will look at two mathematical approaches, one which is published in IEC 79 and the other used in UK. It may be noted that these are to be used by experienced persons as applicable and verified with practical experimentation data available to the extent possible.

Release of gas and vapor

This mathematical approach is based upon a combination of fluid dynamics, kinetic theory of gases and practical measurements.

The mass release of gas from an orifice can be calculated as given below,

$$G = 0.006aP\left(\frac{M}{T}\right)^{0.5} \text{ kg/s}$$

Where

G = mass release, kg/s

a = cross-sectional area of leak, m^2

P = upstream pressure, N/m^2

M = molecular weight

T = absolute temperature of released gas, K.

The equation as above is valid if upstream absolute pressure exceeds 2×10^5 N/m^2. Below this pressure the effect of atmospheric pressure is significant and the diffusion effect need to be considered.

The volume of released gas V,

$$V = \frac{V_0 G T}{T_0 M}$$

Where

V_0 = molar volume, m^3/kg

T_0 = melting point of ice, 273 K.

The formula then becomes,

$$V = \frac{0.082GT}{M}$$

The distance 'X', at which the gas or vapor due to mixing with air while traversing falls below LEL, can be calculated as below,

$$X = 2.1 \times 10^3 \left[\frac{G}{E^2 M^{1.5} T^{0.5}}\right]^{0.5}$$

Where

E = lower explosive limit (LEL), %.

It may be noted that the above formula is valid as long as jet velocity is sufficiently high as compared to air/wind speed.

In case the difference between the two reduces then the following formula which is based on experimentation can be used,

$$X = 10.8\left(\frac{GT}{ME}\right)^{0.55}$$

Typical progression of gas release

Now let us look at how a typical gas or vapor release looks like. As the gas is released, a rapid expansion and fall in pressure takes place, shown in Figure 3.1 by 'a'. The lines 'b' indicate the probable way the envelope of explosive and flammable gases will develop during mixing stage. The curve 'c' indicates the likelihood of area till which place the hazardous area will extend.

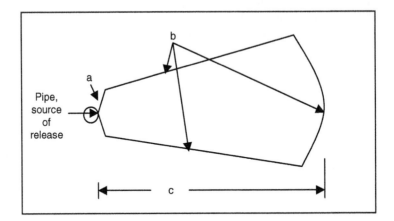

Figure 3.1
Geometry of gas release formed in hazardous area

Examples of gas and vapor release

Example 1

Leakage from a pipe flange gasket,

Type of Gas	Ethylene
Molar weight	28
Ambient temperature	22 °C
Size of orifice	$4 \times 10^{-5} \text{ m}^2$
The pressure in pipe	$4 \times 10^5 \text{ N/m}^2$
LEL at the given ambient	2.7%

The released mass will be

$$G = 0.006aP\left(\frac{M}{T}\right)^{0.5} \text{ kg/s} = 0.03 \text{ kg/s}$$

The distance will be

$$X = 2.1 \times 10^3 \left(\frac{G}{E^2 M^{1.5} T^{0.5}}\right)^{0.5} = 2.7 \text{ m}$$

The hazardous area will be a sphere extending 2.7 m from the source of release in all directions.

Obstruction to the gas and vapor release

In case the path of release of gas or vapor is obstructed by the presence of a barrier or an object before the distance X is determined as above, then the volume need to be calculated as given below,

$$B = 100 - L\left[\frac{(100 - \text{LEL})}{X}\right] \cdots \%$$

Where
    B = % gas/vapor in air
    L = distance to obstruction
    X = distance to LEL.

Thus knowing the volume ratio the new effective molecular weight is calculated as given below,

$$M_{(\text{mixture})} = \left[\frac{M_{(\text{gas})} \times \%_{(\text{gas})}}{100}\right] + \left[\frac{M_{(\text{air})} \times \%_{(\text{air})}}{100}\right]$$

$$V_{(\text{gas})} = \frac{0.082GT}{M_{(\text{gas})}}$$

$$V_{(\text{mixture})} = V_{(\text{gas})} \times \left[\frac{100}{\% \text{ gas in mixture}}\right] \text{m}^3$$

$$G_{(\text{mixture})} = V_{(\text{mixture})} \times \frac{12.19\, M_{(\text{mixture})}}{T} \text{ kg}$$

$$LEL_{(\text{mixture})} = LEL_{(\text{gas})} \times \left[\frac{100}{\% \text{ gas in mixture}}\right] \%$$

At the point of obstruction the new values are calculated of the mixture and new dispersion distance found out using the formula for X (Figure 3.2).

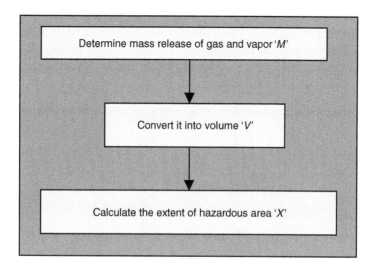

Figure 3.2
Summary of gas and vapor release calculations

Release of liquid below its atmospheric boiling point

As in the case of gas and vapor, in case of liquid below boiling point will take the form of jet or mist from an orifice like opening, depending on the pressure. The formula to calculate the release is,

$$G = 1.13a \, [\sigma_1(P - 10^5)]^{0.5}$$

Where

    G = mass release, kg/s

    P = upstream pressure, N/m^2

    σ_1 = liquid density at atmospheric conditions, kg/m^2.

The velocity of jet can be calculated based on

$$v = v \cos\phi (2gh + v^2 \sin^2 \phi)^{0.5} + \frac{v \sin\phi}{g}$$

The horizontal distance up to which extent of hazardous area will be present can be found out by

$$X = \left(\frac{2v^2 h}{g} \right)^{0.5}$$

Where

    v = velocity, m/s

    g = gravity acceleration (9.81 m/s^2)

    h = height of initial release, m.

Release of liquid above its atmospheric boiling point

In this outdoor scenario the liquid at the point of release will evaporate partially after absorbing heat from atmosphere and the rest will fall on the ground based on the trajectory of jet, the pressure, etc. However, after falling on ground the evaporation will take place very rapidly, almost immediately. The release quantity in such cases is calculated as given hereunder,

$$G = 0.8A \, [2 \, \sigma_m (P_1 - P_c)]^{0.5}$$

Where

    A = cross-sectional area of leak, m^2

    σ_m = density of released mixture, kg/m^3

    P_1 = containment pressure, kg/m^2

    P_c = 0.55 × vapor pressure, kg/m^2.

It may please be noted that the vapor-pressure curve for the material in question is required for calculating σ_m with reference to P_c.

The fraction of mass which vaporizes on being released can be calculated as given below,

$$M_g = \frac{(T_1 - T_c)\, C_1}{L}$$

Where

    M_g = fraction of mass release which is vapor
    T_1 = process temperature, K
    T_c = temperature giving P_c, K
    C_1 = heat capacity of liquid, kJ/kg°C
    L = latent heat of vaporization, kJ/kg.

Now the density of the mixture can be calculated as below,

$$\sigma_m = \frac{1}{\left[\left(\dfrac{M_g}{\sigma_v} \right) + \left\{ \dfrac{(1 - M_g)}{(\sigma_1)} \right\} \right]}$$

Where

    σ_m = density of released mixture, kg/m^2
    σ_v = density of vapor at T_1
    σ_1 = density of liquid at T_1.

From above, the velocity of release can be calculated and then the formulae given for release from jet in the previous section can be applied.

Assessment and estimation of hypothetical volume 'V_z'

This approach is based on IEC 79.

The ideal formula for estimating a hypothetical volume V_z of potential hazardous and explosive atmosphere around the source of release is

$$V_z = \frac{\left(\dfrac{dV}{dt} \right)_{min}}{C}$$

Where
  C = number of fresh air changes per unit time, s^{-1}.

In an enclosed space the value of C is

$$\frac{\dfrac{dV_{tot}}{dt}}{V_0}$$

Where
  dV_{tot}/dt = the total flow rate of fresh air
      V_0 = the total volume being ventilated.

In open space this value is 0.03/s based on a average wind speed of 0.5 m/s.

$(dV/dt)_{min}$ is the minimum volumetric flow rate of fresh air (volume per time, m^3/s)

The theoretical minimum ventilation flow rate can be obtained as given hereunder,

$$\left(\frac{dV}{dt}\right)_{min} = \frac{\left(\frac{dG}{dt}\right)_{max}}{k \times LEL} \times \frac{T}{293}$$

$(dG/dt)_{max}$ = the maximum rate of release at source (mass per time, kg/s)

$\quad LEL$ = the lower explosive limit (mass per volume, kg/m^3)

$\quad k$ = a safety factor applied to the LEL; typically

$\quad k = 0.25$ (continuous and primary grades of release)

$\quad k = 0.5$ (secondary grades of release)

$\quad T$ = the ambient temperature (in Kelvin).

The formula for V_z, as given above, would hold for an instantaneous and homogeneous mixing at the source of release, given ideal flow conditions of the fresh air. In practice, such ideal situations will generally not be found, for example because of possible impediments to the air flow, resulting in badly ventilated parts of the area. Thus, the effective air exchange at the source of release will be lower than that given by C in formula for enclosed space, leading to an increased volume V_2. By introducing an additional correction (quality) factor, f to formula above, one obtains:

$$\frac{f \times \left(\frac{dV}{dt}\right)}{C} \sim n$$

where f denotes the efficiency of the ventilation in terms of its effectiveness in diluting the explosive atmosphere, with f ranging from

f = one (ideal situation) to, typically

f = 5 (impeded air flow).

The volume V_z represents the volume over which the mean concentration of flammable gas or vapor will be either 0.25 or 0.5 times the LEL, depending on the value of the safety factor, k used in formula. This means that, at the extremities of the hypothetical volume estimated, the concentration of gas or vapor will be significantly below the LEL, i.e. the hypothetical volume where the concentration is above the LEL would be less than V_2.

Estimation of persistence time *t*

The time (f) required for the average concentration to fall from an initial value X_0 to the LEL times k after the release has stopped can be estimated from:

$$t = \frac{-f}{C} \ln \frac{LEL \times k}{X_0}$$

Where X_0 is the initial concentration of the flammable substance measured in the same units as the LEL, i.e. % vol or kg/m^3. Somewhere in the explosive atmosphere, the concentration of the flammable matter may be 100% vol (in general only in the very close vicinity of the release source). However, when calculating t, the proper value for X_0 to be taken depends on the particular case, considering among others the affected volume as well as the frequency and the duration of the release, and for most practical cases it seems reasonable to take a concentration above *LEL* for X_0.

Examples of calculations to ascertain the degree of ventilation

The example of toluene gas for the three cases have been taken followed by an example for leakage from compressor seals. These are illustrative in nature and based on IEC 60079.

Example 1

Characteristics of release	
Flammable material	Toluene vapor
Source of release	Flange
Lower explosion limit (LEL)	0.046 kg/m^3 (1.2% vol)
Grade of release	Continuous
Safety factor, k	0.25
Release rate, $(dG/dt)_{max}$	2.8×10^{-10} kg/s
Ventilation characteristics	
Indoor situation	
Number of air changes, C	1/h (2.8×10^{-4}/s)
Quality factor, f	5
Ambient temperature, T	20 °C (293 K)
Temperature coefficient, (T/293 K)	1

Minimum volumetric flow rate of fresh air

$$\left(\frac{dV}{dt}\right)_{min} = \frac{\left(\frac{dG}{dt}\right)_{max}}{k \times LEL} \times \frac{T}{293} = \frac{2.8 \times 10^{-10}}{0.25 \times 0.046} \times \frac{293}{293} = 2.4 \times 10^{-8} \text{ m}^3/\text{s}$$

Evaluation of hypothetical volume V_z

$$V_z = \frac{f \times \left(\frac{dV}{dt}\right)_{min}}{C} = \frac{5 \times 2.4 \times 10^{-8}}{2.8 \times 10^{-4}} = 4.3 \times 10^{-4} \text{m}^3$$

Time of persistence

This is not applicable to a continuous release.

Conclusion

The hypothetical volume V_z is reduced to a negligible value. The degree of ventilation is considered as high with regard to the source.

Example 2

Characteristics of release	
Flammable material	Toluene vapor
Source of release	Failure of flange
Lower explosion limit (LEL)	0.046 kg/m^3 $(1.2\%$ vol$)$
Grade of release	Secondary
Safety factor, k	0.5
Release rate $(dG/dt)_{max}$	2.8×10^{-6} kg/s
Ventilation characteristics	
Indoor situation	
Number of air changes, C	1/h $(2.8 \times 10^{-4}$/s$)$
Quality factor, f	5
Ambient temperature, T	20 °C (293 K)
Temperature coefficient $(T/293$ K$)$	1

Minimum volumetric flow rate of fresh air

$$\left(\frac{dV}{dt}\right)_{min} = \frac{\left(\dfrac{dG}{dt}\right)_{max}}{k \times LEL} \times \frac{T}{293} = \frac{2.8 \times 10^{-10}}{0.5 \times 0.046} \times \frac{293}{293} = 1.2 \times 10^{-8} \, \text{m}^3/\text{s}$$

Evaluation of hypothetical volume V_z

$$V_z = \frac{f \times \left(\dfrac{dV}{dt}\right)_{min}}{C} = \frac{5 \times 1.2 \times 10^{-4}}{2.8 \times 10^{-4}} = 2.2 \, \text{m}^3$$

Time of persistence

$$t = \frac{-f}{C} \ln \frac{LEL \times k}{X_0} = \frac{-5}{1} \ln \frac{1.2 \times 0.5}{100} = 25.6 \, \text{h}$$

Conclusion

The hypothetical volume V_z is significant but can be controlled.

The degree of ventilation is considered medium with regard to the source on this basis. However any release would persist and the concept of Zone 2 may not be met.

Example 3

Characteristics of release	
Flammable material	Toluene vapor
Source of release	Failure of flange
Lower explosion limit (LEL)	0.046 kg/m³ (1.2% vol)
Grade of release	Secondary
Safety factor, k	0.5
Release rate $(dG/dt)_{max}$	6×10^{-4} kg/s
Ventilation characteristics	
Indoor situation	
Number of air changes, C	12/h (3.33×10^{-3}/s)
Quality factor, f	2
Ambient temperature, T	20 °C (293 K)
Temperature coefficient (T/293 K)	1

Minimum volumetric flow rate of fresh air

$$\left(\frac{dV}{dt}\right)_{min} = \frac{\left(\dfrac{dG}{dt}\right)_{max}}{k \times LEL} = \frac{T}{293} = \frac{6 \times 10^{-4}}{0.5 \times 0.046} = \frac{293}{293} = 26 \times 10^{-3} \text{ m}^3/\text{s}$$

Evaluation of hypothetical volume V_z

$$V_z = f \times \frac{\left(\dfrac{dV}{dt}\right)_{min}}{C} = \frac{2 \times 26 \times 10^{-3}}{3.3 \times 10^{-3}} = 15.7 \text{ m}^3$$

Time of persistence

$$t = \frac{-f}{C} \ln \frac{LEL \times k}{X_0} = \frac{-2}{12} \ln \frac{1.2 \times 0.5}{100} = 0.85 \text{ h (51 min)}$$

Conclusion

The hypothetical volume V_z is significant but can be controlled.

The degree of ventilation is considered medium with regard to the source on this basis. Based on the persistence time the concept of Zone 2 would be met.

Example 4

Characteristics of release	
Flammable material	Propane gas
Source of release	Compressor seals
Lower explosion limit (*LEL*)	0.039 kg/m³ (2.1% vol)
Grade of release	Secondary
Safety factor, k	0.5
Release rate (d*G*/d*t*)$_{max}$	0.02 kg/s
Ventilation characteristics	
Indoor situation	
Number of air changes, C	2/h (5.6×10^{-4}/s)
Quality factor, f	5
Ambient temperature, T	20 °C (293 K)
Temperature coefficient (*T*/293 K)	1

Minimum volumetric flow rate of fresh air

$$\left(\frac{dV}{dt}\right)_{min} = \frac{\left(\dfrac{dG}{dt}\right)_{max}}{k \times LEL} \cdot \frac{T}{293} = \frac{0.02}{0.5 \times 0.039} \cdot \frac{293}{293} = 1.02\,\text{m}^3/\text{s}$$

Evaluation of hypothetical volume V_z

$$V_z = \frac{f \times \left(\dfrac{dV}{dt}\right)_{min}}{C} = \frac{5 \times 1.02}{5.6 \times 10^{-4}} = 9200\,\text{m}^3$$

Time of persistence

$$t = -\frac{f}{C} \ln \frac{LEL \times k}{X_0} = \frac{-5}{2} \ln \frac{2.1 \times 0.5}{100} = 11.4\,\text{h}$$

Conclusion

The hypothetical volume V_s is significant and for a room (10 m × 15 m × 6 m) will extend beyond the physical boundaries of the room and will persist.

The degree of ventilation is considered low with regard to the source.

3.9 Area classification – dust

The principles involved are similar to those used to classify plants handling flammable gases, vapors or liquids but it should be recognized that the behavior of flammable dusts is not as predictable as that of flammable gases or vapors.

When dust is released into the atmosphere it disperses in the air as a cloud and may become widely spread by air movement. A cloud of flammable dust within its flammable range can be ignited, and flash-fire, in a confined space, explode. A light dust remains in suspension longer than a heavy dust but both eventually settle and lie dormant in a layer on exposed surfaces. Should these surfaces be hot, or should another source of ignition be present, a layer of flammable dust is a constant risk and may be ignited producing a fire whose severity is dependent on the burning characteristics of the bulk material.

Some flammable dusts in layer form when ignited, have the ability to propagate combustion by flame or smoldering, the latter particularly when the dust is present in bulk. In some cases when the ignition source is removed, combustion of the dust layer ceases. In other cases, it continues and the dust layer is said to train fire.

A layer of flammable dust can be disturbed to form a dust cloud, which may spread and eventually settle again to form another layer. This cycle can be repeated from time to time. Should a small explosion occur, layers of flammable dust over a large area could be disturbed to form a cloud. This, on ignition can create a secondary explosion and/or fire of considerably greater damage-potential than the small primary explosion. In a plant handling flammable dust, a high standard of 'housekeeping' is therefore essential.

The problems associated with dust layers and dust clouds are complex. The data that may be relevant in the case of ignition of dust layers and clouds are ignition temperature, MIE and thermal stability and, in addition, in the case of dust clouds, lower flammable limit. For the latter, typical values are in the range 0.01–0.06 kg of a flammable dust dispersed in each m^3 of air and such a concentration is clearly visible.

It should be noted, however, that these parameters do not provide a direct measurement of the sensitivity of the dust to ignition and are subject to certain limitations.

In the case of dust layers, the data are affected by the thickness of the layer, the temperature of the surface on which the dust rests, and that of the immediate surroundings. For example, the ignition temperature may fall as the thickness of the layer increases. It will vary should the dust be exposed to hot surfaces for a period and although in many cases the ignition temperature will rise when the dust degrades; there are other instances when it will fall.

In the case of dust clouds, the data may be affected by solvent content, presence of additives and by the particle size distribution in the cloud. For example, for a given material, a cloud of dust with a higher proportion of smaller particles is likely to have a lower ignition temperature than one with a predominance of large particles. It is therefore essential to obtain expert advice to provide such data and to ensure that it is relevant to the particular dust in the form in which it will occur in the plant.

3.9.1   Classification method

Because a cloud of flammable dust can arise not only as a result of release of dust from plant equipment but also from disturbances of deposits of dust around the plant, and because the area of spread of dust clouds cannot readily be quantified, it is recommended that a generalized method of classification based on judgment and experience should be used. Additionally, however, in certain cases it may be appropriate to use the techniques of the source of hazard method.

In carrying out an area classification, it is necessary to:

- Identify those parts of the plant where flammable dust can exist including, where appropriate, the interior of process equipment.
- Assess the likelihood of occurrence of a flammable atmosphere (taking into account the general level of 'housekeeping' which will be maintained in the plant) thereby establishing the appropriate zonal classification.
- Delineate the boundaries of the zones taking into account the effect of likely air movement.

In assessing the area classification of a plant, the influence of the classification of adjacent plants must be taken into account.

The classification should be carried out in accordance with the following criteria:

Zone 20

This classification will apply where a flammable atmosphere is continuously present or present for long periods.

Typical examples of Zone 20 are the spaces within such equipment as powder-conveying systems, cyclones and filters containing flammable dust.

Zone 21

This classification will apply where a flammable atmosphere is likely to occur in a normal operation.

A concentration of flammable dust above the lower flammable limit is likely to form an atmosphere through which it is difficult to see and in which it is almost impossible to work. In normal operation, such a situation is unlikely to arise over large areas of the plant. Nevertheless there are certain manual and mechanical operations (which need to be identified positively) which will give rise to a local flammable atmosphere, but this is not likely to spread more than say 1 m beyond the immediate area. In plants handling flammable dusts, therefore, Zone 21 classification will usually apply only to small areas. However, in a room, which contains a number of local Zones 21, it may be more practical to give the whole room this classification.

Typical examples of Zone 21 are the areas local to bag-filling and emptying points and powder-handling equipment, from which release is likely to occur in normal operation in sufficient quantity to produce a cloud of flammable dust. Another example is the space within a container where a cloud of flammable dust may exist from time to time.

Zone 22

This classification will apply where a flammable atmosphere is not likely to occur in normal operation, and if it occurs it will only exist for a short time.

Typical examples of Zone 22 are the areas around powder-handling equipment from which release of flammable dust is not likely to occur during normal operation of the plant equipment, and areas around those classified Zone 21, should there be a likelihood of an abnormal release of flammable dust extending beyond the boundary of the local Zone 21.

3.9.2 Non-hazardous areas

These areas are self-evident once the hazardous areas have been classified.

In exceptional circumstances, a non-hazardous area may be achieved by a very high level of 'housekeeping' and/or an efficient extraction system.

Areas, in which pipes or ducts containing flammable dusts are installed without joints, or with strong joints designed not to leak, may be considered non-hazardous, provided there is a negligible risk of them being damaged.

3.9.3 Surface temperatures

Irrespective of the zonal classification, the temperature of all surfaces on which a flammable dust can settle shall be below the ignition temperature of the dust in layer form. Similarly, the temperature of all surfaces with which a cloud of flammable dust can come into contact shall be below the ignition temperature of the dust in cloud form.

In some exceptional cases where, for process reasons, a surface temperature, which is higher than the ignition temperature of the dust concerned, is required, the apparatus shall be designed and maintained to prevent accumulation of dust on such hot surfaces and the plants shall be operated and maintained to prevent the formation of dust clouds.

3.10 Classification procedure

The first step is to assess the likelihood of explosive and flammable atmosphere existing, in accordance with the definitions of Zone 0, Zone 1 and Zone 2. Once the likely frequency and duration of release (and hence the grade of release), the release rate, concentration, velocity, ventilation and other factors which affect the type and/or extent of the zone have been determined, there is then a firm basis on which to determine the likely presence of an explosive gas atmosphere in the surrounding areas. This approach therefore requires detailed consideration to be given to each item of the process equipment which contains a flammable material, and which could therefore be a source of release.

In particular, Zone 0 or Zone 1 areas should be minimized in number and extent by design or suitable operating procedures. In other words, plants and installations should be mainly Zone 2 or non-hazardous. Where release of flammable material is unavoidable, process equipment items should be limited to those which give secondary grade releases or, failing this (that is where primary or continuous grade releases are unavoidable), the releases should be of very limited quantity and rate. In carrying out area classification, these principles should receive prime consideration. Where necessary, the design, operation and location of process equipment should ensure that, even when it is operating abnormally, the amount of flammable material released into the atmosphere is minimized, so as to reduce the extent of the hazardous area.

Plants in which flammable dusts are present, as well as flammable gases, vapors or liquids, shall be classified separately for both types of risks (Figure 3.3).

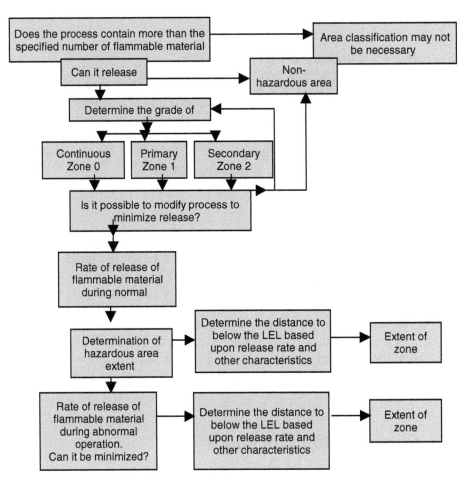

Figure 3.3
Classification process ... defined

3.11 Responsibility and personnel involved

Who should be associated with/during this classification procedure?

The responsibility for the area classification of a plant rests jointly with the engineering, process, safety and other departments responsible for the design and operation of the plant concerned. Representatives of all the interested parties, including the electrical engineer, must agree on the classification and record their findings.

The usual approach is for the senior management to appoint an area classification team, which is responsible to the senior management. It usually comprises expertise from the disciplines of mechanical, electrical, civil, instrumentation, process control, chemical, scientific and production engineering, which will need to agree on the risks. In addition, generally an attempt is made to involve plant operating personnel and site execution personnel at the stage of designs itself.

In the case of operating plants, responsibility for area classification rests with the works manager.

Once a plant has been classified and all necessary records made, it is important that no modification to equipment or operating procedures is made without discussion with those responsible for the area classification. Unauthorized action may invalidate the area classification. It is necessary to ensure that all equipment affecting the area classification, which has been subjected to maintenance, is carefully checked during and after re-assembly to ensure that the integrity of the original design, as it affects safety, has been maintained before it is returned to service.

3.12 Documentation

The task will involve the completion of location drawings, which will show how the areas are classified. This should correspond to markings and area designations on the plant. Documentation should record the philosophy adopted by the team and the details of how they have approached each possible hazard. Such documentation forms part of the 'plant safety documentation'.

It is recommended that area classification is undertaken in such a way that the various steps, which lead to the final area classification, are properly documented.

The results of the area classification study and any subsequent alterations to it shall be placed on record.

Those properties which are relevant to area classification of all process materials used in the plant should be listed and should include flashpoint, boiling point, ignition temperature, vapor pressure, vapor density, explosive limits, gas group and temperature class.

As indicated above, the information has to be comprehensive and necessarily include the following in addition to any other information felt necessary by user,

- Appropriate area classification drawings giving location details of electrical installation, gas and vapor dispersion characteristics and calculations.
- Recommendations from relevant codes and standards. Certification and approval documents for approved apparatus and equipment, giving their sub-grouping, surface temperature class, any special requirements, etc.
- A study of ventilation characteristics in relation to flammable material release parameters so that the effectiveness of the ventilation can be evaluated. Any specific purging requirement to be placed on record.
- Sub-grouping and surface temperature information for all non-certified apparatus installed. Its use and any special installation instructions.
- Details of special cable connections if any. This should also include cabling checks requirement for multi-core cable in intrinsically safe installation.
- Any safe shutdown procedures/features for apparatus and its control gear.

3.12.1 Drawings, data sheets and tables

Area classification documents should include plans and elevations, as appropriate, which show both, i.e. type and extent of zones, ignition temperature and hence temperature class and gas group.

Given in Figure 3.4 is a single line diagram of general requirement of 'hazardous area verification documentation'.

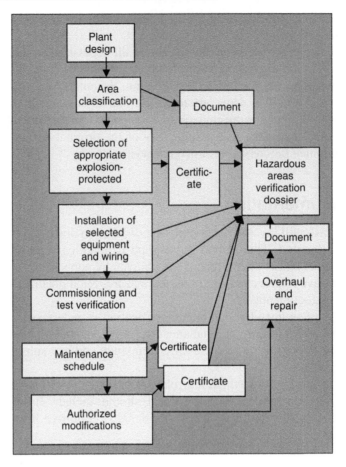

Figure 3.4
Major documentation requirement for hazardous area

3.13   Policy and guidelines for implementation

The plant owner must be able to demonstrate that a safe working environment has been provided. He can only do this if all the possible risks are considered and appropriate controls are in place to minimize the risks to acceptable levels. Knowing how likely vapors may be present, what and where they may be, will enable him to choose equipment that has been assessed for that duty. The design, operation and other factors of an individual plant will force the plant management to generate working practices and procedures. These must also take into account the hazards on the plant, in order for the procedures to minimize risk.

The policy on area classification, including modifications consequent upon changed conditions, shall be embodied in a formal system of which the principal features shall be:

(i) Records of area classification, updated when appropriate, shall be kept and shall be available for use by inspection and maintenance personnel. They shall preferably be in the form of drawings showing the area classification, the apparatus-group and temperature class, and in the case of dust risks, the properties of the dust.

(ii) A system shall be instituted for the control and recording of all plant/process changes or modifications. A requirement of the system shall be the assessment of the effect of any such changes on area classification and on the apparatus group and temperature class of apparatus in use; where dust risks are involved, the effect of any changes in the properties of the dusts shall be assessed. The system shall ensure that the electrical apparatus is either still suitable for the new conditions or, if necessary, changed to make it so.

(iii) Also, where a 'control of modification procedure' is already established in a works or plant, any reviews called for under that procedure should include area classification and the dependencies noted above.

(iv) The decisions made must be documented and published so that all interested parties can select, install and maintain equipment in accordance with this statement.

(v) A review of the area classification shall be carried out at intervals, which should not exceed, say 2 years.

(vi) New chemicals or processes introduced into the plant may change these conditions. Periodic review therefore will be necessary to ensure that changes have been accommodated and communicated to those affected.

The review may be based on checking that,

- The effect of all plant/process changes or modification on area classification (see (ii) above) have been accounted for, and
- The operation and integrity of all process equipment is compatible with the area classification.

The date and outcome of any review shall be recorded.

When the area classification has been changed, account shall be taken of the effect on the apparatus group, the temperature class and the methods of safeguarding of installed apparatus.

Under certain shutdown or other special conditions, there may be no risk of a flammable atmosphere occurring for a temporary period. If this can be guaranteed by the authority responsible (and a certificate issued to this effect), the temporary use of ordinary types of electrical apparatus is permissible. Such apparatus and all temporary connections thereto must be withdrawn from the area before the expiry of the certificate, and precautions must be taken to ensure that they cannot inadvertently be used when no certificate is in force. Conversely there may be occasions when the risk of a flammable atmosphere occurring is temporarily increased so as to require special precautions such as, for example, temporary mechanical ventilation or, in the extreme, the disconnection of electrical supplies.

Those responsible for the area classification must ensure that the locations of fixed ignition sources (e.g. furnaces, flames and high-temperature surfaces), and the areas to which sporadic sources of ignition (e.g. vehicles) will have access, are not at variance with the agreed area classification. Where conflict exists, it shall be resolved by consultation with those concerned with plant layout and design.

Some typical points to be considered and their implications are listed as a guide in the following table:

Subject	Consideration
Sources of release	Any joint in pipe-work must be a potential source of release: minimizes number of joints. Continuously welded pipe preferred. Piping to instruments where there are drain cocks: How often are drain cocks used?
Process pressures	The distance from the source of release that vapor will travel if released. This governs the extent of the zoning. Higher plant pressure will have larger zone areas

(Continued)

Subject	Consideration
Vapor density	If low, then vapor will rise, making roof spaces a potentially hazardous area If high, then vapor will fall and may be trapped in pipe trenches
Upper and lower flammability limits	What is the likelihood of dilution of release? Would installing fans reduce below the LFL? Dilution more likely than concentration
Prevailing winds	Use this to ease leaked vapor dispersion
Plant layout	Wider spacing of vessels may be preferred. How high should bund walls be?
Plant operating temperature	What are flashpoint and ignition temperatures of flammable materials used?
Environmental conditions	Ambient temperature may affect T rating of apparatus
Grouping of vapors	Suitability of Ex protection method

3.13.1 Guidelines with respect to sources of release of flammable gas and vapor in process plants

Leakage from pipe or equipment joints

Another source of leakages which should be closely examined is

- Pipe joints
- Equipment joints.

The quantity of leakage of gas and vapor from the joints depends on

- The type of joint face and
- Gasket used.

The gaskets can be

- A compressed-asbestos fiber (CAF) gasket. In this case the blow-out of a section of gasket between adjacent bolts is always a possibility. It is reasonable to assume an orifice 25×1.6 mm for such a blow-out.
- A metal-clad or spirally supported gasket with backing ring, or a joint is used with a trapped gasket or ring, a blow-out is virtually impossible. The orifice through which a leak can occur is reduced to something like 0.05 mm over a length of 50 mm (i.e. to about one-tenth of the cross-sectional area possible with a simple joint and gasket).

In many situations the hazardous area arising from a joint leak from the use of metal-clad gasket of this size is insignificant in comparison with other likely leakages.

This leads us to examine what type of joints needs to be looked at. The answer is, the joints which are subjected to sharp changes in temperature are the most likely ones to leak, and this cause of leakage is of particular significance on hydrogenation units. It may, however, be countered by the use of trapped or solid metal ring gaskets with, in severe cases, extended bolts and sleeves. In general, with suitably designed equipment, joints will leak only under *abnormal* conditions and hence only *Zone 2* classification is required.

The liquid or gas or vapor leaking from a joint will be in the form of a jet and full area of influence need to be classified based on the exit velocity, pressure and time taken to shut-off the source. All possible directions of travel need to be classified. The area is a narrow strip fanning out on either sides of a joint located in a vertical plane, and extending all around a joint located in a horizontal plane. This applies irrespective of the height of the joint above the ground, but will be accentuated in the case of a joint in an elevated position. At ground level the area may be further extended by spread of the liquid over the ground. The horizontal distance traveled by the liquid can be found by calculation from the velocity at which it escapes (obtained from the standard equation for flow of an incompressible fluid through a flooded orifice) and from the height of the source of escape above the ground, adjustment then being made to the calculated figure to make allowance for air resistance and the break up of the jet as it falls.

The rate of escape and vapor-generation from a leak of flashing liquid or *liquefied flammable gas*, and also the horizontal distance required for dispersal to the *lower flammable limit* of the vapor when it is released at or near ground level, have been calculated on the bases of flashing-flow and turbulent diffusion in the surrounding air (assuming a wind speed of 8 km/h), respectively. As in the case of a leak of these materials from a pump gland or seal, the vertical extent of the *Zone 2* is 3 m above a joint situated at or near ground level, falling to 3 m above ground level at the horizontal limit of the zone.

For a leak of flashing liquid from a joint in an elevated position the horizontal extent of the *Zone 2* is only 70% of that for the same leak at or near ground level, since the vapor diffuses more quickly at a height above the ground than at ground level.

An escape of gas through a joint leak will be at sonic velocity and the rate of escape can be calculated as already described for leakage from compressors.

3.13.2   Sample-point discharge to atmosphere

While sampling the liquid in open the following need to be taken care of:

- The temperature is to be low enough to ensure no loss of 'light ends' from the sample.
- A tundish should be provided below the sample-point to collect purgings from the sample line for delivery into a sealed drain or a closed container.

This will avoid spillage and negligible escape of vapor. Only under *abnormal* circumstances (e.g. insufficient cooling of the sample) will there be a significant *flammable atmosphere*; hence only *Zone 2* classification is required. The extent of the *Zone 2* is estimated as 1.5 m all around the sample-point.

In the case of *liquefied flammable gases* sampling should always be, into a closed system with provision for venting to a safe place.

Thus a *flammable atmosphere* will only occur around the sample-point, that too under *abnormal* circumstances (e.g. leakage of the sampling equipment or inadvertent opening of a sample-point before the equipment is connected). In the worst case the sample-point will be left fully open discharging to atmosphere and, to limit the quantity which can escape in this way, it is recommended that all *liquefied flammable gas* sample-points be limited to a 6 mm bore.

The rate of escape has been calculated on the basis of flashing-flow and the extent of the resultant *Zone 2* on the basis of turbulent diffusion of the vapor cloud produced in the surrounding air, assuming a wind speed of 8 km/h.

3.13.3   Drain-point discharge to atmosphere

These are some of the most neglected areas of the plant but frequented by operators and should be closely examined. The classification of the area round a drain-point discharge to atmosphere depends upon,

- Frequency of usage. Like, during regular operation, once per shift or so, on or only during start-up/shutdown or while preparing plant for maintenance work
- The nature of the material drained
- Whether drain-point discharges downward or otherwise
- Whether drained liquid is at near ambient temperature or not. If the hydrocarbon is a liquid under ambient conditions the amount of vapor generated from an escape through the drain-point is small, and in *normal* operation there is no significant *flammable atmosphere* round the drain-point
- If the hydrocarbon is a *liquefied flammable gas*, however, vapor will be generated in the first place by adiabatic flash-off on reduction of pressure to atmospheric, and then by evaporation of the remaining cold liquid as it picks up heat from the surroundings.

Normally, in a process, the drain operator draws off water from the bottom layer. During this the following can happen. As soon as the operator sees the hydrocarbon being drained he shuts the valve. Thus during each operation some flammable material escapes to atmosphere and hence the area needs to be classified as Zone 1. If the operator fails to shut the drain valve at the right time hydrocarbon may be discharged continuously, and the *hazardous area* round the drain-point thereby increased, but this should occur only under *abnormal* circumstances in which case the additional *hazardous area* should be classified *Zone 2*.

Because of the large clouds of vapor, which can be generated from escapes of flashing liquids, it is recommended that all drain-points through which such liquids can escape should be limited to 19 mm bore and the extents of the zones be calculated on this basis.

A drain-point which is used only infrequently (e.g. once a year) for the purpose of preparing equipment for maintenance work will give rise only to a *Zone 2*, the extent of which should be assessed on the basis of the nature of the material and the amount which is likely to be left inadvertently in the plant and which will escape when the drain is opened.

3.14   Area classification examples

Some examples of area classification problems are now considered. These attempt to illustrate the typical situations for which common practice is applied. To provide a basis for assessing the extent of classified areas, studies are commissioned by the owners based on the process parameters and the ways in which release of flammable gases, vapors and liquids can be expected to occur during month-in, month-out running of equipment handling such materials.

The organizations then develop guidelines based on the sample calculations discussed in this chapter for methods to calculate the size of gas or vapor clouds so formed and the distances necessary for their dispersal and dilution with air, in well-ventilated situations, to below their lower flammable limit. The object would be to establish a physical model of the escape from each source of hazard and to use it to derive figures for the extents of

Zones 1 and 2, either from practical observations and/or calculations or, in the limit, on the basis of personal judgment.

Described below are some such case studies to illustrate typical installations and area classification principles based on various standards (like IEC 60079).

3.14.1   Tank area classification

In Figure 3.5, a fixed roof tank with vents is filled and emptied. If the vapor given off were heavier than air, then vapor would remain present above the surface of the liquid, if the flashpoint were lower than the ambient temperature of the tank and surrounding environment.

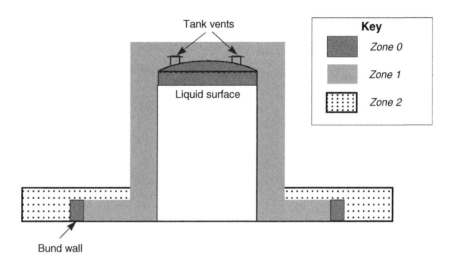

Figure 3.5

When the tank is emptying, air is sucked in the vents and mixes with the vapor to form a flammable mixture. Inside of the tank is normally classified as Zone 0 because the air/gas mixture may be present for long periods. Note that the zone is not concerned with concentration levels at this stage.

When the tank is filling the vapor is pushed out of the vents and falls down the side of the tank. Turbulence in the tank during filling may allow more vapor to be liberated. For a distance around the tank, vapor will appear, albeit diluted by the external air. At the base of the tank, vapor will collect up to the height of the bund wall around the tank, which helps to contain liquid spillage but tends to hamper the dispersal of any vapor. These areas are typically designated Zone 1 because the gas/air mixture is likely to occur in normal operation.

For a further distance around the tank, at the base level, and for a distance above the ground, vapor is only likely to appear if there is a major spillage, which is considered abnormal and so is designated Zone 2. The prevailing wind will influence the size of the surrounding Zone 2.

It is always necessary to consider realistic operational conditions and not just static conditions. If the pumping rate into the tank was increased this may result in more vapor being pushed out and the Zone 1 area may be increased. Prevailing winds and the proximity of other tanks would affect the dispersal or concentration levels, and a judgment needs to be made on this issue.

Every area in a plant must be given a recognizable category of risk that is to be observed during all operations concerning electrical apparatus. Each area would have a zone, apparatus group and a T rating assigned to it. A typical drawing of a tank farm with a pipe trench would look similar to that shown in Figure 3.6.

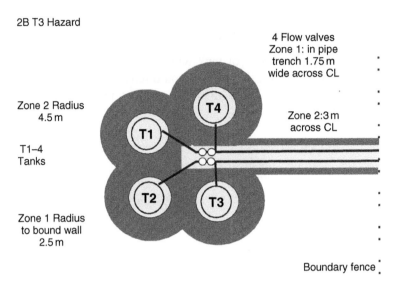

Figure 3.6
Tank farm area classification

The points to observe are the scale or dimensions of the drawing and the way that the areas become additive and are classified to take account of the installation. The pipe trench also has flow instruments installed on the pipe and valves to take the product off to the tank. This poses an increased risk. The vicinity is therefore designated Zone 1 (Figure 3.7).

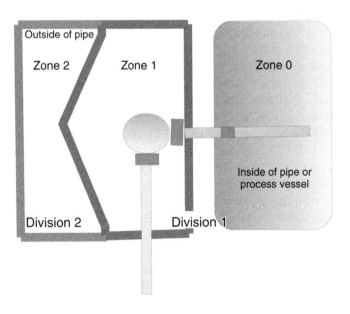

Figure 3.7
Thermocouple addition changes area classification

In the above example, a continuously welded pipe has a thermocouple installed. Whereas before no external gas hazard need have been considered, the mechanism of the thread is considered a source of potential weakness. It may result in the classification of the immediate area as a Zone 1 and an extended Zone 2. The distances that zones extend are a function of the pressure in the pipe, the considered likelihood of failure of the thread and the assembly's susceptibility to external damage, etc.

3.14.2 Leakage from pump glands and seals

The centrifugal/reciprocating pumps have gland packing, mechanical seal or stuffing box which are wetted with the liquid handled, and as a result there is some escape of liquid or vapor to the atmosphere. In mechanical seals the leakage and maintenance is much less and hence should be preferred choice, particularly for pumps handling *hot flammable liquids* or *liquefied flammable gases*.

When a flammable liquid below its atmospheric pressure boiling point is being handled and the glands or mechanical seals are well maintained then the liquid or the vapor which escapes is so small in quantity that it has no significant effect on the surrounding atmosphere. Under these *normal* conditions therefore the extent of the *hazardous area* around either the packed gland or the mechanical seal is too limited to be of practical significance for area classification. It is concluded therefore that there is no significant *Zone 1* around either a well-designed and maintained packed gland or a mechanical seal of a pump handling a *flammable liquid* below its atmospheric pressure boiling point.

With a pump handling either a flammable liquid at or above its atmospheric pressure boiling point or a liquefied flammable gas it is assumed that a mechanical seal will be used for the reasons already stated. When handling these materials the removal of frictional heat generated at the rubbing faces (an essential condition for the successful operation of any mechanical seal) is more difficult to achieve than with less-volatile liquids or with liquids at or near ambient temperature. There is therefore an increased likelihood of a seal leak existing with these materials. Further, with a flammable liquid handled at an elevated temperature the extent of vaporization of the liquid which escapes from the seal is greater than with a liquid at a lower temperature, and when the temperature exceeds the boiling point of the liquid at atmospheric pressure a considerable part of the escape flashes-off as vapor. Similarly, extensive flash-off of the escaping liquid also occurs with liquefied flammable gas. In these cases, therefore, a local hazardous area may exist in normal operation in the immediate vicinity of the seal. It is concluded therefore that it is prudent to have a local Zone 1 (in most cases a sphere of 0.3 m radius) around a well-designed and maintained mechanical seal of a pump handling either a flammable liquid at or above its atmospheric pressure boiling point or a liquefied flammable gas.

Deterioration of the gland packing or failure of the seal in service gives rise to increased leakage. The extent of vaporization of *flammable liquid* below its atmospheric pressure boiling point is still very small but the liquid is likely to splash over the pump assembly and surrounding ground. This causes a *flammable atmosphere* in the vicinity of the area so-wetted. The extent of the *hazardous area* is estimated to be, as a maximum, 3 m beyond the edge of the wetted area. This area is *Zone 2* since the gland deterioration or seal failure is an *abnormal* occurrence. In the case, either of *flammable liquids* handled at temperatures above their atmospheric pressure boiling point or of *liquefied flammable gases* the increased leakage due to seal failure necessitates an increase in the extent of the area classified by the provision of a *Zone 2* outside the areas already classified *Zone 1*.

Figures 3.8 and 3.9 give an illustrative example of how the area housing pumps are to be classified.

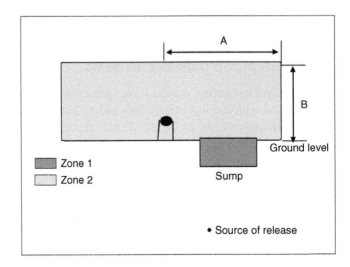

Figure 3.8
Pump – indoor installation

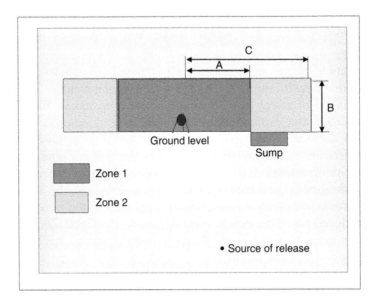

Figure 3.9

Pump mounted – outdoor installation

A normal industrial pump mounted at ground level, situated outdoors, pumping flammable liquid.

Principal Factors which Influence the Type and Extent of Zones		
Plant and process		
Ventilation		
Type	Natural	Artificial
Degree	Medium	High
Availability	Poor	Fair
Source of release	*Grade of release*	
Pump seal	Primary and secondary	
Product		
Flashpoint	Below process and ambient temperature	
Vapor density	Greater than air	

Taking into account relevant parameters, the following are typical values, which will be obtained for a pump having a capacity of 50 m³/h and operating at a low pressure:

A = 3 m horizontally from source of release,
B = 1 m from ground level and up to 1 m above the source of release.

A normal industrial pump mounted at ground level, situated indoors, pumping flammable liquid.

Principal Factors which Influence the Type and Extent of Zones	
Plant and process	
Ventilation	
Type	Artificial
Degree	Medium
Availability	Fair
Source of release	*Grade of release*
Pump seal (packed gland) and pool at floor level	Primary and secondary
Product	
Flashpoint	Below process and ambient temperature
Vapor density	Greater than air

Taking into account relevant parameters, the following are typical values, which will be obtained for a pump having a capacity of 50 m³/h and operating at a low pressure:

A = 1.5 m horizontally from source of release,
B = 1 m from ground level and up to 1 m above the source of release,
C = 3 m horizontally from source of release.

3.14.3   Relief valve and blow-off discharge to atmosphere

The discharges from relief or safety valves are to be carried out in such away so as to,

- Dilute it to below its *lower flammable limit* in air before it reaches a source of ignition
- Limit its threshold value of toxicity or tolerable smell before it reaches ground level or a working platform above ground level.

The first of these two conditions is particularly relevant to area classification, although the second may also have some bearing on the positioning of electrical equipment from the point of view of maintenance work.

Gases lighter than air can be discharged from a point above plant buildings and structures and be allowed to disperse by diffusion in the atmosphere, except that with gases of very high hydrogen content it is preferable to discharge them at high velocity (to overcome back-diffusion of air into the vent stack); the gases are then diluted with air entrained by jet-mixing.

A discharge from a relief valve, or an emergency blow-off of a gas or vapor heavier than air, may only be put direct to atmosphere at a velocity which is sufficient to ensure dilution to the *lower flammable limit* with air entrained by jet-mixing. If this condition cannot be met the discharge should be put to a flare system.

In general a relief valve will discharge only under *abnormal* conditions and the classification around the point of discharge to atmosphere is *Zone 2*, although to allow for small leakages past the valve *Zone 1* is proposed for 1 m around the discharge pipe tip. If, however, because of a special process feature a particular relief valve is expected to discharge frequently to the atmosphere, the whole classification may become *Zone 1*. The extent of the classified area around the point of discharge to the atmosphere from a relief valve or an emergency blow-off is, in general, the area required for dilution to the *lower flammable limit* by air entrained by jet-mixing (Figure 3.10).

Pressure breathing valve in the open air, from process vessel:

Principal Factors which Influence the Type and Extent of Zones	
Plant and process	
Ventilation	
Type	Natural
Degree	Medium
Availability	Fair
Source of release	*Grade of release*
Outlet from valve	Primary
Product	
Gasoline gas density	Greater than air

Taking into account relevant parameters, the following are typical values, which will be obtained for a valve where the opening pressure of the valve is approx. 1.5 bar.

A = 3 m all directions from source of release,
B = 5 m in all directions from source of release.

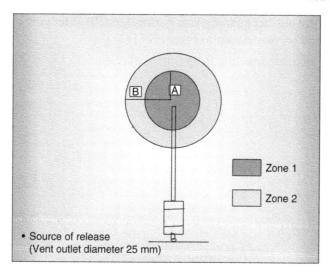

Figure 3.10

3.14.4 Leakage from valve glands

The valve glands are another source of leakage of flammable liquids, gases and vapor. In a plant such leakages or drips are generally guided toward a drain point from where they are led off to a sealed drainage system.

Hence the *flammable atmosphere* at or near ambient temperature exists only close to the liquid surface, and the classified area is below the leaking valve and in the immediate vicinity of the wetted area of paving. Whether the classification is *Zone 1* or *2* depends upon the frequency of the leakage and the number of valves installed in close proximity to each other.

It is considered unlikely that leakage of gas from a valve gland will be large enough to justify more than a nominal classified area – say a sphere of 1 m radius. Again the frequency of leakage and number of valves in close proximity determine whether the classification is *Zone 1* or *2*.

Control valve installed in a closed process pipe-work system conveying flammable gas:

Principal Factors which Influence the Type and Extent of Zones	
Plant and process	
Ventilation	
Type	Natural
Degree	Medium
Availability	Fair
Source of release	*Grade of release*
Valve shaft seal	Primary
Product	
Gas	Propane
Gas density	Greater than air

Taking into account relevant parameters, the following are typical values, which will be obtained for this example (Figure 3.11).

A = 1 m in all direction from source of release.

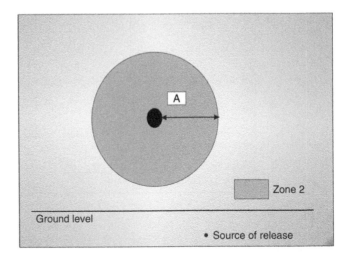

Figure 3.11

3.14.5 Process-mixing vessel – area classification

In a fixed process-mixing vessel situated indoors, being opened regularly for operational reasons, the liquids are piped into and out of the vessel through all-welded pipe-work flanged at the vessel:

Principal Factors which Influence the Type and Extent of Zones	
Plant and process	
Ventilation	
Type	Artificial
Degree	Low inside the vessel. Medium outside the vessel
Availability	Fair
Source of release	*Grade of release*
Liquid surface within the vessel	Continuous
The opening in the vessel	Primary
Spillage or leakage of liquid close to vessel	Secondary
Product	
Flashpoint	Below process and ambient temperature
Vapor density	Greater than air

Taking into account relevant parameters, the following are typical values, which will be obtained for this example (Figure 3.12):

A = 1 m horizontally from source of release,
B = 1 m above source of release,

C = 1 m horizontally,
D = 2 m horizontally,
E = 1 m above the ground.

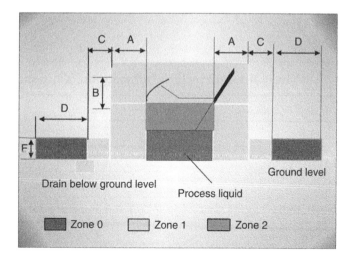

Figure 3.12

3.14.6   Vapor escape from an oil–water separator

Oil collected in the sealed drainage system of a plant area and then discharged to an oil/water separator generally covers the whole range of volatility of materials handled on the plant, but is not rich in light constituents since the oil 'weathers', at least to some extent, and the light constituents escape through the vents of the drainage system. The oil on the water is still flammable, however, and the separator area and the area immediately around it to a distance of 3 m and to a height of 1 m above ground level should be *Zone 1*.

In exceptional cases there can be a large spillage of light material, although this should only occur as the result of some *abnormal* occurrence on the plant. To cover this contingency a further area to a horizontal distance of 7.5 m around the *Zone 1* (including Zone 1 area) and to a total height of 3 m should be *Zone 2*.

It is considered prudent to have a *Zone 0* between the surface of the liquid and the ground level. The proposed extents of *Zones 0*, *1* and *2* are shown in Figure 3.13.

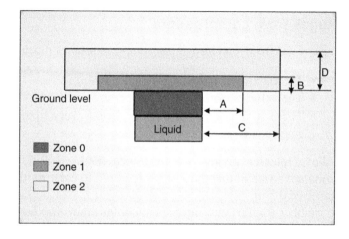

Figure 3.13

The extents quoted are related to a typical plant oil/water separator dealing with spillage of a variety of materials from a plant area. Smaller distances may apply to a small separator dealing with a single material of known composition.

Oil/water gravity separator, situated outdoors, open to the atmosphere, in a petroleum refinery:

Principal Factors which Influence the Type and Extent of Zones	
Plant and process	
Ventilation	
Type	Natural
Degree	Medium
Availability	Poor
Source of release	*Grade of release*
Liquid surface	Continuous
Process disturbance	Secondary
Product	
Flashpoint	Below process and ambient temperature
Vapor density	Greater than air

Taking into account relevant parameters, the following are typical values, which will be obtained for this example:

A = 3 m horizontally from the separator,
B = 1 m above ground level,
C = 7.5 m horizontally,
D = 3 m above ground level.

3.14.7   Leakages from compressors

Leakage of gas from a reciprocating compressor takes place during *normal* operation,

- Through the glands
- The valve-pocket cover-joints
- There is an escape for a short period every time the snubber vessel and inter-stage cooler drains are operated.

The above releases giving rise to a local *Zone 1*. But the quantities during normal operation are so small that the Zone 1 required to cater to them is only a nominal one, say a distance of 0.5 m.

The snubber (buffer) and inter-stage cooler drains are used to run-off accumulations of liquid, either into an open tundish device to the oil/water drain or via a knock-out pot vented to a safe place. With the open discharge the amount of gas escaping in *normal* operation is small, since the drain is closed as soon as all the liquid has run-off and gas appears, and a *Zone 1* of 0.5 m all around the drain discharge is considered to be adequate.

In the case of centrifugal compressors there is no leakage from the seal itself during *normal* operation but the escape routes of gas and vapor are,

- Some oil leaks inward into the machine and outward to the atmosphere. This gets contaminated with process gas
- Oil–vapor emission from breathers on the bearing housings
- Leakage from gas reference pressure piping joints.

To cover all such eventualities, generally, it is recommended that a Zone 1 of 0.5 m can be marked. Generally it is noticed that during normal operation the seal oil pots are to be opened to check for leakage rate, etc. A zone of 1 m may be marked around seal pots and open drains.

The gases which are lighter than air, like hydrogen, tend to rise on release but velocities may be too low to promote turbulent diffusion of the gas in the surrounding atmosphere. This retards the dilution process of the escaped gas or vapor. In compressor houses there are hot air currents generated due to heat dissipation from the machinery and these rise up toward the roof. These help to diffuse the released gas and vapor. Hence, the whole space under the compressor together with further spaces around the roof-ventilators and outside the line of the open walls of the compressor-house; the proposed extent of the Zone 2 is shown in Figure 3.14.

With gases heavier than air (e.g. cold ethylene and propylene handled in refrigeration compressors) a major escape tends to flow downward from the source of escape and then over the compressor-house floor, giving rise to a *Zone 2* the extent of which need to be determined based on,

- The pressure being handled – low or high
- The air flow velocity in the vicinity.

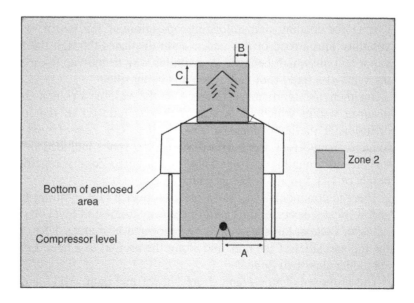

Figure 3.14

An illustrative classification is given for hydrogen compressor building which is open at ground level:

Principal Factors which Influence the Type and Extent of Zones	
Plant and process	
Ventilation	
Type	Natural
Degree	Medium
Availability	Good
Source of release	*Grade of release*
Compressor seals, valves and flanges	Secondary
Close to the compressor	
Product	
Gas	Hydrogen
Gas density	Lighter than air

Taking into account relevant parameters, the following are typical values, which will be obtained for this example:

A = 3 m horizontally from source of release,
B = 1 m horizontally from ventilation openings,
C = 1 m above ventilation openings.

3.14.8 Discharge to atmosphere during loading/off-loading operations

During the loading and unloading operation of the petrol or other liquids of similar volatility into a road or rail tank wagon displaces through the filling-hatch a mixture of vapor and air which behaves in a similar way to the vapor–air mixture expelled through the vents of a fixed-roof tank when it is being filled.

But there is significant difference in the volume of such displaced vapor. Thus the distance within which the concentration is reduced to the *lower flammable limit* by diffusion in the atmosphere is also smaller. The *Zone 1* next to the tank shell can be reduced from 3 to 1.5 m but in this case the space included in the vertical projection of this area to ground level should also be classified *Zone 1*, together with a further area to a height of 1–1.5 m above the ground.

There is commonly weather-protection over a tanker-filling bay in the form of a roof, and vertical sides extending part of the way to ground level. To allow for *abnormal* rates of vapor escape, the space under the roof and also areas 1.5 m outside the *Zone 1* round the top and sides of the tank and along the ground to a height of 1 m above ground level should be classified *Zone 2*.

The proposed *Zones 1* and *2* are shown in Figure 3.11, and apply to the loading or off-loading of liquids with *flashpoints* below 32 °C (90 °F) and liquids with higher *flashpoints* where the liquid temperature may rise above the *flashpoint*.

Loading and off-loading of *liquefied flammable gases* is done through closed pipe systems and the material left in flexes at the end of a movement is either vented to a safe place, displaced back to storage by the use of inert gas or contained by means of self-sealing couplings. In *normal* operation, therefore, escape of the gas to atmosphere in the vicinity of the loading operation is either non-existent or trivial in quantity, but in view of the volatility of the material it is recommended that there should be a nominal *Zone 1*, as shown in Figure 3.11 for a liquid tanker installation. The worst contingency during loading and off-loading operations is the failure of the liquid flex connection. On the basis that the whole content of the flex is discharged almost instantaneously, but that further escape of *liquefied flammable gas* from the loading line or the tank wagon is prevented by excess-flow devices, a *Zone 2* extending 1–1.5 m horizontally and to a height of 1 m above the ground is recommended. Where excess-flow devices cannot be guaranteed, greater distances may be necessary.

Tank car filling installation, situated outdoors, for gasoline, top filling:

Principal Factors which Influence the Type and Extent of Zones	
Plant and process	
Ventilation	
Type	Natural
Degree	Medium
Availability	Poor
Source of release	*Grade of release*
Opening in the tank roof	Primary
Spillage or leakage at ground level	Secondary
Product	
Flashpoint	Below process and ambient temperature
Vapor density	Greater than air

Taking into account relevant parameters, the following are typical values, which will be obtained for this example (Figure 3.15).

A = 1.5 m horizontally from source of release,
B = horizontally to island (gantry) boundary,
C = 1.5 m above source of release,
D = 1 m above ground level,
E = 4.5 m horizontally from drainage channel,
F = 1.5 m horizontally from Zone 1,
G = 1 m above Zone 1.

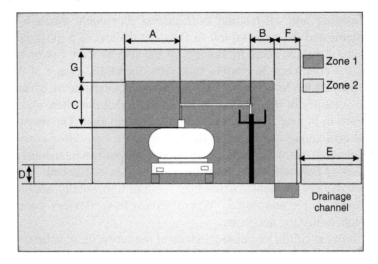

Figure 3.15

3.14.9 Hazardous area classification in a mixing room

This example shows one way of using the individual examples from above individual cases.

Taking into account relevant parameters, the following are typical values, which are obtained for this example:

A = 2 m
B = 4 m
C = 3 m
D = 1.5 m.

If the room is small, it is better that the whole room is classified as Zone 2 (Figure 3.16).

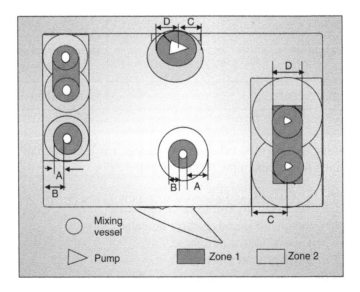

Figure 3.16
Mixing room in a factory

3.14.10   Cyclones and bag filters

Figure 3.17 shows situation where a cyclone is separating the granules from the dust of the same material in a process. The bag filter and cyclone separator are housed in separate rooms. In this case the product stream after the separation of dust can be classified as non-hazardous.

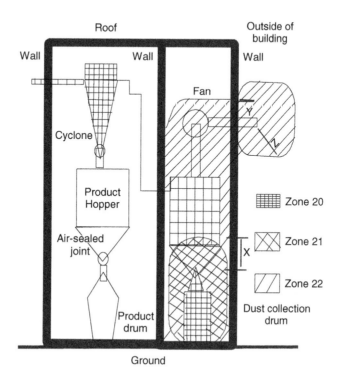

Figure 3.17

The Zone 20 areas are,

- Interior of the cyclone as the dust is present
- Interior of bag filter as dust is present.

As the system is operating under slight vacuum there would not be any leakage from pipe-work or flanges, thus they can be ignored.

The Zone 21 areas are,

- A distance of 1 m above and around the discharge nozzle as it is a primary source of release. This is possible as dust is discharged in normal operation in the bag.

The Zone 22 areas are,

- Dust layers can be formed (even after good house keeping) due to settlement of dust and hence this zone is likely to extend in whole of the room/building.

In Figure 3.17,

X = 1.5 m
Y = 1.5 m
Z = 2 m.

3.14.11   Bag emptying station – classification

Classification applicable to Figure 3.18 is as follows:

- *Zone 20*:   Inside the hopper, because an explosive dust/air mixture is present frequently or even continuously during normal operation, and layers of dust of uncontrollable and excessive thickness are formed or are present.
- *Zone 21*:   The open manhole is a primary grade source of release; it is likely that an explosive dust/air mixture occurs. Hence the classification around this manhole will be Zone 21 with a width of 1.5 m from its edge and down to the floor.
- *Zone 22*:   Because of outside effects, the Zone 22 area will be limited to whole area of room if it is not too large.

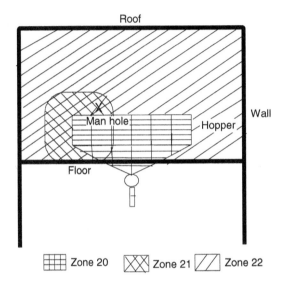

Figure 3.18

All distances are only for the sake of this explanation. In practice, other distances are possible.

Additional protection measures such as explosion venting or explosion isolation, weather-proofing, etc. may be necessary.

3.15   Summary of area classification

Area classification can be summarized as the multiplication of two levels of probability equating to an acceptable level of safety.

This also has implications of increased cost and greater difficulty in maintenance within the area. For example, it is impractical for personnel to have uncontrolled access into an area classified as Zone 0.

If the plant was globally declared Zone 2 then there may be safety implications because higher areas of risk may not be adequately catered for. Working procedures would also be more relaxed increasing the overall risk. Conversely, taking the approach that all areas are Zone 0 is impractical because of the increased cost of Ex-protected equipment and the operational difficulties related to Zone 0. A sensible balance must be agreed and implemented (Figure 3.19).

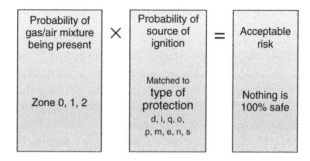

Figure 3.19
Acceptable risk

3.16   A case study

Let us have a bird's eye view of the ESSO accident mentioned at the beginning.

Facts

The explosion occurred when hot lean oil came into contact with equipment that had been operating well below normal temperatures. This caused the brittle metal to fracture and release massive amounts of hydrocarbon gas, which came into contact with an ignition source, causing a violent explosion. The equipment had reached such low temperatures because the supply of lean oil had ceased for over 3 h due to the failure of two pumps.

History

Gas Plant 1 was 30 years old; it was the oldest of the three units operating at Longford. Submissions made to the inquiry suggested that Exxon had to decide, based on profit considerations, whether to update the equipment, work it into the ground or abandon the Longford plant altogether.

Findings

The Commission stated that –

> *a HAZOP study would, without doubt, have revealed the factors that contributed to the fatal accident.*

The Commission quoted from ESSO's own documents showing that in 1994 ESSO deliberately limited a major aspect of its safety efforts, its periodic risk assessment (PRA), on the pretext that a HAZOP study was pending. The Commission concluded:

> *Accordingly, the 1994 PRA was directed away from process-related hazards and concentrated on hazards caused by mechanical equipment failure and operator error. Scenarios addressing the consequences of 'low temperatures', 'high level' and 'no flow' were not used.*

As part of restructuring its operations, ESSO transferred all on-site engineers to Melbourne. This affected the running of the plant, yet ESSO made no analysis.
The Commission found that:

> *The relocation of engineers qualified as a permanent change to operating practices requiring risk assessment and evaluation before implementation in conformity with*

ESSO's management of change philosophy. Yet such relocation was implemented without any such assessment ever taking place.

In addition, the Kennett State Liberal government limited the Commission's terms of reference in order to bury he government's own role in scrapping health and safety measures and replacing them by corporate 'self-regulation'. The Victorian Trades Hall Council, supported by ESSO and the Labour Party, asked the Commission to extend its inquiry but their application was rejected.

Nevertheless, the Commission's final report implies that the Kennett government must also bear responsibility for the disaster.

Had ESSO been required to submit a safety case (safety report), with respect to its facilities at Longford before 25 September 1998, it is likely that it would have identified the very hazards which were in evidence on that day, hazards which a proper HAZOP study of GP1 would also have identified.

We need to ponder on the questions –

Do you agree with Commission's conclusions based on the facts given?

State reasons for your agreements or disagreements.

Conclusion

In conclusion it may be stated that the question of possible risks of explosion must be addressed at the early stages of planning a new facility.

When classifying potentially explosive areas, the influence of natural or artificial ventilation must be considered in addition to the levels of flammable materials being released.

Furthermore, the classification figures relating to explosion technology must be determined for the flammable materials being used. Only then a decision can be reached on the division of potentially explosive areas into zones and the selection of suitable apparatus. IEC 60 079-14 (DIN 60 079-14) applies to the installation of electrical apparatus in potentially explosive areas Group II.

4

Design philosophy and selection of equipment/apparatus

4.1   General

It is worth mentioning that using electricity in potentially explosive areas has been a concern since the beginning of the twentieth century, when these areas were encountered in mines. To avoid the danger of explosions, protective regulations in the form of laws, specifications and standards have been developed in most countries and are aimed at ensuring that a high level of safety is observed. Due to growing international economic links, extensive progress has been made in harmonizing the regulations for explosion protection. The 9/94 EC Directive has created the conditions for a complete harmonization in the European Union (EU). However, worldwide there is still much to be done in this area. The aim of this course is to provide both experts and interested laymen with an overview in the field of explosion protection; in conjunction with electrical apparatus and installations, it does not replace the study of the relevant statutory regulations and applicable standards.

In mining, miners underground have always lived under the threat of firedamp explosions. Herein lies the origin of explosion protection, which has been consistently developed in industrialized countries and now provides a high level of safety.

In the preceding chapters, different facets of the classification of the area made hazardous by the possible presence of explosives atmospheres of air mixed with gas, vapor, mist or dust has been discussed. The knowledge acquired until now will make it possible:

- To classify the areas as hazardous and non-hazardous
- To determine the severity and its classification
- To the extent it will extend.

These hazardous areas can vary greatly in size, based on explosive fuel, or use of equipment. The explosive atmosphere may contain fumes, dust or fibers. Hence, it becomes imperative to look into the constructional details of electrical equipment to be installed in hazardous areas, in order to minimize the risk of fire or explosion. It is to be noted that despite how well the construction is, there always remains a residual risk, however small it may be. As a general principle, it must always be borne in mind that:

Any system, no matter how well it is engineered, is ultimately dependent upon human reliability to actually work.

Hence, as a cardinal principle:

Electrical apparatus, including electrically operated process instruments, should not be installed in a hazardous area when it is practical and economic to site it elsewhere. The yardstick that the benefit should be recognizable is to be strictly applied for locating any electrical equipment in hazardous area.

Electrical installations in hazardous areas involve high initial capital expenditure on methods of safeguarding and continuing high inspection and maintenance costs relative to comparable installations in non-hazardous areas. Electrical and instrument engineers have, therefore, a special responsibility, if need be by influencing basic plant layout and design, to ensure that the need for electrical apparatus in hazardous areas is kept to a minimum.

Where the installation of electrical apparatus in hazardous areas is unavoidable, special methods of safeguarding must be adopted to avoid danger. The following sections define the methods of safeguarding available and describe briefly such matters as certification, conditions of use and other special requirements pertaining to their application. In all cases it is essential to refer for full information to the appropriate standards and codes of practice, in particular to the various parts of SABS/IEC 60079 and SABS/IEC 61241 – series of standards for electrical apparatus in hazardous atmospheres.

4.2 Risks ... history

The use of electrical equipment in hazardous environment always leaves an element of residual 'risk', however small it may be. Minimizing the risk requires the correct selection of equipment, with known safe operating characteristics being matched to a given hazard, with known safe working limits.

Much research has been done in the past by private companies (such as the Imperial Chemical Industries in the UK), by electrical research establishments belonging to government organizations and insurance company associations. These organizations have all created documents, written with the benefit of experience in the use of electrical equipment in hazardous situations. The risks and hazards of an installation do become safety issues that require definition and communication to other personnel and organizations associated with the installation. Contractors and other installers must have basic information on the hazards of a plant to fulfill their responsibility of correctly observing precautions.

4.2.1 Gas, vapor and mist risks

In early times the protection measures were developed for the coal mining industry to mitigate the loss against gas clouds (fire damp) encountered in the underground mines, where generally men worked. There was a pre-supposition that the source of gas would be isolated once detected. It was rare that the gas cloud (firedamp) engulfed men before they were evacuated from the mine and the equipment switched off. But, still keeping in view a remote chance of that happening, a very high safety factor or high level of protection was adopted, as the probability of escape or switching off electrical equipment was relatively low, once the incident occurred.

These levels of risks have been associated with Zone 1 of the surface industry (above ground) and these norms applied for electrical equipment. Hence, these are the oldest norms available. As a result these standards have been developed extensively.

To begin with, the surface industry did not consider the installation of equipment in Zone 0, as no need was felt to install any electrical equipment in this zone. It was much later with the rapid advancement in instrumentation technology and the advent of a high level of automation in the process industry that a demand has been made to place sensors, probes and electrical equipment in this type of zone. These are basically placed inside process vessels to monitor and track process parameters so as to have high level of accuracy in automation to get high level of performance efficiencies. Thus, the standards too have been developed for sensors for Zone 0 applications.

Zone 2 is a surface industry phenomenon. Initially it was presumed that the high-quality industrial equipment, which did not spark or get hot in normal working conditions, was good enough for Zone 2. However, the latest guidelines belie this thinking and much work is underway internationally for developing these standards.

Thus, we have seen that standards for Zone 1 equipment are much more developed than those for Zone 0 or Zone 2. Much is being done to correct this anomaly.

4.2.2 Dust risks

Historically no requirement was felt to place equipment in the area when dust was present. About 70% of dusts involved in industrial processes are flammable but these are not covered by IEC 60079, which was for some time the mainstay for hazardous area protection. The new IEC 61241 series standards now address the requirements for dust explosive atmospheres. Dusts are recognized as a hazard but the properties of dusts change the way that they are thought of in classification terms.

Dust explosions are said to be more frequent than for gases and are generally reported in industries such as,

- Coal mining
- Grain handling and processing
- Metal powder manufacturing.

Dust explosions occur when a sufficient amount of finely divided combustible material is suspended in an atmosphere that supports combustion in the presence of an ignition source of appropriate energy. Dust explosion can occur even while handling a seemingly harmless material. Almost 70% of the industrial dust (fine grain size material) being present in various process industries can be regarded as presenting a dust explosion risk under certain circumstances. Some examples of type of dust or powders encountered in industry are:

- Coal dust
- Grain dust
- Malt fines
- Wheat flour
- Sawdust
- Wood fines
- Milk powder
- Some variety of spices
- Coffee and tea fines
- Sugar

- Powdered soup
- Some organic pharmaceutical powders
- Sulfur dust
- Polyethylene powder
- Polypropylene powder
- ABS powder
- Aluminum powder
- Manganese powder.

Properties of dusts differ from gases in that energy levels required to cause ignition are higher, of the order of millijoules rather than microjoules and the ignition temperature is generally lower. Concentrations are not directly comparable with those of vapors because the behavior of dusts tends to settle rather than to disperse to low- or high-collecting points.

The damage generally is more self-contained and for this reason, the hazard is not so rigorously protected. Higher IP-rated enclosures are sometimes used as the only precaution.

Dusts differ from gases in that they settle and accumulate over time. As stated earlier it takes much more spark energy to ignite a dust cloud than for a gas so the most likely cause of ignition is build up of heat. This means that heat may be prevented from being dissipated and the dust is assisting in creating its own potential source of ignition. The heat required to ignite dust is generally lower than for gases. Once ignited, the shock wave produced tends to dislodge other dust accumulations and so the explosion burning time is lengthened.

Some of the processes where powders as described above are handled always have inherent explosion risks. These are further compounded if the material is handled pneumatically in ducts when compared to its handling by belt conveyor or screw conveyor systems. Generally most risky zones are,

- Filter units
- Cyclones
- Air handling units
- Milling units
- Driers
- Sizing and classification equipment
- Conveying systems like bucket elevators, duct conveyors, etc.

The standardization work for equipment in dust atmosphere is of recent origin and is continuously being updated.

4.3   Classification concepts

Explosion protection is safety technology at its highest level. A series of pioneering developments and innovative designs have enabled various companies to exploit new and more economical and safer possibilities and use fields of application to maximize the gains. Broad specialist knowledge and well-founded understanding of explosion protection are of benefit to all. The use of high-quality moulded plastics for explosion-protected switchgear in particular contributed to this development.

The classification system specifies the following concepts:

- Area classification
- Apparatus grouping
- Temperature classification
- Methods of protection.

Having understood the '*Area Classification System*' concept we will discuss and try to understand the other three concepts.

The different methods of protection available must also be considered at this stage. The suitability of certain methods is only acceptable for some levels of risk owing to the considered integrity of the method of protection against the severity of the risk.

The concept of classification has been developed in order to be clear about the hazards involved and the precautions necessary.

4.4   Equipment . . . a definition

The term 'equipment' is defined as:

> *any item which contains or constitutes a potential ignition source and which requires special measures to be incorporated in its design and/or its installation in order to prevent the ignition source from initiating an explosion in the surrounding atmosphere.*

Also included in the term 'equipment' are safety or control devices installed outside the hazardous area but having an explosion-protection function. 'Protective systems' are defined as items that prevent an explosion that has been initiated from spreading or causing damage. They include flame arresters, quenching systems, pressure-relief panels and fast-acting shut-off valves.

Equipment is classified according to:

- The maximum (spark) energy it can produce
- Its maximum surface temperature.

For equipment to be considered safe for use on a plant, the categories of classification on the equipment must be the same or 'better' than that required by the plant.

The statement of the categories to be encountered on a plant (by its owner or user) provides any potential supplier of equipment (or contractor) with a clear understanding of the limits that must not be exceeded if the equipment is to be operated safely. The suppliers may have their equipment tested for conformance with these limits and then gain official approval or certification for those categories met. In this way, the system helps to match plant and equipment together in a safe way.

The nature of the explosive atmosphere	This relates to the characteristics of the gas(es) or dusts likely to be encountered. In order not to ignite it, there is a maximum spark or flame energy level as well as a maximum surface temperature that must not be exceeded
The probability that an explosive mixture is present	This requires an assessment of where and when gas or dust could accumulate in sufficient concentrations and/or layers, respectively
The maximum energy that can be produced	This relates to the 'apparatus group' of the equipment. This must be matched to the nature of the explosive atmosphere that the equipment may encounter
The maximum surface temperature	This relates to the 'temperature class' of the equipment. This must be matched to the nature of the explosive atmosphere that the equipment may encounter

Flammable materials are categorized by their defined ignition characteristics. These categories are easy to compare with formally assessed or 'certified' apparatus to be used when in expected contact with the given flammable materials.

Matching the marking of equipment to the grouping system in which the plant is considered safe requires the explosive mixtures to be placed into categories that are easy to compare.

It would be uneconomical and sometimes not even possible to design all explosion-protected electrical apparatus in such way that it always meets the maximum safety requirements, regardless of the use in each case. For this reason, the equipment is classified into groups and temperature classes in accordance with the properties of the explosive atmosphere for which it is intended. First of all a differentiation is made between two groups of equipment:

1. *Group I*:   Electrical apparatus for mining.
2. *Group II*:   Electrical apparatus for all remaining potentially explosive atmospheres.

4.5   Apparatus grouping

The 'apparatus grouping' concept originally emerged during the development of the surface industries when a greater range of chemical processes began to use flammable materials.

The mining industry recognized and took precautions against the risk of flammability but only had one gas to contend with, methane. The methane purity varied and so tests were devised to ascertain which purity of the methane was the easiest to ignite. Thus was found the form known as firedamp owing to its relatively high hydrogen concentration. This became the gas used in the testing of electrical equipment to be used in mining. Equipment was only ever tested for the ignition of firedamp on the basis that if the equipment could not ignite this gas then it could not ignite any other form that was more difficult to ignite.

Many different gases or vapors could be in use in industries above ground (referred to as surface industries). Each gas or vapor has its own characteristics and originally equipment was tested for the specific gas it was likely to encounter when in service. It would therefore be impractical to individually type-test apparatus to determine in which gases it was safe or unsafe.

Consequently, a grouping system was developed. Electrical equipment for explosive gas atmosphere is now grouped according to the type of industry in which they are used:

Group I	Electrical equipment for underground industries (mining) susceptible to methane
Group II	Electrical equipment for all places with explosive gas atmosphere other than mining, i.e. surface industries

4.5.1   Group I

For mines where flammable gases other than methane may normally and naturally occur, the electrical equipment shall be constructed in accordance with the requirements for Group I, but shall be submitted to the tests prescribed for the appropriate Group II explosive mixtures and marked accordingly.

Group I gas considered in mining is methane:

Apparatus Group	Representative Gas	Energy Band
I	Methane	200 μJ

4.5.2   Group II sub-division

Electrical equipment of Group II is subdivided according to the nature of the explosive gas atmosphere for which it is intended as follows:

Prescribed sub-division   Within Group II, there are three sub-divisions. These are relating to the energy required to cause ignition. Typical gases are chosen as the 'representative gases' for each group and are the ones used in equipment testing. The energy band is expressed in microjoules.

Apparatus Group	Representative Gas	Energy Band (μJ)
IIA	Propane	>180
IIB	Ethylene	>60
IIC	Hydrogen	>20

This is best explained by considering the emission of energy from 'electrical apparatus'. The term 'apparatus' is used now, as it is the recognized and defined name used in the standards to describe electrical equipment that has undergone testing. If the energy emitted from apparatus is tested and proven to be less than a particular group value then it is acceptable for use in that group.

It is also acceptable for use in any other group with a higher value. Thus, apparatus safe for hydrogen is usable in any gas because hydrogen is the most easily ignitable gas. However, apparatus suitable for ethylene (IIB) is not considered safe in hydrogen and is not permitted.

During the testing of electrical apparatus, energy emission levels are measured or assessed, and an apparatus group is assigned depending on the maximum level put out by the apparatus.

This group is also classified based on maximum experimental safe gap (MESG) and minimum igniting current as explained later in the chapter.

Temperature classification   For all types of protection, the temperature classes T_1–T_6 correspond to the classification of electrical equipment according to its maximum surface temperature.

Particular explosive atmosphere   The electrical equipment may be tested for a particular explosive atmosphere and marked accordingly.

Group IIA

Atmospheres containing acetone, ammonia, ethyl alcohol, gasoline, methane, propane, or flammable liquid-produced vapor, or combustible liquid-produced vapor mixed with air that may burn or explode, having either a maximum experiment safe gap (MESG) value greater than 0.90 mm or minimum igniting current ratio (MIC ratio) greater than 0.80.

Group IIB

Atmospheres containing acetaldehyde, ethylene, or flammable liquid-produced vapor, or combustible liquid-produced vapor mixed with air that may burn or explode, having either a maximum experiment safe gap (MESG) value greater than 0.50 mm and less than or equal to 0.90 mm, or minimum igniting current ratio (MIC ratio) greater than 0.45 and less than or equal to 0.80. This is best explained by considering the emission of energy from 'electrical apparatus'.

Group IIC

Atmospheres containing acetylene, hydrogen, carbon disulfide or flammable liquid-produced vapor or combustible liquid-produced vapor mixed with air that may burn or explode, having either a maximum experiment safe gap (MESG) value less than or equal to 0.50 mm or minimum igniting current ratio (MIC ratio) less than or equal to 0.45.

The grouping was originally referred to, in older standards, as gas grouping because the gases were initially tested to form the groups. Now it is known as apparatus grouping because of the clarity of its application. The tests are always performed using the most easily ignitable mixture of the representative gas with air.

4.6 Surface temperature classification

The ignition temperature, i.e. the temperature at which an ignition could occur, for example due to a hot surface of the apparatus, is dependent on the type of existing gases or vapors. This ignition temperature is influenced by several factors and is thus dependent on the stipulated testing order. Depending on the measuring system the results can thus differ in the various countries. Further information regarding materials, not stated can be found in the respective guidelines and literature.

The maximum temperature of the exposed surface of electrical apparatus must always be lower than the ignition temperature of the dust, gas or vapor mixture, where it is to be used.

4.6.1 Group I: Electrical equipment

- Where coal dust can form a layer … 150 °C.
- For internal surfaces, if the above risk is avoided, for k instance by sealing against the ingress of dust … 450 °C.

4.6.2 Group II: Electrical equipment

In order to be able to mark and select electrical apparatus simply in regard to its maximum surface temperature, there are several temperature classes. The gases can be classified into the temperature classes according to their ignition temperature, whereby the maximum surface temperature of the respective class must be lower than the ignition temperature of the corresponding gases.

The standards define temperature classification as:

Temperature class (T class or rating)	One of six values of temperature allocated to electrical apparatus derived from a system of classification according to the maximum surface temperature of the apparatus

The ignition temperature of gases and vapors are in no way related to the ease of ignition by energy. Ignition temperature has to be a completely separate consideration. In scientific circles, it is still not fully understood why this should be the case.

The temperature classification system requires that the maximum surface temperature of the apparatus is measured or assessed. The value must fall in between two temperature classes T ratings in the following list. The lower of the two is the rating given to the apparatus.

The following is the temperature classification as defined by standards,

T Class	Maximum Surface Temperature (°C)
T_1	450
T_2	300
T_3	200
T_4	135
T_5	100
T_6	85

It may be noted that above temperature classification is based on an ambient temperature of −20 to +40 °C.

Gases and vapors are assessed in a similar way. The ignition temperature of the most easily ignitable mixture is determined and compared to the list. The higher value of temperature class is chosen as that for the mixture. The classification of mixtures of gases or vapors with air according to their MESGs and minimum igniting currents, and ignition temperature of gases and vapors are listed in the SABS/IEC 60079-12 and SABS/IEC 60079-20 standards. This was originally taken from the codes of practice BS 5345 Part 1.

As an example, apparatus awarded T_4 may be used in the presence of T_4-rated gases. This means that the apparatus cannot reach a temperature above 135 °C. The T rating applied to the gas or vapor means that it is the highest it can safely encounter.

The values in the list are based on an ambient temperature of 40 °C. The T rating applied to the apparatus is in effect a temperature rise from ambient. T_6, for example, is a rise of 45 °C up to a limit value of 85 °C. In the cases where the ambient temperature is higher on a plant, special consideration must be given and the T rating may be adjusted.

Note that it is the maximum 'surface temperature' that is of interest. Components within the apparatus may get hotter than the T rating given to the apparatus. If the flammable vapor cannot come into contact with the component at an elevated temperature because of the construction, then it need not be considered. Only the surfaces of the apparatus, which come into contact with the vapor, are of concern.

In some cases, temperature ratings on the apparatus cannot be measured for assessment and can only be awarded as a result of experience or justified calculation. Small surface areas are permitted to be at a higher temperature than large ones and so some dispensation is given to small components in some circumstances. Precise and detailed guidance on this subject, specific to apparatus designers, is given in various relevant parts of the SABS and IEC 79 Standards, depending on the protection method employed.

Figure 4.1 gives a comparison of some common gases and vapors with a temperature scale on which the T-rating values are placed.

Note that the gas hydrogen, in the most sensitive apparatus group of IIC has an extremely high ignition temperature. This gas is used for cooling in some industrial applications because of its high thermal conductivity. This demonstrates how diverse the properties of flammable materials can be.

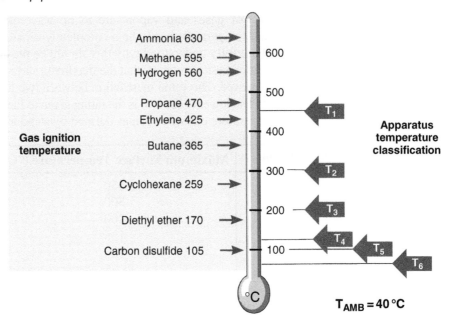

Figure 4.1
Temperature rating system

The following is an illustrative list of gases and vapors with respect to temperature classifications, and equipment or apparatus to be used in such atmospheres need to be marked suitably.

Temperature Class	Limiting Temperature (°C)	Gases and Vapors Against Which Protection is Afforded
T_1	450	Chlorobenzene, ammonia, carbon monoxide, coal gas, hydrogen, water gas
T_2	300	Acetone, ethane, ethyl chloride, xylene, benzene (pure), methane, methanol, naphthalene, propane, toluene, ethyl acetate, ethanol, ethylene, cyclohexanone, iso-amylacetate, 1,4-dioxan, *n*-butane, *n*-butyl alcohol, acetic acid, buta-1,3-diene, vinyl acetate, ethyl benzene
T_3	200	Acetylene, *n*-propyl alcohol, petrol, crude oil, turpentine, cyclohexane
T_4	135	*n*-Tetradecane, tetrahydrofuran, trichlorosilane, ethyl glycol, *n*-hexane, *n*-heptane, *n*-nonane acetaldehyde, ethyl ether
T_5	100	No gas or vapor specified as yet
T_6	85	Carbon disulfide

4.6.3 The American temperature classification system

In the American system, the temperature classification is also defined with sub-classification of the T ratings for apparatus into closer divisions. The system places the onus on the user to ensure that this T rating is below the ignition temperature of the gas or vapor.

T Class	°C
T_1	450
T_2	300
T_2A	280
T_2B	260
T_2C	230
T_2D	215
T_3	200
T_3A	180
T_3B	165
T_3C	160
T_4	135
T_4A	120
T_5	100
T_6	85

4.6.4 Comparison of temperature classification based on national and international codes

The following table will give an idea how this classification is applied worldwide,

IEC/CENELEC Australia		Japan (RIIS-TR-79-1)		USA (NEC 1984)		Min. Ignition Temp. (°C)
Class	Maximum Surface Temp. (°C)	Class	Maximum Surface Temp. (°C)	Class	Maximum Surface Temp. (°C)	
T_1	450	G1	360	T_1	450	450
T_2	300	G2	240	T_2	300	300
				T_2A	280	280
				T_2B	260	260
				T_2C	230	230
				T_2D	215	215
T_3	200	G3	160	T_3	200	200
				T_3A	180	180
				T_3B	165	165
				T_3C	160	160
T_4	135	G4	110	T_4	135	135
				T_4A	120	120
T_5	100	G5	80	T_5	100	100
T_6	85	G6	70	T_6	85	85

(Source: Nordland Lighting)

4.7   Concepts and techniques of explosion protection

Electrical equipment can be designed, manufactured and operated in that if they are in a hazardous area they will not contribute to causing an explosion in several ways. Today, there are three basic approaches to providing explosion protection to electrical circuits in hazardous location. They are as follows:

- Explosion confinement
- Ignition source isolation
- Energy-release limitation.

Understanding these techniques starts with the combustion triangle described in Chapter 1. This triangle illustrates the three basic ingredients necessary for combustion to occur. In hazardous locations, the fuel source can be in the form of flammable vapors, liquids, gases, combustible dust or fibers, with the oxidizer being oxygen in the surrounding air. With these two ingredients present in their most easily ignitable concentration and the introduction of sufficient electrical or thermal energy, ignition will result. Either confining or preventing this combustion from occurring can achieve a means of explosion protection.

4.7.1   Explosion confinement

A common method of providing explosion protection allows the three basic ingredients described above to coexist and potentially ignite; yet is confined within an enclosure strong enough to withstand the explosion. This technique is known as explosion proofing in North America and flame proofing internationally. All joints of the enclosure are designed in such a manner that the resulting flame, sparks or hot gases are sufficiently cooled before reaching the outside atmosphere. In addition, all external surfaces must be kept below the auto ignition temperature for the specific gas the enclosure will be exposed to.

Although a popular technique, this technology does pose many drawbacks. Since the enclosures must contain an explosion, they are bulky, heavy and difficult to install. All wiring entering and exiting this enclosure must also be placed in a hardened conduit system, which requires special seals and fittings installed according to strict regulations. This entire system must be inspected frequently to ensure integrity. Loose bolts on the lid of an enclosure effectively eliminate any explosion protection for the system.

4.7.2   Ignition source isolation

The second method of explosion protection is based upon the isolation of the ignition source or energy ingredient, from the fuel/air mixture. Although there are many acceptable methods to choose from, the most popular is purging (internationally referred to as pressurization). This method reduces the concentration of the fuel/air mixture initially inside the housing to an acceptably safe level. By maintaining a high pressure inside the housing any electrical device effectively becomes isolated from the surrounding atmosphere.

Other ignition source isolation techniques include oil immersion, sealing, encapsulation, sand, powder and inert gas filling.

4.7.3   **Energy release limitation**

This form of explosion protection permits the energy source to exist within the fuel/air mixture but by design limits the amount of electrical and thermal energy, which could be released to levels that are incapable of causing ignition.

The most widely used forms of explosion protection, which utilize this technique, are non-sparking and IS. While both share a common foundation, they do differ greatly in many aspects. These differences deal mainly with the application of safety factors. Each does have its merits, which is why it is not uncommon to see the two techniques used together.

It should also be noted that non-sparking is not recognized outside of the United States and Canada while IS is accepted worldwide. It is generally regarded as the safest form of explosion protection. To date, no explosion can be attributed to an intrinsically safe system.

Non-sparking equipment (Ex 'n') is allowed in South Africa in Zone 2. Limitation of electromagnetic energy is required in South Africa for equipments used in mines. It is required by the regulations. The use of light metal alloys in enclosures are not allowed for IS equipment. Clause 8.1 of IEC 60079-0 is applicable to IS equipment. Am 1 of the 1998 edition of 79-0 also addresses this for Group II equipment.

By definition, IS is an explosion-protection technique applied to electrical equipment and wiring intended for installation in hazardous locations. The technique is based upon limiting both electrical and thermal energy to levels which are incapable of igniting a hazardous mixture which is present in its most easily ignitable concentration. These levels have been established through extensive laboratory testing and are available in both table and graph, called ignition curves, form. These can be found in all relevant IS standards including ANSI/UL 913, FM 3610 and CSA 22.2 No. 157.

When intrinsically safe systems are discussed as being safe under normal or abnormal conditions, it is inferred that regardless of a circuit's condition, power levels will not be of sufficient magnitude to ignite a specific hazardous mixture. When discussing an electrical measurement or control circuit in a hazardous location, abnormal conditions, commonly referred to as fault conditions, would generally be considered to be those circumstances in which the circuit has failed in an unsafe manner creating the risk of explosion. With IS, these fault conditions are considered normal and expected. Fault conditions may be opening, shorting or grounding of field wiring as well as the application of higher voltages than were intended for the circuit. Each would collectively increase the potential of igniting a flammable or combustible mixture.

In general, a circuit can only be made intrinsically safe if it requires less than 1 W of power. It is for this reason that IS is limited to measurement and control circuits. An example is a standard 4/20 mA transmitter or LED indicator light. Any circuit powered by higher energy levels requires the use of other explosion-protection techniques.

Intrinsic safety reduces the risk of ignition by electrical apparatus or connecting wiring in hazardous locations. Requirements for an intrinsically safe system do not include reducing risk of explosion related to mechanical or electrostatic sparking, chemical action, radio waves or lightning strikes. Protection against such events should be employed as well.

4.8   Methods of explosion protection

When electrical equipment is to be located in a hazardous area it must be designed, manufactured and certified for that purpose. There are several methods of protection available and these are based upon the various protection techniques.

The zonal classification of the hazardous area that the equipment is to be located in will, or partially, determine the equipment's method of protection. This electrical equipment is known as 'explosion-protected' equipment, the symbol being Ex, or if to CENELEC standards EEx, followed by the letter designating the mode of protection. Care must be taken not to confuse the term 'explosion protected' with the North American term of 'explosion proof' used to describe their hazardous area equipment. Each technique of protection is assigned a code letter depicting the type of protection.

4.8.1   Exclusion of the explosive atmosphere (criterion a)

This is when the gas/air or vapor/air mixture is prevented into coming into contact with components or equipment that could cause ignition.

The following methods satisfy this requirement:

Pressurized (Ex 'p')   If clean dry air or an inert gas is pumped into an enclosure housing electrical equipment and a positive pressure is maintained at 50 Pa with respect to the surrounding atmosphere then flammable gas or vapor will be excluded. However, if the air supply to the enclosure fails then the electrical supply to it must shutdown. There are a few variations to this system and it is popular for control panels manufactured in North America. This method is generally suitable for Zones 1 and 2.

Purged (Ex 'pl')   Similar to Ex 'p' except that an air flow or inert gas flow is maintained in an enclosure to ensure that there is no build up or presence of a flammable gas or vapor. A positive pressure will be maintained in the enclosure or area but the air or inert gas will be continually released at a lower outlet pressure than inlet, thus purging the enclosure or area (Figure 4.2).

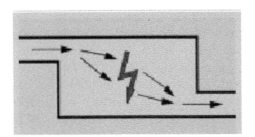

Figure 4.2
Principles of purging Ex 'p'

Ventilated (Ex 'v')   Used in large areas to dilute flammable gas or vapor to well below LEL and to reduce the temperature of electrical equipment by airflow passing over the equipment. In this system air is fed into a potentially hazardous area increasing the proportion of air in the atmosphere thus decreasing the gas/air or vapor/air mixture to less than the LEL.

Encapsulation (Ex 'm')   The main requirement for encapsulation is that the apparatus to be protected is encapsulated in resin with at least 3 mm of resin between it and the surface. In this system, the flammable gas or vapor cannot come into contact with arcs or sparks due to the operation of the apparatus (Figure 4.3). Used mainly for items not readily repairable. This method is generally suitable for Zones 1 and 2.

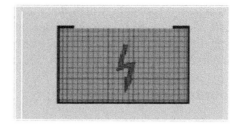

Figure 4.3
Principle of encapsulation Ex 'm'

Oil immersion (Ex 'o')   Not a popular method as the integrity of the system depends upon the presence and suitable depth of mineral insulating oil. The apparatus must be covered by at least 25 mm of oil (Figure 4.4). This method is generally suitable for Zone 1. This Ex technique of oil immersion is not allowed in South Africa.

Figure 4.4
Principle of oil immersion Ex 'o'

4.8.2   Prevention of sparking (criterion b)

In selecting components or equipment that will not provide a source of ignition when in normal use.

Increased safety (Ex 'e')   Perhaps the most widely used method of protection. The design and manufacture of this equipment assures safety against ignition through ensuring that the temperature of the equipment will not become excessive, and that the incidence of arcs and sparks in normal service is prevented. This is achieved by the use of high-integrity insulation, temperature de-rating of insulating materials, enhanced clearance distances, control of maximum temperatures, impact test requirements for the enclosure, and protection against the ingress of moisture and solids. In the case of Ex 'e' motors, design measures are taken to ensure that frictional sparks caused by the rotor against the stator will not occur and the temperature rating of the windings is de-rated (Figure 4.5). An IP rating of IP 54 is standard. This method is generally suitable for Zones 1 and 2.

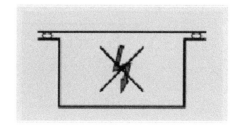

Figure 4.5
Principle of increased safety Ex 'e'

Non-sparking (Ex 'n') Apparatus which does not produce arcs, sparks or hot surfaces in normal operation is considered within this scope, and this system of protection is common amongst three-phase induction motors used in hazardous areas. This system relies on sound construction, and its IP rating is normally IP 54. This method is generally suitable for Zone 2.

4.8.3 Explosion containment (criterion c)

If a gas/air or vapor/air mixture manages to enter an enclosure that contains electrical equipment and that mixture is ignited then this enclosure must be robust enough to contain the explosion and ensure that the escaping products of the explosion do not cause ignition outside of the enclosure.

Flameproof (Ex 'd') This method is also widely used but tends to be less popular than Ex 'e' equipment. The basis of Ex 'd' equipment is that ordinary non-certified electrical apparatus such as relays, switches, terminal blocks, etc. are located in an enclosure and that if a gas/air or vapor/air mixture enters the enclosure in sufficient quantities and ignites then the enclosure will contain the effects of the ignition. The products of that ignition will be cooled by passing out through the flange between the enclosure's lid and body known as the flame path, therefore not being able to further ignite flammable gas or vapor outside of the enclosure. The larger the flameproof enclosure the greater the amount of flammable gas or vapor able to accumulate in it and the greater the flame path width required. The flame path gap must be kept to a minimum in order to inhibit the ingress of flammable gas or vapor, and that is known as the MESG. The MESG is in part determined by the internal volume of the enclosure and the gas grouping of the enclosure so that IIC gases have a smaller MESG than that of IIA and IIB gases for the same size of enclosure (Figure 4.6). This method is generally suitable for Zones 1 and 2.

Figure 4.6
Principle of Ex 'd'

Sand-filled (Ex 'q') Also known as powder-filled. The equipment enclosure is filled with finely granulated quartz that will suppress any arcs or hot surfaces and stop ignition outside of the enclosure (Figure 4.7).

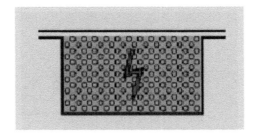

Figure 4.7
Principle of Ex 'q'

4.8.4   Energy limitation (criterion d)

The amount of energy into a hazardous area can be limited so that there is insufficient energy allowed into the circuit to cause ignition.

Intrinsically safe (Ex 'ia' and Ex 'ib')   This is a common type of protection where the required limitations of voltage and current allow its use. With IS, there is always the need for a certified interface unit such as a Zener barrier or galvanic coupler to couple the supply from the safe area to the intrinsically safe equipment in the hazardous area. Some types of equipment termed 'simple' apparatus may be non-certified for use in a hazardous area but are still supplied through a certified interface. Simple apparatus is defined in IEC 60079-11: 1999 as 1.5 V, 100 mA and 25 mW. It does not store energy, e.g. LEDs, switch contacts and resistors. Zener barriers perform their safety function by limiting current to the hazardous area via an infallible current-limiting resistor (CLR) at voltage defined by a Zener clamp. Zener barriers come with either an Ex 'a' or Ex 'b' certification the difference being that Ex 'a' equipment will perform its IS function with two faults and Ex 'b' equipment with one fault. The galvanic coupler uses a different safety philosophy than Zener barriers including the use of a transformer isolated supply. These are becoming popular due to, amongst other reasons, there being no need to provide an earth as is required by Zener barrier systems (Figure 4.8).

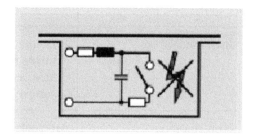

Figure 4.8
Principle of IS Ex 'i'

4.8.5   Special situations

Special protection (Ex 's')   This is not a common mode of protection but can be utilized to certify equipment that does not comply with any of the recognized methods of protection after extensive third-party testing.

4.9   Typical applications of methods of protection

Typical applications of each method are given in the table below. In general they can be described as best suited to either high electrical power applications or those where energy levels are or can be maintained at significantly lower levels. Thus, it should be clear that not all methods could be used for all applications. Part of the equipment selection process may be to determine which method of protection is the most suitable for a given application in a given set of circumstances.

The table below lists the Ex protection methods, codes and standards they conform to. The description here outlines the essential aims of the technique to provide protection and is subsequently discussed in more detail.

Type of Protection in Accordance with IEC or EN	Method/Technique	Application
'p' IEC 60079-3 EN 50016	Pressurization, continuous dilution and pressurized rooms	Switchgear and control cabinets, analyzers, large motors
'q' IEC 60079-5 EN 50017	Powder/sand filling	Transformers, capacitors, terminal boxes for heating conductors
'o' IEC 60079-6 EN 50015	Oil immersion	Transformers, starting resistors
'm' IEC 60079-18 EN 50028	Encapsulation	Switchgear with small capacity, control and signaling units, display units, sensors
'e' IEC 60079-7 EN 50019	Increased safety	Terminal and connection boxes, control boxes for installing Ex components (which have a different type of protection), squirrel cage motors, light fittings
'n' IEC 60079-15 EN 50021	Non-sparking	All electrical apparatus for Zone 2, less suitable for switchgear and control gear
'd' IEC 60079-1 EN 50018	Flameproof enclosure	Switchgear and control gear and indicating equipment, control systems, motors, transformers, heating equipment, light fittings
'i' IEC 60079-11 EN 50020	Intrinsically safe apparatus or system	Measurement and control technology, communication technology, sensors, actuators
's'	Special protection	To be approved for each application

4.10   Mixed techniques

The new directives give a lot of freedom to designer and manufacturers to innovate such that most optimized solutions are implemented both in terms of cost and safety.

Keeping this in view it has become quite common for more than one explosion-protection technique to be used on a single item of apparatus. Such as,

- A flameproof motor with an increased safety terminal box.
- Flameproof apparatus is often designed to also meet DIP requirements by including gaskets in such a way that they do not affect the flame paths.
- Apparatus commonly referred to as increased safety (Ex 'e') is actually a combination of explosion-protection techniques (Figure 4.9).
- A typical increased safety fluorescent luminaire may incorporate the following explosion-protection techniques (Figure 4.10):

  - *Ex 'e'* : Increased safety for the housing.
  - *Ex 'm'* : Encapsulation for the ballast (ballast may be Ex 'dm').
  - *Ex 'q'* : Powder-filling for the capacitor (capacitor may be Ex 's').
  - *Ex 'de'*: Flameproof, in part, for single-pin tube holders.
  - *Ex 'de'*. Interlocking isolation switch (when fitted).
  - *Ex 'e'* : Increased safety for the terminals.

The explosion-protection designation is therefore Ex 'emdq', as shown in Figure 4.9.

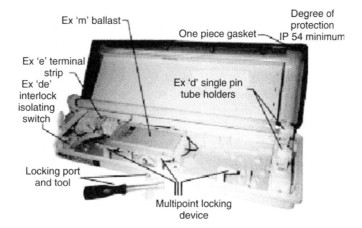

Figure 4.9
Typical fluorescent luminaire of the increased safety type (combining several explosion-protection techniques)

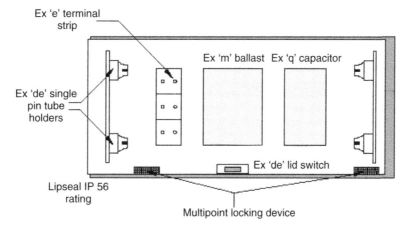

Figure 4.10
Single-line diagram of a typical fluorescent luminaire of the increased safety type

4.11 Dust explosion-protection methods

As with gas, the first priority should be to remove the source of ignition away from the area where there is probability of formation of dust clouds. In reality this may not be always possible to achieve either removal of ignition source nor obviate the possibility of removal of explosive atmosphere caused due to dust or powders.

Generally three methods are widely used in industry,

1. Principle of operating under inert atmosphere
2. Principle of containment by
3. Principle of ingress protection.

The method of operation of process under inert atmosphere implies that the concentration of oxygen is monitored and it is limited so as to prevent the explosion. Thus it is essential that the limiting oxygen concentration (LOC) of the dust or powder be known to apply this protection method. The integrity of the nitrogen or carbon or carbon dioxide blanket is also of utmost importance. Hence the installation of reliable oxygen measurement instrumentation is essential.

In case the consequences of a dust explosion are excessively dire then prevention measures for explosion containment methods need to be deployed. This is achieved by manufacturing the process equipment in such a way that it is strong enough to withstand the maximum pressure of the explosion. This is an expensive option and can be implemented in small operations.

Explosion relief panels are also sometime engineered into the equipment. These relief devices release the pressure if it builds up excessively so that other equipment does not suffer extensive damage.

Another method widely used till date in industry is to deploy an IP class of enclosure to keep the dust atmosphere away from electrical ignition sources. The Ingress Protection Code IEC 60529, AS 1939 *Degrees of protection provided by enclosures for electrical apparatus (IP Code)*, specifies the degree of protection provided by enclosures to levels of solids and water as given hereunder,

1. Contact with, or approach to live parts and against the contact with moving parts (other than smooth rotating parts and the like) inside the enclosure, and protection against the ingress of solid foreign objects.
2. Harmful effects due to the ingress of liquid or water.

These are designated such that, the characteristic letters IP (for 'ingress protection') is followed by two numerals ('the characteristic numerals').

1. The first numeral indicates the degree of protection described in (1).
2. The second numeral the degree of protection described in (2).

In some cases reference may be made to only one numeral the other is replaced with an 'X', e.g. IP 3X or IP X3.

The IP code can be optionally extended by an additional letter A, B, C or D (after the second numeral) where the protection of persons against access to hazardous parts is higher than that indicated by the first numeral, or where the first numeral is unspecified.

Note that IP X6 D, protected against driving rain, has been omitted from the 1990 edition of AS 1939.

Generally, standards refer to the class of protection as given here and manufacturers like to have it indicated on their apparatus or equipment, based on the verification testing done by notified bodies. It is fairly common practice for manufacturers who want their apparatus to be used in the open to seek an IP classification, with IP 65 being the most common.

The scope of protection for IP protection classes with reference to the first and second characteristic numerals is as follows:

Code First Digit	Physical Protection and Foreign Solid Body Ingress Protection	Code Second Digit	Protection against Ingress of Liquid
0	No protection against ingress of solid foreign bodies No protection of persons against contact with live or moving parts inside the enclosure	0	No protection
1	Protection against inadvertent or accidental contact with live or moving parts inside the enclosure by a large surface of the human body, for example a hand, but not protections against deliberate access to such parts Protection against ingress of large solid foreign bodies (50 mm diameter)	1	Protection against drops of condensed water Drops of condensed water falling vertically on the enclosure shall have no harmful effect
2	Protection against contact with live or moving parts inside the enclosure by fingers Protection against ingress of large solid foreign bodies (12.5 mm diameter)	2	Protection against drops of liquid Drops of falling liquid shall have no harmful effect when the enclosure is tilted at any angle up to 15° from the vertical
3	Protection against contact by objects of thickness greater than 2.5 mm Protection against contact from tools	3	Protection against rain Protection against water spray at an angle up to 60°
4	Protection against contact by objects of thickness greater than 1 mm Protection against contact with a wire	4	Protection against splashing liquid Protection against water spray from all directions
5	Complete protection against contact with live or moving parts plus harmful deposits of dust. The deposit of dust may not be fully prevented but sufficient amount of dust will not be allowed to enter so as to have a harmful effect on operation of equipment Protection against contact with a wire	5	Protection against water jets
6	Complete protection against contact and ingress of dust, i.e. it shall be dust-tight	6	Protection against conditions on ships decks Protection against strong water jets
		7	Protection against immersion in water
		8	Protection against indefinite immersion in water

It may please be noted that 'water' is used as the test for liquid ingress. Enclosures may suffer a greater degree of penetration from other liquids such as solvents where corrosion or reactions may also result. Assessment of ingress by other liquids shall therefore be determined by testing.

The example of how labeling is done is illustrated hereunder,

IP 67 = Ingress protection – dust-tight and protected against immersion
IP 56 = Ingress protection – dust-protected and protected against heavy sea.

These enclosures are also used in conjunction with other methods of protection where flammable liquids are likely to be present giving rise to flammable atmosphere.

4.12 Selection of explosion-protection technique for safeguarding

The certainty of protection working correctly is partly to do with the integrity of each explosion-protection technique but is also concerned with other installation-related topics such as fault tolerance and degree of maintenance necessary.

No one method of explosion protection is ideal or can be described as better than the other without reference to the risk perception and HAZOP studies done for the plant. The selection of electrical apparatus for hazardous areas will depend on the nature of the hazardous area. It is a case of selecting the technique best suited to economically and safely solve the particular application in the zone.

The techniques should not all be considered as alternatives. Each has its advantages and disadvantages and it is difficult to make direct comparisons. Where there is a choice the plant designer must base the selection primarily on what method of protection is permitted in the zones encountered by an application. Only then can the other benefits offered by each technique be considered. Conversely, disadvantages may dissuade the use of certain techniques in certain situations.

A few guidelines are described hereunder so as to help the designer to make an informed choice.

Gas and vapor risks

> *Zone 0*   Intrinsically safe system and special Category 'Ia' provided sparking contacts are additionally protected.

Special protection Ex 's'. Ex 'ia' or Ex 's' are two options available for selection of apparatus for Zone 0 areas.

Ex 's' use provides some possibilities for innovative use of appropriate apparatus. While Ex 's' apparatus can be used in Zone 0, there are no specified requirements that would automatically make the apparatus suitable for such use. It is generally dependent on the manufacturer or user to demonstrate its suitability.

For example, some low-powered apparatus complying with Ex 'm' (encapsulation) has been certified 'Zone 0 Ex "s"'.

Another approach, which has been considered by some designers of the plant, is that of duplicated techniques for use in Zone 0, e.g. a pressurized flameproof enclosure (Ex 'pd'). This might make it possible to use higher-powered apparatus that would not normally be permitted in a Zone 0 area.

However, it would not allow for one of the biggest advantages of using intrinsically safe apparatus, which is that maintenance can be carried out 'live' without the need to 'gas free' the area.

IEC is now developing a standard for apparatus suitable for Zone 0.

Zone 1   Flameproof enclosure 'd'
Intrinsically safe system Category 'Ia' or 'Ib'
'Approved' apparatus and type of protection 's'
Pressurizing or purging 'p'
Type of protection 'e'
Type of protection 'm'
Type of protection 'q' and 'o'.

Zone 11 areas represent the most common applications and provide a much larger range of options. The two techniques for gas atmospheres, which have dominated the scene generally, are:

* *Ex 'd' (flameproof)*: Particularly where power levels are reasonably high.
* *Ex 'i' (intrinsic safety)*: Because of its nature such apparatus must be restricted to low-power applications.

Pressurized (in accordance with the requirements for Zone 11) and ventilation techniques can also be used but they tend to be for a particular installation, and the apparatus is only infrequently submitted for certification.

Ex 'e' (increased safety) has emerged as a popular technique, in particular for luminaries and motors. It has a distinct advantage over flameproof apparatus in that it is generally cheaper.

Ex 'm' (encapsulated apparatus) has been used for many years for items such as solenoids.

Ex 's' is acceptable, provided it is in accordance with the requirements for Zone 1 (or 0).

Certified Ex 'q' (powder-filling) and Ex 'o' (oil-immersion) apparatus are included in the acceptable techniques for Zone 1 areas.

Zone 2   Any method of safeguarding suitable for Zones 11 and 12
Type of protection 'n'
Type of protection 'e'
Non-sparking apparatus
Totally enclosed apparatus.

Ex 'n' (non-sparking) is permitted only in Zone 2 areas and is the most common technique for such areas. Typical apparatus include luminaries and motors for high-powered apparatus. For low-power applications, apparatus that are intrinsically safe when used with barriers can often be used as Ex 'n' without the barriers, but, the apparatus must be certified for use in Zone 2.

Dust risks

Most dusts require higher ignition energy than gases, millijoules rather than microjoules, and so the energy levels associated with intrinsically safe apparatus have a considerable level of safety. Common dusts such as wheat-flour, paper, rice, wheat, wood, sucrose, aspirin, coal and rubber all have minimum cloud ignition energy much higher than 1 mJ.

The relevant standard for the selection of apparatus for Zones 20, 21 and 22 is

AS/NZS 61241.1.2 *Electrical apparatus for use in the presence of combustible dust. Part 1.2: Electrical apparatus protected by enclosures and surface temperature limitation – Selection, installation and maintenance.*

Zone 20   Intrinsically safe system Category 'Ia', provided sparking contacts are additionally protected.

Encapsulation Ex 'm' – to be duly certified for use in Zone 20

Pressurizing Ex 'p' – to be duly certified for use in Zone 20.

Encapsulated apparatus (Ex 'm') complying with AS 2431 or IEC 60079-18, provided the apparatus is installed in accordance with AS/NZS 2381.1.

Pressurized rooms or enclosures (Ex 'p') complying with the requirements for dust hazardous areas, specified in AS 2380.4.

Intrinsically safe apparatus (Ex 'i') complying with AS 2380.7 or IEC 60079.11, provided that the following conditions are satisfied:

- The apparatus complies with either Ex 'ia' or Ex 'ib'.
- The apparatus is IIC.
- The minimum dust cloud ignition energy to which the apparatus will be exposed is higher than 1 mJ.
- Associated safe area apparatus is not installed in Zones 20, 21 and 22 areas, unless protected by an appropriate protection technique.
- The apparatus is either encapsulated or protected by an enclosure complying with at least the degree of protection IP 5X given in AS 1939.
- The apparatus is installed in accordance with the requirements of AS 2381.7.

Apparatus intended for use in Zone 20 shall be verified by the manufacturers as suitable for use in Zone 20, with particular reference to the layer depth and all the characteristics of the material(s) being used.

Zone 21   Intrinsically safe system Category 'Ia' or 'Ib'

Pressurizing or purging 'p'

Encapsulation Ex 'm'

Enclosures having degree of protection IP 54 except that if the dust is likely to train fire or is electrically conducting IP 65 may be required.

All protection that are used in Zone 20. The intrinsically safe, pressurizing and encapsulation protection can be used but need to be certified for Zone 21.

Zone 22   Intrinsically safe system Category 'Ia' or 'Ib'

Pressurizing or purging 'p'

Enclosures having degree of protection IP 54.

The Australian/New Zealand Standard for pressurized apparatus for use in Zones 20, 21 and 22 areas is AS 2380.4.

Combined dust and gas and vapor risks

Gas and Vapor Risk	Dust Risk	
Zone 0 Zone 0 Zone 0 Zone 1 Zone 2	Zone 20 Zone 21 Zone 22	Intrinsically safe system Category Ia, provided sparking contacts are additionally protected

Gas and Vapor Risk	Dust Risk	
Zone 1 Zone 1	Zone 21 Zone 22	Intrinsically safe system Category Ia or Ib Pressurizing or purging Any other method of safeguarding suitable for Zone 1 gas and vapor risks, provided the enclosure is to degree of protection IP 54, except that if the dust is likely to train fire or is electrically conducting an enclosure to IP 65 may be required for the Zone 21 dust risk
Zone 2 Zone 2	Zone 21 Zone 22	Intrinsically safe system Category Ia or Ib Pressurizing or purging Any other method of safeguarding suitable for Zone 2 gas and vapor risks, provided the enclosure is to degree of protection IP 54, except that if the dust is likely to train fire or is electrically conducting an enclosure to IP 65 may be required for the Zone 21 dust risk

The methods listed above are discussed in more detail in later sections. The zone of use is discussed in more detail after each technique has been described subsequently. The different levels of integrity of each protection method makes some less acceptable or suitable for higher-risk situations.

The recognized methods of explosion protection might be applied to electrical equipment in order for it to be certified.

4.13   Conclusion

We have so far seen that only explosion-protected equipment may be used in areas in which a dangerous, explosive atmosphere may still be expected, despite the implementation of primary explosion-protection measures.

The certainty of protection working correctly has partly to do with the integrity of each explosion-protection technique but is also concerned with other installation-related topics such as fault tolerance and degree of maintenance necessary.

Safety in potentially explosive areas can only be guaranteed by a close and effective working relationship amongst all parties involved.

The operator is responsible for the safety of his equipment. It is his duty to judge where there is a risk of explosion and then divide areas into zones accordingly. He must ensure that the equipment is installed in accordance with regulations and is tested before initial use. The equipment must be kept in a fit state by regular inspection and maintenance.

The manufacturers of explosion-protected apparatus are responsible for routine testing, certification and documentation, and are required to ensure that each device manufactured complies with the design tested.

The installer must observe the installation requirements, and select and install the electric apparatus correctly for its intended use.

Till now we have studied the explosion protection and responsibility of operators. In the next few chapters we will look into the details of the type of explosion-protected equipment to be manufactured and the use to which it can be put to. Then we will follow that up with installation-related discussions.

5

Protection concept 'd'

5.1 General

Explosion protection is safety technology at its highest level. A series of pioneering developments and innovative designs have enabled various companies to exploit new, more economical and safer possibilities and apply to many different fields of application to maximize the gains. Broad specialist knowledge and a well-founded understanding of explosion protection are of benefit to all. In this chapter, we shall discuss one of the oldest and widely used protection concepts, flameproof *Ex 'd'*.

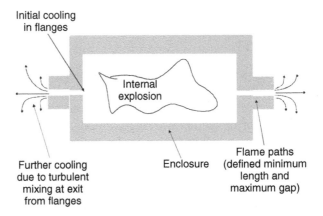

Figure 5.1
Principle of operation of a flameproof enclosure

The flameproof enclosure is one of the oldest of the protection concepts being considered in various parts of the world as suitable for Zones 1 and 2 areas, dating back to before World War II (Figure 5.1). This owes its origin to the mining industry and was developed in parallel in the UK and Germany. Dr Ing Carl Beyling (a German who developed this concept in 1908) was the person who was honored by the Institution of Mining Engineers, UK in 1938. It is interesting to note that the letter 'd' also is of German origin.

The concept of the 'Flameproof enclosure' followed the use of electricity in coal mines in Europe. In those early times (beginning of twentieth century), it was DC motors and other DC-operated electrical equipments which were in use. These produced highly ignition-capable sparking in normal operation. While the ideal way would have been to completely exclude the explosive gases from coming in contact with this sparking, it was not possible from a practical point of view. This was due to the unreliability of the equipment at that time, which needed to be frequently opened for maintenance and thus breaking any seal with the attendant difficulty of remaking it with confidence. This, no doubt, is still, to a degree, the case as it is still considered that to make and break a seal of the desired quality in the field with confidence is generally not possible. Added to this, the rotating shaft and rods also make sealing permanently and under all conditions a nearly impossible task.

On accepting the fact that an explosive atmosphere cannot be permanently kept out nor sparking be eliminated under all conditions, it became imperative to look for the means to 'contain' the result of such internal explosions as and when they occur, so as to minimize the risk for the persons present in the vicinity. Thus, an enclosure sturdy enough needed to be designed so as to contain the forces of such internal explosion without any damage to itself.

It has also been discovered that the flame front caused by the explosion could be rendered harmless by passing it through a small gap (if the gap size could be guaranteed to have a specific maximum dimension during the explosion). The apparatus so enclosed would not constitute an unacceptable risk, even if the explosive atmosphere did penetrate its enclosure and was ignited internally. Thus the concept of flameproof enclosure is born.

5.2 Definitions

This method is also widely used along with the equally popular Ex 'e' concept.

The basis of Ex 'd' equipment is that ordinary non-certified electrical apparatus such as relays, switches, terminal blocks, etc. are located in an enclosure. If a gas/air or vapor/air mixture enters the enclosure in sufficient quantities and ignites, the enclosure will contain the effects of the ignition.

This method is generally suitable for *Zones 1 and 2*.

Flameproof enclosure

The term 'FLP' (mnemonic for Flameproof) originates from the mining industry where it was first used. The term flameproof is applied to enclosures of electrical apparatus certified by a certified testing body as having been examined, type-tested where necessary, and found to comply with the SABS/IEC 60079-1 or AS 2380.2 or BS/EN 50018 (1995). The enclosure must be strong enough to withstand the stresses of internal ignition.

A *flameproof* enclosure is defined in the standards as: 'An enclosure for electrical apparatus that will withstand an internal explosion of the flammable gas or vapor which may enter it without suffering damage and without communicating the internal flammation to the external flammable gas or vapor for which it is designed, through any joints or structural openings in the enclosure.'

A *flameproof* enclosure is designed to withstand the pressure of an internal explosion; it is not necessary therefore to provide openings for pressure relief. Where there is a joint, however, or where a spindle or shaft passes through the enclosure, the products of the explosion can escape. It should be understood that the aim of a flameproof enclosure is not necessarily the total avoidance of any gaps in an enclosure. The misconception that it should be *'gas-tight'* is misplaced. The principle recognizes that some openings are unavoidable in practice and so restricts itself to requiring that the size of such openings should not exceed the safe limit above which the nature of the escaping flame is such as to ignite a specified flammable atmosphere. On the other hand, it is not the aim to require joints to be deliberately spaced to give an opening.

Flameproof joint

The place where the corresponding surfaces of the different parts of a flameproof enclosure come together, where the flame or products of combustion may be transmitted from the inside to the outside of the enclosure.

Length of flame path

The shortest path traversed by a flame through a joint from the inside to the outside of an enclosure.

Gap

The distance between the corresponding surfaces of a flameproof joint after the electrical equipment has been assembled.

Pressure piling

A condition of rise in pressure resulting from ignition of pre-compressed gases in compartments or sub-divisions other than those in which ignition was initiated and which may lead to a higher maximum pressure than would otherwise be expected.

Maximum experimental safe gap (MESG)

Any path, which the flame or hot gases may take, needs to be of sufficient length and constriction to cool the products of the explosion so as to prevent ignition of a flammable atmosphere external to the enclosure.

The products of the ignition will be cooled by passing out through the flange between the enclosure's lid and body known as the flame path, therefore not being able to further ignite flammable gas or vapor outside of the enclosure. The larger the flameproof enclosure, the greater the amount of flammable gas or vapor able to accumulate in it and the greater the flame path width required. The flame path gap must be kept to a minimum in order to inhibit the ingress of flammable gas or vapor and that is known as the MESG. The MESG is in part determined by the internal volume of the enclosure and the gas grouping of the enclosure so that IIC gases have a smaller MESG than for IIA and IIB gases, for the same size of enclosure. The dimensions of these flame paths are critical and are specified in SABS/IEC 60079-1.

Other tests for flame energy levels were devised from Davy and Faraday's original work.

An alternative test is the eight-liter capacity bronze sphere, which is made of two halves and joined by a flange of one-inch path length. The sphere is enclosed in a tank of eight cubic feet in capacity filled with the most easily ignitable mixture of a test gas. The distance between the flanges of the two halves of the sphere, known as the gap, is varied in a series of tests with each gas. A substantial spark is introduced into the inside of the sphere. The size of the gap is increased until the point is found at which sufficient flame energy is released from the sphere, through the gap, to ignite the surrounding gas. The object is to find the maximum gap size that will not transmit ignition. This is said to be the 'maximum experimental safe gap'.

To add a margin of safety the MESG was reduced to a safe working gap.

It was realized that the sensitivity MESG and the minimum ignition energy, MIE values for any given gas correlated. This is illustrated in the table below:

Representative Gas	Ignition Temp. (°C)	MESG	Safe Working Gap	Minimum Ignition Energy (μJ)
Propane	466	0.016″	0.016″	180
Ethylene	425	0.004″	0.008″	60
Hydrogen	560	0.001″	0.004″	20

Note that the ignition temperatures do not correlate to MESG or MIE. Temperature must be considered separately.

5.3 Certification in brief

Flameproof enclosures cannot yet be certified independently of their contents but this situation may change for small enclosures because of current work.

Alteration to the position of the internal components is not permitted because conditions may be created inadvertently which will lead to pressure piling. No modification, addition or deletion to the enclosure or its internal components shall be made without the written permission of the certifying authority. Such permission shall be obtained through the manufacturer of the apparatus, unless it can be verified that such change does not invalidate the certification.

It should be noted that a flameproof enclosure is not tested for its ability to withstand the effects of an internal electrical fault.

5.4 Construction requirements

The Ex 'd' method of protection relies on the mechanical strength of an enclosure to withstand an internal explosion. It must not allow the explosion to propagate to a flammable atmosphere surrounding the enclosure.

The enclosure itself is tested to perform in a controlled way with specified devices inside the enclosure.

Some of the other details that are covered by the standards are:

- Minimum surface roughness (machined surfaces).
- Inspection access to measure the gap between flanged joints – usually with feeler gages.
- Holes in joint surfaces.
- Joints, including, class of fit, pitch, minimum number of threads engaged.
- Gaskets and 'O' rings.
- Operating rods and spindles.
- Shafts and bearings.
- Breathing and draining devices.
- *Fasteners*: Especially the requirement that all holes must be blind holes, so that if a bolt is removed there will not be direct access into the flameproof enclosure. Minimum wall thickness – 3 mm or 1/3 the diameter of the hole – whichever is greater. Minimum length of thread engaged and the depth of the hole to ensure the bolt does not 'bottom out' before it is fully engaged.
- Mechanical strength, pressure piling.
- Cable glands and cable connection techniques.

- *Terminals for external connections*: The terminals must be generously sized and of sufficient cross-section for the required current-carrying capacity. They should also be positively located, and constructed in such a way that the conductors cannot loosen or be damaged by the clamping action.
- Acceptable methods for internal connections or joints.
- *Clearances*: Minimum clearance distances are specified between conducting parts at different potentials, which vary depending on the working voltage.
- *Creepage distances*: The grading of different insulating materials is covered, as well as minimum creepage distances for different working voltages.
- Surface profiles for insulating materials.
- Insulating materials including thermal stability, temperature rating, strength and requirements for varnishes.
- *Windings*: This section covers the minimum wire sizes, impregnation techniques and test voltage for different wire sizes.
- *Limiting temperature*: In general the 'T' rating must not be exceeded by any part of the equipment within the TE time during normal operation or starting, in order to prevent ignition of an explosive atmosphere.
- *Overtemperature protection*: The windings have to be protected to ensure the limiting temperature cannot be exceeded in service. This is usually performed by a suitable current-dependent device. This device needs to be selected with great care and the calibration may need to be checked, to ensure that it will operate as required, i.e. it must disconnect the motor from the supply within the TE time, if the motor is subjected to a locked rotor situation.
- Internal wiring.
- *Degree of protection*: Minimum IP ratings. For motors with bare live parts this is IP 54.
- Maximum values for TE time for different starting current ratios. The TE time cannot be less than 5 s and the I_a/I_n ratio cannot exceed 10.
- Limiting temperatures for insulated windings for different classes of insulation.
- Requirements for impact tests.

Additional requirements for rotating electrical machines

- Clearance for internal fans
- Minimum radial clearances (or air gap) between the rotor and stator at different speeds
- Special requirements for cage rotors.

Cable entries

It is necessary, at the time of ordering, to specify the number of cable entries and the size requirement of cable glands for use with a flameproof enclosure. These have to be machine-cut by the manufacturer and this operation shall not be carried out on site.

- Any enclosure must have the ability to be opened up in such a way as to allow the apparatus to be installed inside it. It is a natural consequence that in closing the box, there may be a 'gap' between the mating surfaces of the lid and the body of the enclosure. The mating surfaces must be of certain critical dimensions that, if ignition occurs inside the box, the resultant flame cannot come out through the path left by the mating surfaces. The flame path must have a defined and purposely constructed length. This dimension is stated in the construction standards.

- The standards say that there must be no intentional gap, but if there is a gap, it must not be larger than certain dimensions listed in the standards. Figure 5.2 shows a typical Ex 'd' enclosure.
- An intentional gap is not required in order to release the pressure of any explosion. However if an explosion occurs, it is likely that the force will try to raise the lid of the box and expose a gap. The gap in this case is known as the flame path and is again a consequence of the combination of a lid forced to rise by the pressure of an explosion. It is not intentional. The size of the gap must not exceed that stated in the standards in order for the arrangement to remain safe or within accepted safety margins.

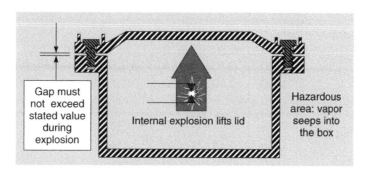

Figure 5.2
Ex 'd' flameproof enclosure

The resultant flame from an internal explosion causes a pressure rise in the box. Consequentially, the force on the lid causes the bolts to stretch and a gap may or may not occur between the lid and the box flange. Many tons of force is experienced under these conditions.

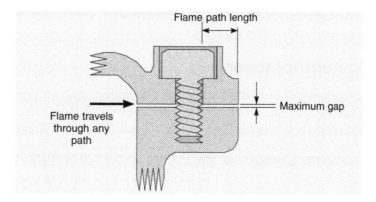

Figure 5.3
Ex 'd' flame path

If the lid does lift and a flame front can begin to travel through the resultant gap the maximum size and minimum length of the path open to the flame is critical to its ability to quench the combustion energy. This energy is dissipated by various methods as it travels through the 'flame path' and eventually the flame is extinguished through the cooling effect of the energy dissipation that takes place (Figure 5.3).

Minimum width of joint and maximum gap for I, IIA, IIB enclosures
AS 2380.2 TABLE 1.1

Width of Joint L (mm)	Maximum Gap for Volume, V (cm³), (mm)								
	$V < 100$			$100 < V < 2000$			$V > 2000$		
	I	IIA	IIB	I	IIA	IIB	I	IIA	IIB
Flanged and spigot									
joints	0.30	0.30	0.20	–	–	–	–	–	–
6 < L < 9.5	0.30	0.30	0.20	–	–	–	–	–	–
9.5 < L < 12.5	0.40	0.30	0.20	0.40	0.30	0.20	0.40	0.20	0.15
12.5 < L < 25	0.50	0.50	0.20	0.50	0.40	0.20	0.50	0.40	0.20
25 < L									
Operating rods and									
spindles	0.30	0.30	0.20	–	–	–	–	–	–
6 < L < 12.5	0.40	0.30	0.20	0.40	0.30	0.20	0.40	0.20	0.15
12.5 < L < 25	0.50	0.40	0.20	0.50	0.40	0.20	0.50	0.40	0.20
25 < L									
Shafts with rolling brgs									
6 < L > 12.5	0.45	0.45	0.30	–	–	–	–	–	–
12.5 < L < 25	0.60	0.50	0.40	0.60	0.45	0.30	0.60	0.30	0.20
25 < L > 40	0.75	0.60	0.45	0.75	0.60	0.40	0.75	0.60	0.30
40 < L	0.75	0.75	0.60	0.75	0.75	0.45	0.75	0.75	0.40

For rods, spindles and shafts, the gap is the maximum diametric clearance. Usually if equipment is to be used for Groups I, IIA and IIB, it will be designed and built to meet the requirements of a IIB enclosure, which exceeds the requirements of IIA and I. Separate tables apply for Group IIC with much tighter tolerances and gaps. As IIC requirements are not common, standard equipment is not usually produced to IIC to cover the requirements of the other groups. Equipment for IIC requirements is usually produced only specifically when required.

5.5   Flameproof theory

It is not possible or desirable to produce equipment with no gaps between the different parts, especially with an electric motor with a revolving shaft. The allowable gap between the different parts is accepted, but strictly defined. The 'gaps' help to relieve the pressure inside the enclosure caused by an explosion. Experimental testing is used to establish the 'maximum experimental safe gap' for different gases and gas mixtures. This data has been used to establish AS 2380.2 table 1.1 (refer clause 4). It can be seen that the wider the joint the larger the allowable gap, and that gases that explode more violently require a smaller gap.

The standard that covers the testing to establish the 'maximum experimental safe gap' is IEC 60079-1A (Appendix D).

In South Africa all cable glands have to comply with SABS 808 or SABS 1213. These requirements are not the same as 60079-0 and 60079-1. The SABS standards take preference. This is illustrative of a case where more stringent conditions apply.

In Australia the cable glands have to comply to AS 1828 and until this standard is around (likely to be withdrawn) users will have to follow it in addition to what is mentioned in AS/NZS 60079.0.

Flame path dimensions for joint surfaces interrupted by holes

Sl No.	Joint Width (L) (in mm)	Length of Flame Path between Enclosure and Exterior (l) (in mm)
1.	< 12.5	6
2.	≥ 12.5 < 25	8
3.	≥ 25	9

Explosion proof vs flameproof

Americans refer to 'explosion proof', while the UK and IEC refer to 'flameproof' motors or equipment. According to the IEC definitions, this is an Ex 'd' piece of equipment. Ex 'd' equipment is designed to contain an internal explosion to escape between the 'flame paths', but cool any flame in the hot gases so that no flames escape from the enclosure to ignite any external flammable gases – hence 'flameproof'.

Although Ex 'e' equipment is designed to be used in a Zone 1 area, it cannot be described as 'explosion proof' or 'flameproof', as it will not contain an explosion if one did occur. Ex 'e' equipment is manufactured to an approved 'explosion-proof technique'.

5.6   Other general requirements for explosive atmospheres

Fasteners

Fasteners can only be undone by the use of a special tool, if the parts they are securing are necessary to achieve a type of protection (Ex 'd', Ex 'e', etc.). For Ex 'd' equipment this is usually achieved by using socket head cap screws with a counter bore, or hexagon head bolts with a counter bore or shroud.

Metallic enclosures

The amount of aluminum or magnesium in external alloys is limited by their properties, especially for Group I applications. This is because coal dust and aluminum under ideal conditions can create spontaneous combustion at temperatures as low as 150 °C (refer clause 5.1.1 of IEC 60079-0).

AS 2380.1 specifies that the such enclosures for electrical equipment in Group I shall contain:

- Not more than 6% in total of magnesium and titanium together
- Not more than 15% in total of aluminum, magnesium or titanium, singly or in combination, except as stated above.

For Group I equipment, it is usual for all external parts to be cast iron or steel.

Other requirements these enclosures have to meet

- External temperatures during normal operation – 'T' ratings
- IP ratings
- Usual electric tests including a high-voltage test at $2 \times$ the operating voltage plus 1500 V (2330 V for 415 V equipment)

- Impact tests
- Marking, including information required on nameplates.

Drilling and tapping suitable holes to install spark plugs pressure transducers and gas pipes modify the equipment.

5.7   Testing

Type tests are carried out on all Ex 'd' equipment before it can be approved. These tests are performed in two stages.

Pressure test

Tests of the ability of the enclosure to withstand pressure, and the rate of rise of pressure, developed inside the enclosure during an explosion – the 'reference pressure'. The enclosure is then overpressure tested to 1.5 times the 'reference pressure', usually with compressed air or water, to ensure a factor of safety. During manufacture all equipment that is of a steel fabricated construction must be routine tested at a static pressure of 1.5 times the 'reference pressure' to ensure the integrity of the welding. If equipment is of cast or molded construction, it can be exempted from the requirement for routine pressure testing if the sample passes a four times overpressure test.

Flame test

These tests are done to ensure that the enclosure is flameproof, i.e. that an internal explosion does not create an external explosion.

Methodology

To determine the explosion 'reference pressure' at least three explosions are performed inside the enclosure using the appropriate mix of gas to suit the gas group the enclosure is designed for. For electric motors, these tests are performed both while the motor is running and also while it is stationary.

The equipment is placed in a test chamber, which is filled with the same explosive gas mixture as is used in the equipment under test. The gas inside the equipment is exploded at least five times to prove that the internal explosions do not ignite the external gas.

The equipment is considered to have passed the tests if no flame transmission has occurred, and the enclosure has not suffered any damage or permanent deformation that may affect its flameproof properties.

Before any explosion testing is performed, copies of the approved drawings are carefully checked to ensure that the design details meet the requirements of the relevant standards. After explosion testing is completed the samples (prototypes) are disassembled and all the component parts are carefully measured to ensure that they have been manufactured to the dimensions and tolerances specified on the approval drawings.

5.8   Grouping and effect of temperature classification

Flameproof enclosures are grouped according to the specified maximum permissible dimensions for gaps between joint surfaces and the surfaces of other openings in the enclosure. The standards quote the maximum permissible dimensions of gaps for the various enclosure groups but, in practice, joints shall be fitted as close as possible and on no account shall the maximum permissible dimensions be exceeded.

All enclosures are marked with the appropriate standard and group reference. For industrial gases the enclosure groups are IIA, IIB and IIC in IEC 60079-1. Because of the similarity of the enclosure group symbols, care shall be taken to ensure that the IEC standard to which they relate is identified.

Enclosures certified for a particular group may be used with gases and vapors appropriate to an enclosure group having larger permissible maximum gap dimensions. For example, a Group IIB enclosure may be used in place of a Group IIA enclosure but not vice versa.

Hot surfaces can ignite explosive atmospheres. To guard against this, all electrical equipment intended for use in a potentially explosive atmosphere is classified according to the maximum surface temperature it will reach in service. This temperature is normally based on a surrounding ambient temperature of 40 °C or 102 °F. This temperature can then be compared to the ignition temperature of the gases, which may come into contact with the equipment, and a judgment reached as to the suitability of the equipment to be used in that area.

Many Ex-type products marketed by leading companies are certified for use in ambient temperatures up to 55 °C.

Enclosures are marked with a temperature class (T_1–T_6 in accordance with standards) and shall not be installed where flammable materials are used which have ignition temperatures below the maximum for that class.

The following table gives 'Suitability of Group II enclosures for use in the presence of specific gases and vapors' and 'ignition temperatures' of these gases and vapors are criteria employed.

Selection Criteria of Electrical Apparatus in Hazardous Vapor or Gas Atmosphere

Ignition temperature of gas or vapor = 170 °C
Gas temp. class (irrespective of grouping) will be T_4

Parameters	Suitable Electrical Apparatus	Unsuitable Electrical Apparatus
Ambient temperature	40 °C	40 °C
Temperature rise	80 K	100 K
Max. surface temperature	120 °C	140 °C
Temperature class	T_4	T_3

Ambient temperature and its effect on Ex 'd' equipment

It is generally assumed that only the temperature classification will be altered, when the hazardous area located electrical equipment is subjected to ambient temperatures that are outside of the certified ambient temperature range. Unfortunately, it may also cause the method of protection to be bypassed. This may be permanent and leave the certified electrical equipment unsuitable for use in a hazardous area.

The problems include:

- Flameproof equipment is designed and certified to contain an internal explosion and prevent the transmission of this into the hazardous area. An aluminum flameproof enclosure must be able to contain the internal explosion over the full range of ambient temperatures. At ambient temperatures below the certified ambient temperature range, the aluminum could suffer from embrittlement. At such low temperatures, the aluminum flameproof enclosure could be shattered by the internal explosion, allowing the explosion to spread into the hazardous area.
- Aluminum has a large coefficient of expansion, so at ambient temperatures above the certified ambient temperature range, the flame path gaps may become sufficiently large to permit the internal explosion to spread into the hazardous area. The expansion of the aluminum may cause the flame path gap to buckle, leaving the flameproof enclosure permanently ignition-capable.
- Cyclic heating of bolts can loosen them, and flameproof bolts are rarely locked. It only takes one loose cover bolt to bypass the method of protection.

5.9   Conditions of use

Flameproof enclosures are primarily intended for use in Zone 1 gas and vapor risks. When used in Zone 2 gas and vapor risks no relaxation of the application, installation or maintenance requirements shall be permitted. Flameproof enclosures must not be used in *Zone 0*. *Flameproof* enclosures may be used in dust risks and combined dust and gas/vapor risks if the additional precautions specified in the above clauses are complied with. An enclosure that is specifically designed as flameproof/weatherproof shall be used where a *flameproof* enclosure is exposed to the weather or installed in wet conditions as available. Weatherproofing is usually achieved in this type of enclosure by gasketed joints, which are additional to and separate from the flame paths. The weatherproofing of other *flameproof* enclosures may be achieved by the use of a suitable grease or flexible non-setting compound in the flame path, provided that chemicals with which they may come into contact do not adversely affect these. Alternatively, tape may be applied to the outside of the joint, provided the requirements laid down in standards are observed. Gaskets shall not be inserted in flame paths.

Where a *flameproof* enclosure is exposed to corrosive conditions, its safety features may be impaired by corrosion of the enclosure, cement or other sealing materials. It shall therefore be suitably protected by, for example, painting external surfaces (including the cement) and the greasing of flanges. Consideration shall also be given to increasing the frequency of maintenance.

The effect of tape and obstacles on flame paths

ERA reports DT 129 and DT 131 show that the tape-wrapping of flanged joints and other openings, or the presence of obstacles near the edges of flanged joints and other openings, may impair the protection afforded by a *flameproof* enclosure. To compensate for these effects the rules set out below, which are applicable in both *Zone 1* and *Zone 2* shall be observed.

Tape (usually a grease-impregnated linen tape)

- *Group IIC enclosures*: Tape shall not be applied to any flanged joint or spindle or shaft gap.
- *Group IIB and IIA enclosures*:

- Tape shall not be applied to spindle or shaft gaps.
- Tape shall only be applied to flanged joints when strictly necessary. When tape is applied, the gap between the joint surfaces shall not exceed 0.1 mm irrespective of the flange width. The tape shall be restricted to one layer enclosing all parts of the flange with a short overlap, and a new tape shall be applied when the existing tape is disturbed.

Obstacles (such as external covers, guards, supports, pipes, structural steelwork, etc.)

Where the obstacle is more than 40 mm away from the edge of a flanged joint or another opening, no special precautions are necessary.

- *Group IIC enclosures*: There shall be no obstacle within 40 mm of a flanged joint or spindle or shaft gap.
- *Group IIB and IIA enclosures*:
  - There shall be no obstacle within 6 mm of a flanged joint or within 40 mm of a spindle or shaft gap.
  - Where an obstacle is between 6 and 40 mm of a flanged joint the gap shall not exceed 0.1 mm, irrespective of the flange width.

Integral obstacles

Many flameproof enclosures have obstacles external to and integral with the enclosure but since they have been tested and certified in this condition no special rules or precautions are necessary.

The effect of grease, other sealants and paint on flanged joints

ERA Report 5191 describes experiments which showed that the presence in a flanged joint of grease or a non-setting jointing compound caused no deterioration in the *flameproof* qualities of the joint. The same result was obtained where the exterior of the flanges was painted, even if the paint filled and bridged the gap. Therefore no special precautions are necessary when these materials are applied, except that aluminum paint should not be used because of the potential danger from a combination of aluminum and rust.

Aluminum flameproof enclosures and cables with aluminum conductors

ERA Report 70-32 considers the risks, which can arise when aluminum is used as a flameproof enclosure material, and when aluminum conductors are used inside *flameproof* enclosures. Until further information is available, the following precautions, based on the ERA report and other information, can be applied:

- Because aluminum *flameproof* enclosures can eject hot aluminum particles under fault conditions, and because of the danger of arcs burning through the enclosure, the use of such enclosures is restricted to circuits protected by a 15 A or smaller fuse.
- Cables with aluminum conductors shall not be used in *flameproof* enclosures unless the possibility of ejecting hot aluminum particles from the enclosures has been minimized by either:

– Using cable-terminating enclosures whose joints are threaded or spigoted or
– Using fully insulated conductors, and using terminals, which are designed to reduce the likelihood of faults and are shrouded by insulation. The compound filling of boxes is one method of meeting this requirement.

5.10   Illustrations of mechanical construction types

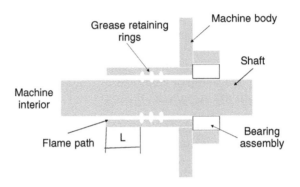

L = length of flameproof joint

Cylindrical joint with grease retaining rings

Figure 5.4
Type of gland for ball or roller bearings

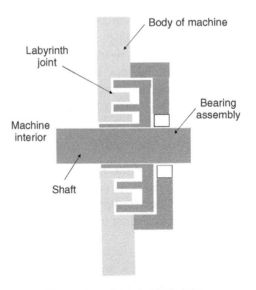

Geometry of a labyrinth joint

Figure 5.5
Geometry of a labyrinth type of gland

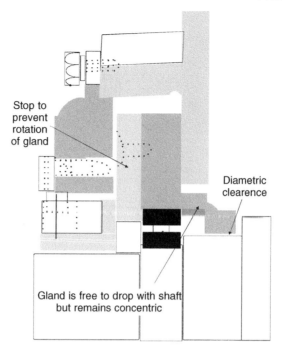

Stop to
prevent
rotation
of gland

Diametric
clearence

Gland is free to drop with shaft
but remains concentric

Figure 5.6
Typical floating gland

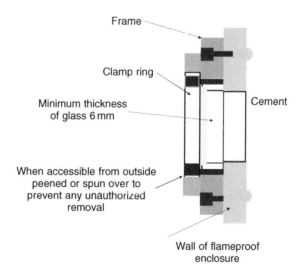

Frame

Clamp ring

Minimum thickness
of glass 6 mm

Cement

When accessible from outside
peened or spun over to
prevent any unauthorized
removal

Wall of flameproof
enclosure

Figure 5.7
Typical cemented window mounting

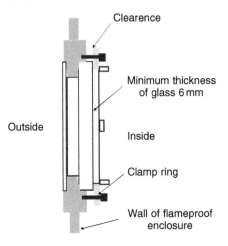

Figure 5.8
Typical non-cemented window mounting

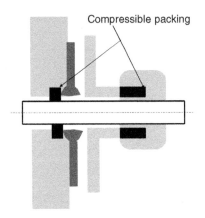

Figure 5.9
Typical gland and clamp entry of flexible armored cable into a terminal box

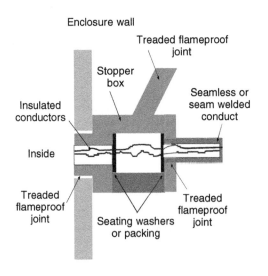

Figure 5.10
Conduit stopping box

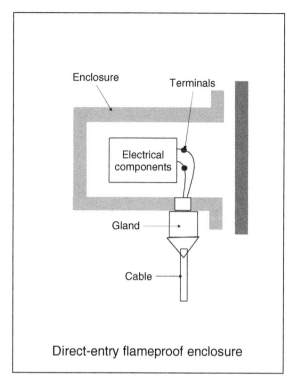

Figure 5.11
Compound filled gland

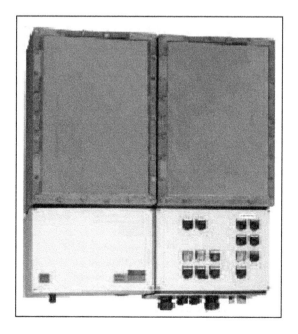

Figure 5.12
Flameproof enclosure for Group IIB
(Courtesy R. STAHL)

(a) Flameproof switch   (b) Flameproof terminal box

Figure 5.13
Typical example for Ex 'd' enclosure

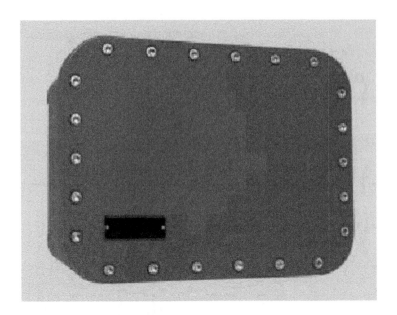

Figure 5.14
Flameproof enclosure with large number of bolts

D4863P

Figure 5.15
Explosion-protected emergency light fittings for fluorescent lamps with sheet metal enclosures

5.11 Summary

To design, test and manufacture flameproof equipment requires a great deal of attention to detail and a high level of safety, which is understandable, as failure will put people's lives at risk. The information above is a brief summary of some of the detail that has to be complied with to meet the requirements of the standards.

User's responsibility

It is the responsibility of the user to install, use and maintain flameproof apparatus in such a way that its safety is preserved. Neglect of the following points may nullify initial flameproofness and although they lie beyond the manufacturer's control, they are brought to the notice of the users of flameproof apparatus as being of paramount importance.

Installation

- Install flameproof apparatus correctly and in full compliance with the regulations of the Department of Mines (or of Labor), as is relevant.
- Prevent the occurrence of short-circuit currents of dangerous values inside the apparatus by providing automatic interruption of the short circuit elsewhere.

Use

- Use apparatus only in atmosphere for which it is certified flameproof.
- Install apparatus so that it is not loaded beyond its rating.
- Do not leave oil-immersed starting resistors in circuit, or operate them more frequently than intended by the design.
- Limit the temperature rise of apparatus such as resistors, space heaters and lighting fittings.
- Remember that flammable gas may originate inside the enclosure of some types of apparatus.

Maintenance

- Prevent clearance gaps from becoming excessive due to wear and from becoming fouled with foreign matter.
- Keep all joint surfaces clean. Slight surface greasing with suitable water-repellent grease is permissible.
- Ensure that all bolts, screws, studs and nuts are present, and are tight and secured against working loose. Replacement bolts, screws, studs and nuts must satisfy the requirements of this specification, or flameproofness will be impaired.

6

Protection concept 'e'

6.1 General

The protection concept of increased safety is one intended for use in Zone 1 and less-hazardous areas. It is generally denoted by adding suffix 'e', i.e. Ex 'e'. This method of protection owes its origin to developments in Germany. 'Erhochte sicherheit', from which the 'e' is taken, means 'increased safety'.

It was originally conceived for high-power equipment, which was rigid or fixed. IEC 79-7 requires the use of good quality materials having well-defined insulation properties and with adequately de-rated mechanical and electrical design specifications.

What does increased safety mean?

Increased safety is a type of protection applied to electrical equipment that does not produce arcs or sparks in normal service and under specified abnormal conditions, in which additional measures are applied so as to give increased security against the possibility of excessive temperature and of the occurrence of arcs and sparks. A series of pioneering developments and innovative designs have enabled various companies to exploit new and more economical and safer possibilities and use fields of application to maximize the gains. Broad specialist knowledge and well-founded understanding of explosion protection are of benefit to all. In this chapter, we shall discuss one of the latest-used protections concept, Increased Safety Ex 'e'.

Although Ex 'e' equipment is designed to be used in a Zone 1 area, it cannot be described as 'explosion proof' or 'flameproof', as it will not contain an explosion if one did occur. Ex 'e' equipment is manufactured to an approved 'explosion-proof technique'.

It is generally coined that Ex 'e' is 'short on name – long on safety'.

6.2 Definitions

Type of protection 'e' is defined in Standard SABS/IEC 60079-7 as:

A method of protection by which additional measures are applied to electrical equipment so as to give increased security against the possibility of excessive temperatures and of the occurrence of arcs and sparks during the service life of the apparatus. It applies only to electrical equipment no parts of which produce arcs or sparks or exceed the limiting temperatures in normal service.

6.2.1　Increased safety

The type of protection applied to electrical equipment that does not produce arcs and sparks in normal service in which additional measures are applied so as to give increased security against the possibility of excessive temperatures and of the occurrence of arcs and sparks.

6.2.2　Limiting temperature

It is the 'maximum permissible temperature' for electrical equipment or a part of electrical equipment and is the lower of the following two temperatures determined by:

- The danger of ignition of an explosive gas atmosphere (T_1–T_6).
- The thermal stability of the materials used.

6.2.3　Initial starting current

The highest rms value of current absorbed by an AC motor while starting from rest (inrush current at zero speed), when supplied at rated voltage and rated frequency.

6.2.4　Starting current ratio I_a/I_n

The ratio between initial starting current I_a and rated current I_n.

6.2.5　Time t_E

The time taken for an AC winding, when carrying the initial starting current I_a to be heated up to the limiting temperature, from the temperature reached in rated service, at the maximum ambient temperature.

6.3　Principles of design for increased safety

Electrical apparatus with type of protection increased safety 'e' is distinguished by the fact that it does not generate any ignitable sparks during normal operation. The aim of this type of protection is to avoid the occurrence of ignitable sparks and thus have a distinctly higher degree of safety compared with conventional electrical apparatus. In addition, the design of such apparatus prevents parts in the interior of the enclosure and on the outer surface of the enclosure from reaching temperatures, which exceed the ignition temperature of any explosive atmosphere that may be present.

The Ex 'e' standard specifically does not permit the inclusion of any discontinuous contact. No switches or switching mechanisms are allowed in this concept of protection. Sparks therefore cannot occur and spark energy does not need to be considered.

The only possible source of ignition to be considered is therefore heat. The same precautions taken to eliminate the possibility of sparking also help to reduce the temperature rise in current-carrying parts. They are made larger than necessary or may be good-quality standard parts, which are suitably de-rated. In this way heat dissipation is increased, so that temperature rise is reduced.

These protection aims are mainly reached by applying the following principles:

- The enclosures are designed in such a way that the entry of moisture and dirt in hazardous quantities is prevented. The IP protection class IP 54 is laid down as the minimum requirement and the enclosures have a mechanical strength that can withstand the typical harsh operating conditions in an industrial plant.

Enclosures must guarantee the minimum protection standard IP 54 even under severe external mechanical forces.

- Internally, the clearance and creepage distances must also be so dimensioned that even under harsh ambient conditions, no short circuits via creepage paths or flashovers can occur.
- The electrical connection terminals are designed in such a way that it is not possible for the cable connected to them to come loose.
- The dimensioning of the apparatus in electrical terms ensures that no inadmissible temperatures can occur inside or on outer parts of the apparatus.

6.4   Certification (components)

Component parts to be included in larger arrangements may be 'component certified' for some flexibility. In an Ex 'e' junction box, for example, the enclosure will be impact-tested. The terminals to be used within will be component-approved.

The main uses of this technique are found in higher-power circuits such as induction motors, fluorescent lighting fittings, junction boxes and terminal housings. The German standards from which this came promote the use of toughened plastic cable sheaths on permanent installations, as opposed to the more expensive steel wire armored cable used elsewhere.

When applied to junction boxes, an Ex 'e' enclosure is given an 'enclosure factor' when certified. This represents the highest number of 'terminal-amps' permitted in the box. Terminals mounted in the box must be component-approved. The total of terminal-amps must be calculated and must be equal to or less than the enclosure factor.

6.5   Construction requirements

The standards permit the construction of apparatus in such a way that during normal operation of the equipment, it is unlikely to become a source of ignition. The rules therefore seek to develop an acceptable level of integrity by considering standard industrial grade equipment and enhancing some aspects of its construction. This is as opposed to the inclusion of specific electrical or mechanical techniques to prevent ignition, which are applied in some of the other methods.

The constructional details covered under this standard are as hereunder.

Enclosures

Enclosures must be designed to be weatherproof and to be impact-resistant. The minimum IP rating is IP 54 before and after a '7 Newton-Meter' drop test has been performed as specified in the standard. In practice, manufacturers often try to achieve IP 65. The enclosure is not designed to withstand external or internal explosion.

Damage to the enclosure must not increase the likelihood of causing short circuits within. Additionally the enclosure material must be resistant to any chemical attack when installed in an adverse environment.

Ex 'e' junction boxes can be certified to provide an 'enclosure factor'. This is the total number of 'terminal-amperes' that can be installed into the box. Terminals are also awarded factor values depending on their size and current-carrying capacity. Thus, an enclosure with a factor of 800 may use 40 off-terminals each with a factor of 20. Some up rating of enclosures is allowed for instrumentation but this severely limits the useful number of terminals permitted (Figure 6.1).

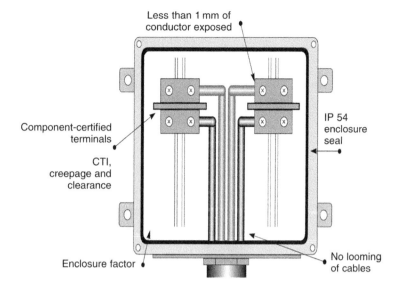

Figure 6.1
Typical enclosure

Terminals for external connections

The terminations and connections are made with a higher degree of security such that in the conditions of use they are unlikely to become loose and spark. Electrical connections are generously sized and of sufficient cross-section for required current-carrying capacity, in order to lower contact resistance and to help dissipate any heat. Terminations for Ex 'e' apparatus are well-insulated and separate with secure clamping arrangements for cables. Anti-vibration designs of assemblies are necessary to be incorporated. They should also be positively located, and constructed in such a way that the conductors cannot come loose or be damaged by the clamping action (Figure 6.2–6.5).

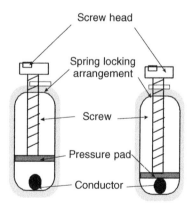

Figure 6.2
Typical single-conductor terminal

Acceptable methods for internal connections or joints and clearances:

- *Clearances*: Minimum clearance distances are specified between conducting parts at different potentials, which vary depending on the working voltage.
- *Creepage distances*: The grading of different insulating materials is covered and minimum creepage distances for different working voltages.

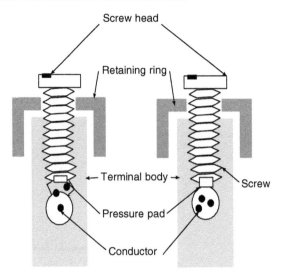

Figure 6.3
Typical multi-conductor terminal

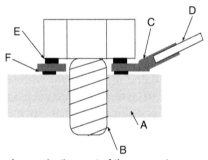

A = conducting part of the apparatus
B = connection screw
C = spade or eylet termination
D = conductor crimped, soldered, welded
        or brazed to termination
E = spring washer to prevent loosening
F = star washer

Figure 6.4
Typical spade/eyelet termination

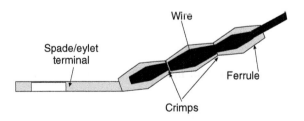

Figure 6.5
Typical crimped connection

Insulation

Insulation needs to be of good quality in design and construction. Insulation properties are tested using a method described in the standard called the Comparative Tracking Index (CTI) test. This gives a measure of the insulator's ability to resist electrical breakdown between adjacent conducting parts over the surface of the insulating material.

- Surface profiles for insulating materials
- Insulating materials including thermal stability, temperature rating, strength and requirements for varnishes.

Windings

This section covers the minimum wire sizes, impregnation techniques and test voltage for different wire sizes.

- *Limiting temperature*: In general the 'T' rating must not be exceeded by any part of the equipment within the t_E time during normal operation, or starting, in order to prevent ignition of an explosive atmosphere.
- *Overtemperature protection*: The windings have to be protected to ensure the limiting temperature cannot be exceeded in service. Usually using a suitable current-dependent device does this. This device needs to be selected with great care and the calibration may need to be checked, to ensure that it will operate as required, i.e. it must disconnect the motor from the supply within the t_E time if the motor is subjected to a locked rotor situation.
- *Internal wiring/layout of conductors*: The risk of localized heating due to power dissipation in conductors is considered and cables are not to be loomed in an enclosure where hot spots could be created. Adequately rated and usually oversized conductors are specified to minimize any heating effect.

Degree of protection

Minimum IP ratings; for motors with bare live parts this is IP 54.

- Maximum values for t_E time for different starting current ratios. The t_E time can not be less than 5 s and the I_a/I_n ratio cannot exceed 10.
- Limiting temperatures for insulated windings for different classes of insulation.
- Requirements for impact tests.
- Requirements for cable glands for use with Ex 'e' equipment.

Effect of ambient temperature

Generally it is assumed that only the temperature classification will be altered, when the hazardous area located electrical equipment is subjected to ambient temperatures that are outside of the certified ambient temperature range. But this need to be examined more closely as, it may also cause the method of protection to be bypassed. This may be permanent and leave the certified electrical equipment unsuitable for use in a hazardous area.

The problems include:

- Increased safety equipment is designed and certified to prevent ignition of the flammable atmosphere by avoiding sparking and ignition-capable hot spots. Subjecting increased safety equipment to ambient temperatures above the certified ambient temperature range may cause the materials to distort and the electrical insulation to breakdown, which may lead to surface tracking. This could cause ignition of the flammable atmosphere. Increased safety equipment is not designed to contain an internal explosion.

- Ambient temperatures above the certified ambient temperature range may also cause the gap between the cover and the enclosure to increase, this may reduce the minimum ingress protection to a level below what could be certified.

- At ambient temperatures below the certified ambient temperature range, the material may become brittle and may not be able to pass the required impact test.

6.5.1 Additional requirements for rotating electrical machines

If Ex 'e' motors are not suitably protected and terminated then product safety and certification is invalid.

Standard Ex 'e' motors are not designed or tested for arduous duty. It may be possible, however with careful motor selection and additional thermal protection, to obtain certification for arduous duty – but only after testing.

In addition to extreme testing Ex 'e' motors also require:

- Special terminals
- Defined types of internal cable connections
- Increased clearances between conducting parts
- Non-hydroscopic insulation with high-tracking resistance
- High-quality insulation and varnish
- Minimum air gaps – minimum radial clearances (or air gap), between the rotor and stator at different speeds
- Clearance for internal fans.

All of the above plus more has to be documented and submitted to the appropriate testing authority for approval.

These additional measures ensure the integrity of the motor, and account for its increased cost. Where human life and property are concerned, the extra safeguards pay their way many times over. Ex 'e' is an abbreviated way of saying 'protection and peace of mind'.

6.6 Principles of testing

As we have seen that increased safety apparatus calls for high degree of integrity of material and manufacturing its assessment involves checking that the manufacturer has complied with the design parameters of the standard.

Thus it calls that testing program for any type of device should include the following as minimum:

- *Test of creepage distances*: This will involve determining the CTI of the insulating material. This is required to be done to establish the minimum distance, also known as creepage distance, required between the live parts or live parts and ground.

  Special apparatus consisting of application of test voltages and ammonium chloride are applied to the material. Two electrodes spaced 4 mm apart are used to apply the test voltage. This determines the grade of insulating material and hence the resultant properties.

  Ceramic materials are exempted from this test.

- *Temperature-rise test*: This test is carried out so as to determine the temperature class of the apparatus. Unlike flameproof apparatus, all surfaces, including internal surfaces of the apparatus, are considered. Exceptions would be internal components using exclusion or containment techniques, e.g. encapsulation or flame proofing.

  For the apparatus to pass the tests not only itself but also the components that are housed in it should also be within the limiting temperature range as per limits of the insulating material used. These limits are to be in line with industry standards, e.g. cable insulation, or are given in AS 2380.6, e.g. insulation of motor windings.

- *Degree of protection tests*:   As already explained this is an important aspect of this protection and the apparatus has to meet various levels of protection against the ingress of solid objects or water specified, which must be tested in accordance with codes and standards as specified for *Degrees of protection provided by enclosures for electrical apparatus (IP Code)*.

  For enclosures containing live bare parts the minimum IP rating is IP 54. For enclosures containing only insulated parts a minimum IP 44 rating is required.

  Rotating electrical machines may be assigned a lower IP rating.

6.6.1   Principles for testing of motors (certified as Ex 'e')

The motors are required to be extensively tested before being certified. A brief description of the same is hereunder,

- *Electrical strength* is established through high potential test.
- *Rate of rise of temperature during starting* is established through locked rotor tests. Measurement of rate of temperature rise for the stator and rotor during locked rotor. Measurement of locked rotor current, which is measured 5 s after switch on.
- *Temperature rise tests* at full load at rated voltage and frequency for the stator and the rotor.
    1. Experimental determination of temperature rise.
    2. Calculation of temperature rise.
    3. Determination of the t_E time.

The t_E is determined from Figure 6.6; OC is the limiting temperature. OA is the ambient temperature (usually 40 °C). AB is the maximum temperature rise for the rotor or stator during full-load operation. BC is the maximum rate of temperature rise for the rotor or stator during locked rotor. Separate calculations are done for the rotor and for the stator, and the smaller of the two t_E times is taken as the t_E time for the motor.

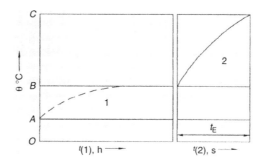

Figure 6.6

It is not unusual for the rotor to be the limiting factor in determining the t_E time. Some motors need to be de-rated from their normal full-load outputs in order to comply with all the requirements of the standards.

6.6.2   A typical test procedure for a motor

A generally followed procedure of 'testing' of a cage motor is described hereunder,

- Record ambient temperature, and continue to record during testing.
- Megger, Hipot of motor stator at ambient temperature.
- Measure winding resistance.
- Subject the motor to locked rotor, at full voltage, if possible. Record locked rotor volts and locked rotor amps at 5 s after switch on, and then switch off. Measure the temperature of the rotor using thermocouples attached to the rotor in a number of locations. Measure the resistance of the stator winding immediately after switch off, and record the time in seconds after switch off. It is recommended to use an automatic recording device such as a data logger. The rotor temperature will continue to increase for a considerable time after switch off as the heat at the center of the rotor conducts to the surface. The stator-winding temperature will start to decrease immediately after switch off. This data is used to determine the rate of temperature rise during locked rotor, and the starting current ratio.
- Run the motor at full load until the external temperatures stabilize. Switch off and measure the winding resistance and the rotor temperature, which again will continue to rise for several minutes after switch off. It can be quite complicated technically to arrange the measurement of the rotor temperature in a number of locations; however, it is usual to start recording after the motor stops. There are a number of ways this can be done. This data is used to determine the temperature rise of the motor.
- The procedure above is considerably simplified, but gives an idea of what is involved. These tests are carried out with extreme care, are quiet complex and usually involve a number of sample motors.

6.6.3   Testing of heating devices

Resistance-heating cables and other heating devices, which are exposed to high temperatures, are subjected to mechanical stresses. The devices or cables to be tested are subjected to being crushed by a steel rod of 6 mm diameter with a 1500 N force applied. They are further subjected to being bent through 90° on the mandrel, as shown in Figure 6.7, at its lowest operating temperature. The device or cable has to pass these tests without any damage to the heater or its insulation. If these are to be used under water then they need to be tested while immersed in water.

6.7   Periodic testing and repair of electrical apparatus

The main points for the periodic testing and repair of the equipment by the user result from the protection aims for type of protection increased safety 'e'.

- *Exterior condition of the enclosure*:
  - The enclosure must be visually checked at regular intervals. When this is done, it must be examined for holes and cracks that could allow the entry of moisture and dirt.

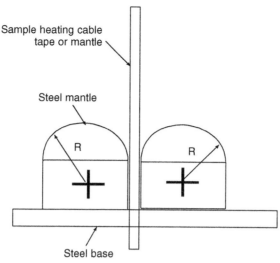

Sample heating cable
tape or mantle

Steel mantle

R

R

Steel base

R = minimum bending radius specified by the
manufacturer

Figure 6.7
Test apparatus for resistance heat devices subject to bending stresses in service

- Further, when the apparatus has been opened up, it must be examined to see if moisture and dirt have managed to penetrate to the interior via the sealed gap. The sealing system must be examined thoroughly. Holes or brittleness in the gasket or visible mechanical damage to the lip of the gasket are not permissible.

- *Interior condition of the enclosure*:

  - The electrical wiring and connections inside the apparatus must be regularly examined for traces of inadmissible large thermal influences. The terminals must be checked regularly for tightness. If screw terminals have to be tightened, this must be done with the torque specified by the manufacturer.
  - Checking to the permitted maximum temperatures in normal operation must be carried out regularly. It can, for example with plugs and sockets, be sufficient just to touch them as a check. If necessary, contact thermometers approved for hazardous areas may be used. Temperature indicators that can be stuck to the enclosure have also proved themselves. A further effective indicator is the change in color of the insulation material resulting from thermal influences.

- *Examination of the condition of cable glands*:

  - The cable glands, together with their cables, must also satisfy the minimum requirements of the IP protection type, i.e. IP 54. In addition to that, they also serve as strain relief for the cables, which are terminated inside the enclosure. Both of these should be ensured by checking the tightness of the cable gland (union nut should sit firmly) and by a visual check of the interface between it and the enclosure, and the interface between the gland, gasket and cable.

– For the case that unused openings are closed by means of stopping plugs, the enclosure has to be checked that it is tight at this joint and the external condition of the stopping plug has to be checked.

- *The main servicing tasks are*:

    – Cleaning of the interior of the enclosure
    – Proper tightening of the electrical screw connections (meeting the specified torque)
    – Replacement of gaskets, if necessary.

When cleaning the outside surfaces of the enclosure, it is essential to take note of any instruction from the manufacturer concerning the necessity of using damp cloth. These instructions are always attached to enclosures that have a high surface resistance and thus conceal the danger of electrostatic charging and discharging when cleaned with a dry cloth.

The possibility of repairing apparatus in this type of protection by the user is better than that in the type of protection flameproof enclosure 'd'.

The jobs for which generally authorizations and the tests are not prescribed are illustrated in the table below.

Description of Activities/Work	Acceptance by Authorized Expert for Increased Safety 'e'
Installation or replacement of terminals according to conformity certificate	Not required
Removal of terminals	Not required
Drilling holes for screw glands to manufacturer's specifications	Not required
Replacement of cover screws and gaskets	Not required
Replacement of type-tested devices	Not required
Replacement of the reflector or the protective glass by original spare parts	Not required
Replacement of type-certified lamp sockets and interlock switches	Not required
Replacement of tungsten lamps or fluorescent lamps of approved type	Not required
Replacement of tungsten and glow lamps by equivalent approved type	Not required
Replacement of ballast accord to conformity certificate	Not required

6.8   Conditions of use

Whilst type of protection 'e' has features in common with type of protection 'n', it is, in many respects, more stringent (e.g. in the case of motors lower temperature rises are specified and special overload protection is required to avoid excessive temperatures under all conditions including stalling).

This type of protection is used mainly for terminal and connection boxes, control boxes for installing Ex-components (which have a different type of protection), squirrel-cage motors, light fittings, etc. (Figure 6.8).

Apparatus with type of protection 'e' may be used in Zone 2 gas and vapor risks with any type of enclosure, which is suitable for the environment, provided it is permitted in the respective equipment standards.

Apparatus with type of protection 'e' may also be used in Zone 1 gas and vapor risks provided that:

- The enclosures of live bare parts and insulated parts are to degrees of protection IP 54 and IP 44 respectively as a minimum, except that where there is a likelihood of harmful gases and vapors entering the enclosure in quantities likely to cause deterioration of the insulation, the enclosure of insulated parts shall also be to IP 54 as a minimum.
- In the case of motors, the methods of control of the rotor and stator winding temperatures are strictly in accordance with the above standards. The devices used for temperature control, whether of the current-dependent or temperature-detector type shall be of high quality and shall be regularly tested.

Apparatus with type of protection 'e' is marked with a temperature class (T_1–T_6) and shall not be installed where flammable materials are used which have ignition temperatures below the maximum for that class.

Although apparatus with *type of protection 'e'* is suitable for use in all gases and vapors, provided account is taken of surface temperature considerations, it is sometimes used in combination with parts which have some other form of protection (e.g. switches which are *flameproof*), in which case attention shall be paid to any gas or vapor grouping of the parts with the other forms of protection.

No modification, addition or deletion shall be made to apparatus with type of protection 'e' without the written permission of the certifying authority (such permission shall be obtained through the manufacturer of the apparatus) unless it can be verified that such change does not invalidate the certification.

Apparatus with type of protection 'e' is suitable for use in dust risks and in combined dust and gas/vapor risks, provided that the additional precautions specified for use in dust atmosphere are complied with. When selecting apparatus special care shall be taken to ensure that the apparatus and its component parts are constructed so as to guard against electrical and mechanical failure in the intended conditions of use. Particular attention shall be given to the need for weather-proofing and protection against corrosion.

Fluorescent light fittings have no moving parts and can be constructed and certified Ex 'e' (Figure 6.9). The starters for this type of light are current-sensitive switches and therefore cannot normally be included. The starter is encapsulated, certified Ex 'm' and given a component approval for use in a given Ex 'e' certified design.

Instrumentation loops are not usefully accommodated within this method. The enclosures are often purchased as junction boxes for use on IS circuits because they are robust and reliable as proven by the Ex 'e' testing. The Ex 'e' certification is not used and should be removed from the box because it can cause confusion over how circuits are protected.

Figure 6.8
Sheet metal emergency light fittings type 6018 explosion-protection marking: II 2 G; EEx e d m IIC T4
(Source: R. Stahl)

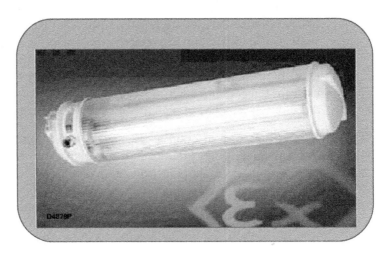

Figure 6.9
The explosion-protected light fitting for extreme conditions, certified Ex 'e'
(Source: R. Stahl)

6.8.1   Ex 'e' motors for surface industry

Induction motors (as opposed to 'commutating' motors) may be designed, as Ex 'e'. Some of the attributes are,

- Physically larger than an equivalent safe area rated device
- Clearances and cooling aspects increased for the security
- Overcurrent trips matched to the motor characteristics used as part of the installation.

These motors need to be wired up with care to ensure the whole system meets the requirements of the standards. If a motor is operated without a suitable overload then the whole installation will no longer meet the requirements of the standard, and would no longer be considered safe.

Ex 'e' motors that are suitable for Zone 1 areas, are generally to be 'approved' for gas Groups IIA, IIB and IIC with a T_3 or suitable temperature rating depending upon use.

These type of motors are generally not explosion-proof, are not suitable for use in a coal mine, and are not approved for use in Group I atmosphere. However, they are popular in petrochemical industries, in spray-painting booths driving extractor fans, etc.

If these motors are to be used with high inertia loads or with frequent stop/starts it may be possible to use a carefully selected inverse time delay overload, and/or additional thermal protection and still maintain the requirements of the standards.

Ex 'e' motors have to be manufactured to high standards, and the prototypes are to be extensively tested (refer clause for testing). A great deal of detailed information has to be documented and submitted to the testing authority before approval is issued, which may help to explain why these motors may appear to be expensive when compared to a standard motor, but the integrity of the product is vital.

6.9   Standards for Ex 'e'

The standards and practices as applicable in various parts of the world are as follows:

- South Africa: The governing standard is SABS IEC 60079-7.
- European Union: CENELEC-EN 50019 (the same as BS 5501: Part 6) is the now accepted standard in Europe.
- United Kingdom: Equipment to the old BS 4683: Part 4 is still manufactured and sold in the UK but new equipment is certified to the European Standard BS EN 50019 or BS 5501-Part 6.
- France: The governing standard is NF EN 50019.
- Germany: The governing standard is VDE 0171 T.6.
- IEC: IEC 60079-7, written in 1990 follows the European Standards and the German Code.
- Australia: The governing standard is AS 2308.6.
- This technique has not been accepted in the USA and Canada up until very recently. The advantages are now being recognized and apparatus is now seen on installations.

In Europe and other parts of the world where Ex 'e' is accepted, use in Group I and Group II; Zones 1 and 2 are permitted.

7

Protection concept 'n'

7.1 General

The most widely used forms of explosion protection, which utilize the technique of energy limitation, are non-sparking and IS. While both share a common foundation, they do differ greatly in many aspects. These differences deal mainly with the application of safety factors. It is for this reason that non-sparking is limited to Zone 2 hazardous locations while IS is acceptable for Zones 0, 1 and 2 locations. Each does have its merits, which is why it is not uncommon to see the two techniques used together. An example is a non-sparking I/O system in a Zone 2 location with an intrinsically safe interface providing connections to the Zone 0 field devices.

It should also be noted that earlier non-sparking was not recognized outside of the United States and Canada but now it is acceptable worldwide in the majority of countries. While IS has been accepted worldwide from an early time, it is generally regarded as the safest form of explosion protection. To date no explosion can be attributed to an intrinsically safe system.

The perspective of the user is that Ex 'n' is a less-costly approach than IS because no interface (e.g. barrier or isolator) is required. It could be argued that the overall installation is less safe with Ex 'n' than with Ex 'i' with only a marginal cost saving. The use of Ex 'n' remains restricted to Zone 2 only. This raises the concern that area classification may be influenced in order to accommodate Ex 'n' apparatus.

7.2 Definitions

Type of protection N is defined in the standard as:

> *A type of protection applied to electrical apparatus' such that, in normal operation, it is not capable of igniting a surrounding explosive atmosphere and a fault capable of causing ignition is not likely to occur.*

The general requirements of such apparatus are that it shall not, in normal operation:

- Produce an arc or spark unless:

  - The operational arc or spark occurs in an *enclosed-break device*

    or

  - The operational arc or spark has insufficient energy to cause ignition of a *flammable atmosphere*

    or

– The operational arc or spark occurs in a *hermetically sealed device.* It is to be noted that sliding contacts are considered to be sparking in normal operation.

- Develop a surface temperature or hot spot capable of causing ignition of an external *flammable atmosphere.*

It is to be noted that this requirement applies to the temperature of internal and external surfaces to which a surrounding atmosphere has access, except internal surfaces within *enclosed-break devices, hermetically sealed devices* or *restricted-breathing enclosures.*

7.3 Principles of design

The 'non-sparking' concept of protection was originally accepted as safe on the basis that manufacturers used good-quality and well-designed industrial equipment with little or no additional requirements. This is if it is operated well within its rating and installed in areas where the risk of contact with a potentially flammable atmosphere were adequately low (Zone 2 only). Under these conditions the UK's Factory Inspectorate issued letters of 'no objection' to the use of such equipment when it first came into service but before any standard of construction was developed.

The Standard BS 4683: Part 3: 1983 in UK, AS 2380.9 in Australia and IEC 60079-15 internationally, eventually emerged to formalize the concept of good design using industrially graded apparatus and this was termed Type 'n' equipment. The standard clarified the method in that it permitted equipment to produce sparks or for the surface temperature of electrical assemblies to rise in temperature but not to levels that could cause ignition.

More recently, the method was re-designated Type 'n' under BS 6941:1988. It has been updated and made more flexible to include the specific needs of instrumentation where circuits with less than 75 V DC were recognized. Discontinuous contacts were permitted, provided the resultant spark could be shown non-sparking. The same curves are now required to assess the safety of Type 'n' circuits. The system is deemed safe in normal operation only, and faults in the equipment or components and their effect on the explosion-protection integrity are not considered. There are some parallels with Ex 'i' in that this technique is 'safe with no faults' and therefore could be likened to an unofficial grade of safety Ex 'ic' if the same logic applied to the IS technique is followed.

7.4 Certification

Type of protection 'n' is the term applied to apparatus certified by *BASEEFA* as having been examined, type-tested where necessary, and found to comply with the requirements of a British Standard or a *BASEEFA* Certification Standard for *type of protection N* applied to a particular kind of apparatus (e.g. BS 5000 Part 16 for motors and *BASEEFA* Certification Standard SFA 3011 for control gear), or with the general British Standard for *type of protection n* (BS 4683 Part 3). The IEC standard applicable is IEC 60079-15.

7.5 Construction requirements

The standards permit the construction of apparatus in such a way that during normal operation of the equipment, it is unlikely to become a source of ignition. The rules therefore seek to develop an acceptable level of integrity by considering standard industrial grade equipment and enhancing some aspects of its construction. This is as opposed to the inclusion of specific electrical or mechanical techniques to prevent ignition, which are applied in some of the other methods.

The case or enclosure of the equipment must also be able to withstand a 7 Nm impact test; the resultant damage must not cause any electrical faults, which would then make the equipment unsafe. Similarly, environmental conditions must not be permitted to defeat safety and therefore the enclosure must be designed for adequate level of ingress protection, i.e. IP 54. Damage to the enclosure because of the test must not degrade the IP rating.

As this protection concept covers different types of apparatus, which in normal operation,

- Are non-sparking
- Produce arcs
- Produce sparks
- Produce hot surfaces.

Because it has to cater to the above requirements, hence a collection of various techniques is to be employed to cater to meet the safety. These are,

- Enclosed-break device.
- Non-incentive component.
- Hermetically sealed device, which in normal operating conditions cannot be opened and have a free internal volume not exceeding $100 \, cm^3$, and be provided with external connections, e.g. flying leads or external terminals.
- Energy-limited apparatus and circuits.
- Restricted breathing enclosure (restricted breathing enclosures are limited to use with gases and vapors whose restricted breathing factor is less than 20, see Table A1 in AS 2380.9 Appendix A for typical breathing factors; acetylene, hydrogen and isoprene have restricted breathing factors higher than 20).

It was originally conceived as a method for handling high-power equipment without the need for enclosures, which were strong enough to withstand an internal explosion. Rotating machines, transformers light fittings, terminal boxes could be certified Type 'n' (Figure 7.1).

If higher levels of heat or sparks were evident in the design of apparatus then extra precautions such as the inclusion of CLRs, sealed or encapsulated devices or small volume Ex 'd'-enclosed contacts were permitted as part of the protection method.

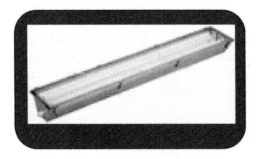

Figure 7.1
Totally enclosed fluorescent fitting – stainless steel, suitable as non-sparking Ex 'n'
(Source: Northland Lighting)

7.5.1   General constructional requirements

These apply to all types of Type 'n' apparatus and address,

- *Environmental protection*:   These circuits are supposed to be carrying incentive energies, and unrestricted entry of environment is assumed to be able to cause internal faults which would allow those energies to be released. Hence, as with other forms of protection exhibiting this phenomenon, the enclosure of the apparatus is required to provide a suitable degree of protection against entry of liquids and solids present in the external environment. For example, IP 54 or IP 44 type of enclosures can be used depending upon whether there are bare or insulated wires present and environment is not clean. These can be further reduced to IP 2X in case of a clean environment. It may be noted that even these types of enclosures are necessary where ingress is possible and could cause an ignition-capable situation.
- *Mechanical strength*:   The apparatus enclosure is required to be proof against mechanical damage and is shown by subjecting it to the mechanical impact test. This test may be carried out with lesser mass (say 0.25 kg) or height (say 0.2 m). The device would have passed the test as long as the IP rating and correct operation of apparatus is ensured, even if enclosure suffers some impact damage.
- *Wiring and connections*:   The internal and external wiring of Type 'n' apparatus need only comply with the requirements for normal industrial apparatus except that additional care needs to be taken to avoid damage by rubbing against metal parts or sharp edges.

    The wiring and connections, internal or external, are to be such that neither hot spot nor looseness does occur. These connections could be

    - Screwed
    - Bolted
    - Crimped
    - Soldered
    - Brazed
    - Welded
    - Pressure type
    - Mechanically wrapped by machine
    - Plugs and socket.

- *Conductor insulation and separation*:   This requires that as with increased safety 'e' and IS 'i', the requirements pertaining to breakdown of such distances that could produce incentive sparking either at the point of breakdown or elsewhere should be maintained for the following conditions:

    - Separation of uninsulated conductors in air (clearance)
    - Uninsulated conductors across a surface of insulating material (creepage in air)
    - Separation of uninsulated conductors across varnished surfaces (creepage under coating) and
    - Separation distances through encapsulation

7.6 Conditions of use

Whilst type of protection 'n' has features in common with type of protection 'e', it is, in many respects, more lenient (e.g. in the case of motors one need not specify the lower temperature rises as done in case of 'e' and need not commission any special overload protection to avoid excessive temperatures under all conditions, including stalling).

Apparatus with *type of protection 'n'* is suitable for use in *Zone 2* gas and vapor risks. The type-tests for normally sparking contacts are carried out in a gas mixture which ensures that apparatus with *type of protection 'n'* is suitable for use in all gases and vapors (including hydrogen and acetylene) when mixed with air, provided account is taken of surface temperature considerations.

Apparatus with *type of protection 'n'* is marked with a temperature class (T_1–T_6 in accordance with standards) and shall not be installed where *flammable materials* are used which have *ignition temperatures* below the maximum for that class.

No modification, addition or deletion shall be made to apparatus with *type of protection 'n'* without the written permission of the certifying authority (such permission shall be obtained through the manufacturer of the apparatus), unless it can be verified that such change does not invalidate the certification.

Apparatus with *type of protection 'n'* is suitable for use in dust risks and in combined dust and gas/vapor risks if the additional precautions specified for protection against dust ingress are complied with.

When selecting apparatus special care should be taken to ensure that the apparatus and its component parts are constructed to guard against electrical and mechanical failure in the intended conditions of use. Particular attention shall be given to the need for weatherproofing and protection against corrosion.

7.6.1 'Division 2 approved' apparatus

Prior to the introduction of apparatus with *type of protection 'n'* apparatus which was considered suitable for use in Division 2 (now *Zone 2*), gas and vapor risks were submitted to *HMFI (Her Majesty's Fire-service Inspectorate)* for examination, BS 4137 forming the basis for this examination. Where the apparatus was found satisfactory *HMFI* issued a letter stating that no objection would be taken to the use of the apparatus in Division 2 areas. The apparatus was therefore commonly referred to as *'Division 2 approved'*.

The conditions of use of this apparatus (including its use in dust risks and in combined dust and gas/vapor risks) are similar to those for apparatus with *type of protection 'n'*, except that as no temperature classification is stipulated, the responsibility for making a judgment on this aspect falls on the user.

7.7 Illustrations

Non-incentive equipment installation

Simply put, if you supply non-sparking equipment with its rated power and use it as per its instruction, it will not produce heat or sparks capable of igniting a hazardous atmosphere. It must not have normally arcing contacts and must not produce heat in excess of 80% of the ignition temperature of the hazardous atmosphere. There is no fixed restriction on the amount of power as long as the preceding statement remains true. The energy level carried by the wiring may far exceed the LEL. Therefore, a Zone 2 wiring method must be used (Figure 7.2).

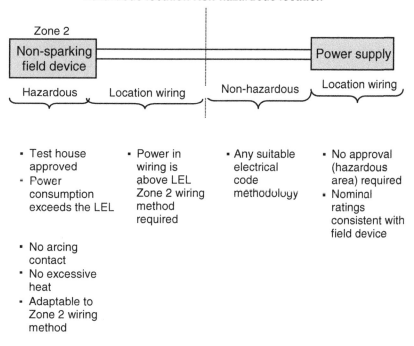

Figure 7.2
Non-sparking equipment installation

Quick disconnects are considered normally arcing if they do not require a tool for quick disconnection.

7.8   Standards for Ex 'n'

The following standards govern this protection concept,

- Australia: AS:NZS 3800:1997.
- South Africa: The governing standard is SABS IEC 60079-15.
- IEC: The standard is known as IEC 60079-15 which is generally based upon BS 6941.
- European Union: CENELEC – EN 50021 governs this protection concept.
- United Kingdom: The governing standard is BS 6941.
- Germany: pr DIN EN 50 021 or pr VDE 0170/0171 T.16 are governing standards for this protection concept.
- France: pr C 23-521 is governing standard for this protection concept.

In the UK, prior to the adoption of BS 6941, the Code of Practice (BS 5345: Part 7) did not give adequate guidance as to how to maintain instrumentation systems that were installed as Type 'n'. Many organizations avoided its use owing to lack of clear rules on whether live working was permitted. This has now been clarified in BS 6941, where, before work is carried out, the maintainer must assess the safety criteria of performing the required work on the system. Procedures must give clear instructions to technicians, with this done this may find favor with users and manufacturers as it substantially reduces cost of the equipment.

8

Protection concept 'i' principles

8.1   Origins of intrinsic safety

The origins of IS date back to the Senghenydd Coal Mine Disaster on 13 October 1913.

During the subsequent investigation, Messrs Wheeler and Thornton, found documents showing that just prior to the explosion occurring, the old 'wet-cell' type of battery, used to power the bell signaling system (shown in Figure 8.1) had been replaced with a more modern 'dry cell' type. In order to find out if this could have been the cause of the incident they set up a test, using equipment they had devised called 'spark test apparatus', to reproduce the conditions in which the circuit of the signaling system would have worked underground.

The test comprised a contact, making and breaking, in an envelope of methane or firedamp, to represent the switching effect of the spade across the signaling wires. The test revealed that the new cells were capable of producing a spark large enough to ignite the gas whereas the old cells could not. The internal resistance of the dry cell was lower than that of the wet cell and so it had a greater current delivery capacity.

Wheeler and Thornton had effectively stumbled upon the technique of energy limiting that is now known as IS.

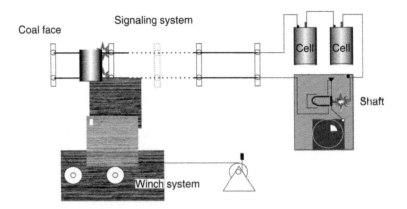

Figure 8.1
The mine signaling system circa 1900

8.2   Principles of IS

The fire triangle (Figure 8.2) analogy can be used to explain the objective of IS. In the presence of the 'most easily ignitable mixture' of a given flammable vapor with air, ignition *cannot* occur if the levels of heat or sparks are insufficient.

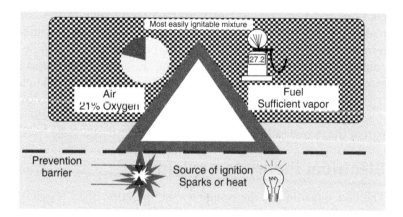

Figure 8.2
The fire triangle

The principle of IS is to ensure that levels of heat or size of sparks that occur in an electrical circuit which comes into contact with a flammable gas, are limited to below those which will cause ignition.

An IS circuit is defined in Standard IEC 60079-11 as:

A circuit in which any spark or thermal effect produced in the condition specified in this International Standard, which include normal operation and specified fault conditions is not capable of causing ignition in a given explosive gas atmosphere.

The standard repeats and qualifies this in where it declares three basic criteria that must be satisfied:

1. Separation from other circuits
2. Temperature classification
3. The inability to cause ignition by sparking.

The definition suggests that sparks and heat are permitted in a circuit under specified fault conditions, but these must never exceed levels that could be incentive. These criteria and the requirement for separation from other circuits provide the high integrity necessary.

The use of the term 'circuit' has important implications. Electrical energy can only produce heat or sparks when electricity is flowing. Since it can only flow in a complete circuit, it is the safety of the circuit that is of concern. The components of the circuit do not pose a threat unless electricity is passing through them, in which case they must be part of a 'circuit'. Heat and sparks occurring in this circuit can be assessed for compliance with the standards for IS.

A 'circuit' can mean any of the following arrangements, increasing in complexity:

- A single cable looped through a hazardous area.
- An assembly of electrical components working together as an electronic device (such as an instrument).
- A number of assemblies can be interconnected in the same circuit.

The circuit may operate with energy levels that are quite safe under normal conditions. Under probable fault conditions acting within or onto the circuit, the circuit must still not be able to emit heat or sparks in sufficient quantities to cause ignition when it encounters a hazardous area. Internal faults and certain external faults must be adequately protected against.

The possible 'faults' are predicted by careful examination of what failure mechanisms could occur. Components are built into the design of the complete circuit in order to maintain energy to known safe levels under these fault conditions. These components are termed 'safety components' and are in-built into the operation of the circuit. To further enhance the integrity, the failures of the 'in-built' components are separately assessed to ensure that if they fail in a specified manner, safety is still maintained.

The dictionary definition of the word 'intrinsic' is given as 'of its nature' or 'in-built'.

Safety, which is achieved by the inclusion of specific components built into a circuit, is part of the nature of the design and it is therefore said to be 'intrinsic'.

8.3   Electrical theory to explain IS

The understanding of the principles and practice of IS requires a working knowledge of simple electrical theory as explained in Chapter 2.

The principles are explained by the application of Ohm's law.

Suppose that a conventional (constant-voltage) power supply provides an open-circuit output voltage, V_{OUT}. In theory, it would be able to deliver from zero up to an infinitely large current.

The current, I_L, supplied to the load will be controlled by the load resistance, R_L, such that:

$$\frac{V_{OUT}}{R_L} = I_L, \quad \text{according to Ohm's law}$$

8.3.1   Power dissipation

The power, P_L (measured in Watts), dissipated in the load resistor R_L is:

$$V_{OUT} \times I_L = P_L, \quad \text{again, according to Ohm's law}$$

In a simple circuit (Figure 8.3), as a result of the power dissipated in the resistance R_L, the temperature rise experienced will depend on its physical properties such as mass, surface area, heat-dissipation properties, the ambient temperature and external cooling effects.

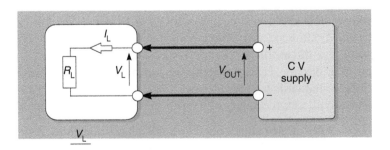

Figure 8.3
Power dissipation in a load R_L

Now consider the effect of taking values between:

R_L = an infinitely high value (where no current is drawn from the supply)
    through to
R_L = an infinitely low value (where the highest current is drawn)

Power dissipation in R_L is inversely proportional to the value of resistance (when the voltage stays constant). In this case, the power dissipated in the R_L increases by the square of the current.

$$\frac{V^2}{R} = \text{Power}, \quad \text{another aspect of Ohm's law}$$

The current delivered from any source of supply is dependent on its 'internal' resistance. A battery has a low internal resistance when charged but this rises as the battery becomes exhausted. Its internal resistance governs the short-circuit current of any simple supply system such as a battery. (A power supply using an electronic regulator is designed to maintain artificially low internal impedance throughout its current supply range.)

In order to limit the total power supply capability of a voltage source, a 'current limiting resistor' (CLR) may be included in the circuit (as artificial internal resistance). This will provide a minimum circuit resistance. The circuit is shown in Figure 8.4.

The effect of changing the value of R_L (in series with a constant R_{CLR}) may now be considered. Taking values between an infinitely high value, open circuit (o/c), through to an infinitely low value, short circuit (s/c), the power dissipated in R_L can be examined.

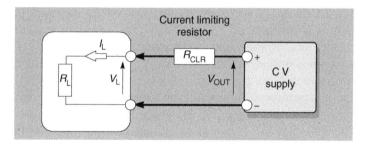

Figure 8.4
Power dissipation in a load with CLR

When R_L is either at either extreme of high values or low values then the power dissipated in R_L is zero. This is because:

- When R_L is very high (open circuit), the circuit current is zero and so $I^2 R_L$ will be zero or
- When R_L is low (short circuit), the current is at a maximum, limited by the value of R_{CLR}. As R_L is zero, $I^2 R_L$ will be zero.

It follows that there must be a value for R_L where the power dissipated in it reaches a maximum value. This is when $R_L = R_{CLR}$ and can be proved with a few simple calculations.

Artificially setting the *minimum* circuit resistance therefore enables control of the maximum power that can be dissipated in loads connected to the supply.

The maximum amount of power dissipated in an IS circuit must not allow a surface temperature rise to occur in conductors or resistive components that come into contact with the gas/air mixture, such that ignition occurs. The amount of power involved will be discussed in more detail later.

8.3.2 Energy

The power dissipation in R_L has been examined during the transition from a low to a high resistance through variable intermediate values.

When the transition is made to happen instantaneously, for example, when a circuit is broken, the rates of change of voltage and current are extremely high. The transition from s/c to o/c, as in the opening of a switch, is known to cause a spark (see Figure 8.5).

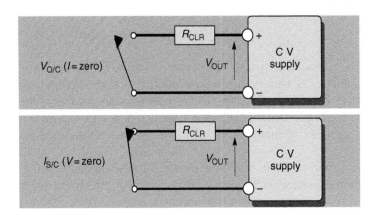

Figure 8.5
Switch action dissipating energy as a spark

A spark dissipates electrical energy. The size of the spark produced at the breaking of a circuit is a function of $V_{O/C}$ and $I_{S/C}$. The speed of transition between the two states precludes any 'power' being dissipated.

It is known that sufficient spark energy will cause the ignition of a flammable atmosphere, but the physics of spark ignition is still not fully understood. The size of the spark is determined empirically and the method of testing a gas for its spark ignition energy is described in the IEC Standard 79.

There are then two techniques that may be used for controlling the size of the spark that which will cause ignition:

1. Reducing the voltage, V_{OUT} and
2. Reducing the current, $I_{S/C}$, by raising the value of R_{CLR}.

8.3.3 Stored energy

In any electrical circuit, the natural effect of inductance, capacitance and resistance will be present.

Stored electrical energy is measured in Joules and is represented by the letter 'E'.

Stored energy may be calculated by the formulae as follows:

- For inductance, $E = \frac{1}{2}LI^2$
  where L is the inductance in Henrys, I is the current flowing through in amperes.
- For capacitance, $E = \frac{1}{2}CV^2$
  where C is the capacitance in Farads, V is the charge on the plates in volts.

Energy in the form of heat is dissipated by power in the resistance of the circuit.

Each and every source of stored energy in an IS circuit must be identified and assessed to ensure that it cannot provide sufficient energy to pose a danger. The stored energy must not be released in an uncontrolled fashion otherwise incentive heat or sparks may be produced. Clearly, where there are many sources in the same circuit then the likelihood of the energy becoming cumulative must also be assessed. It may be necessary to take additional precautions such that the risk of this is minimized to an acceptable level.

The circuit designer must therefore include components, which act to limit power, energy and stored energy in the apparatus. This must be done in a subtle way, which meets the requirements of the standard but does not inhibit or affect the operation or indeed performance of the apparatus. In this way the safety is 'in-built' or intrinsic to the apparatus. There are many techniques that may be used in the internal design of electronic apparatus, but these are closely protected secrets within the IS apparatus manufacturing industry.

8.3.4 Energy limits

The work of Wheeler and Thornton began by investigating the size of voltages and currents (limited by resistance) causing ignition. Empirical studies quantify the maximum values of voltage and current permitted in IS circuits. These cover voltage and current levels, current to inductance levels and voltage to capacitance levels.

As an example, by consulting these voltage – current curves, it will be seen that if a supply of 50 V was connected through a 1000 Ω resistor to allow 50 mA to be made and broken in the presence of the most easily ignitable mixture of hydrogen with air, then the resulting spark would not cause ignition. However, by increasing the current to 60 mA (reducing the resistance to 830 Ω) ignition could be expected. The graphs were plotted to take the worst-case points.

Figure 8.6 shows the ignition curves which are used for determination of voltage/current limits in IS circuits.

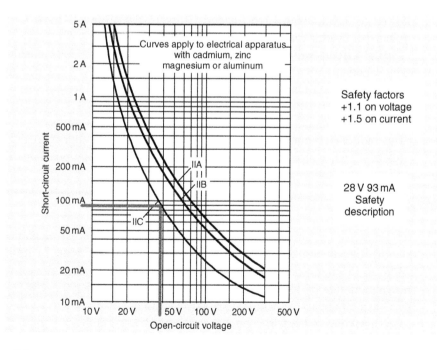

Figure 8.6
The method of determining the permitted current from a given safety description voltage

The use of the curves can be demonstrated by the example set out below:

- Determine the highest open-circuit voltage required: 28 V
- Apply a safety factor of 1.1 (i.e. add 10%)
- Find the intersection of the voltage and the curve
- Read off the current at the intersection
- Apply a factor of safety of 1.5 (i.e. reduce current by 33.3%)
- The value found is the highest short-circuit current.

For circuits operating in IIB and IIA gases or vapors, readings are taken from the appropriate curves.

It will be realized that in order to achieve a short-circuit current of not more than 93 mA from a source of voltage not greater than 28 V, the minimum series resistance, R_{CLR}, value can be calculated by:

$$R_{CLR} = \frac{V_{O/C}}{I_{S/C}} = \frac{28}{0.093} = 300\ \Omega$$

In some interpretations of the standards the 10% additional factor has not been considered necessary, and the extra available power from the circuit has been useful in some applications.

Note that the graphs specifically include the use of aluminum, magnesium, zinc and cadmium, which are known to exacerbate the sparking effect. Since these are in common use in modern electronic circuits, their inclusion in the standards forces the design to use the most onerous conditions, thereby increasing safety margins.

8.3.5   Stored energy limits

There are also maximum permitted values for stored inductance and capacitance against given current and voltage values, respectively. Storage of energy in circuits must not be allowed to accumulate above incentive levels. These levels must be realistic such that circuits can operate and be made safe.

Typical circuit values found on a powered instrument loop can be of the order of 500 µH and 0.1 µF. The working limits of current and voltage due to these values are shown in relation to the IIC (hydrogen) curve in Figure 8.7.

Factors of safety of 1.5 are added to current values and 1.1 added to voltage values, as determined by the design standards for IS apparatus and systems.

The inductance curve for hydrogen, IIC, is usefully approximated to a straight-line value of 40 µJ, which makes calculations for stored energy possible. The same does not apply to capacitance values.

8.3.6   Power limits

Returning to the subject of power dissipation, some experimentation was needed to quantify the amount of power permitted in a circuit as heat. In limiting the power to 1.3 W an IS circuit has been shown to develop a temperature rise of well below 135 °C from an ambient of 40 °C. Therefore the standards state that T4 (135 °C) may be awarded for a piece of apparatus, which dissipates less than 1.3 W.

This provides a load line on the voltage/current graph in Figure 8.7. The combined general limits of power and energy can be visualized in this representation. It can be seen that the choice of voltages and currents revolve around the operating point, which gives the most flexibility to the design of circuits.

These are not absolute limits. Power, energy, voltage and current levels may be increased or varied to suit special applications. Trading off one characteristic for another may do this. High-voltage systems are permitted by introducing higher resistance, thereby limiting the current, and vice versa.

Using the example in 8.3.4, a 28 V 93 mA 300 Ω safety description may be analyzed for maximum power dissipation in order to assess the temperature rise experienced.

The power dissipated in a matched resistance (matched power discussed in Section 8.3.1), that will appear connected across the source of supply is given by the formula:

$$\frac{(V_{O/C} \times I_{S/C})}{4} = \frac{28 \times 0.093}{4} = 0.6503 \text{ mW}$$

Since this value is below the maximum value of 1.3 W dissipation, this satisfies the T_4 rating requirement.

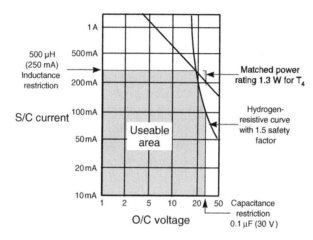

Figure 8.7
The useable area of IS

8.3.7   Component failure and infallibility

Limiting the voltage and current to maximum values is achieved by the specific inclusion of certain electrical components into the design of the complete IS circuit. Such components are required to be carefully chosen for their enhanced reliability. When designed into a circuit, they are said to be 'infallible'. The definition given in the standards is reproduced in Table 8.1.

An infallible component or assembly of components	Components or assemblies of components that are considered as not subject to certain fault modes as specified in this standard. The probability of such a fault mode occurring in service or storage is considered to be so low that they are not to be considered

Table 8.1
Fault definitions

The circuit's IS integrity may be influenced by the reliability of any of the components used and the way they have been designed into the circuit. Some assessment of the heating effect of these components is necessary during the design of the apparatus. It may be that their location and mounting is adjusted to minimize the effect. These components may otherwise affect the reliability of the infallible components. There are two types of faults in IS circuits, as defined in the IS standards and shown in Table 8.2.

A fault	Any defective component, separation, insulation or connection, which is not defined as an infallible component
A countable fault	Any defective component, separation, insulation or connection, which is not defined as an infallible component

Table 8.2
Fault definitions

The countable fault applies to a component, which is so placed in the design of a circuit to provide a limiting function in that circuit. The component is often referred to as a 'safety component'. The standard gives the criteria that these components must meet. These are stringent conditions of acceptable type and de-rating factors that the circuit designer must demonstrate have been met.

The safety components, which are added to the circuit in order to limit voltages and currents, are liable to failure during storage or service. The likelihood of failure must be reduced to an acceptable risk in order to say that it is safe. If the components failed, the circuit could become unsafe. Voltages and currents could rise to incentive levels.

The function of certification of apparatus is an affirmation that the circuit with its components has complied with the requirements of the standard as assessed by a competent authority. The built-in safety of IS forces the examination of all possible faults and their effect on the failure of the components. An IS category is awarded to the apparatus when it has been certified.

8.3.8   Categories of apparatus

The category of IS apparatus refers to the integrity of the preservation of the IS properties in the event of certain specified faults occurring in the components on which safety depends.

There are three recognized categories of IS apparatus in the IEC 79 Standard. These are as follows:

1. *Simple apparatus*: This category needs not consider any fault condition because no fault condition however caused could render the circuit in which it was placed in an unsafe condition. This concept is discussed in detail later on.
2. *Ex 'ib'*: Apparatus that may have one countable fault. Category 'Ib' equipment permits one countable fault to occur, yet the circuit will still remain safe. A second countable fault on an 'ib' circuit may render it unsafe.
3. *Ex 'ia'*: Apparatus that may have two countable faults. If two countable faults can occur but the circuit still limits the design voltage and current then it is awarded Category 'Ia'. A third countable fault occurring could then make the circuit unsafe.

Examples of fault counts will be more easily understood when the implementation of IS is examined in Section 8.5.

8.4   Implementation of IS

To explain the theory of IS and how it is implemented, a simple example will be used. This will demonstrate what has to be considered. Figure 8.8 shows a status-input loop to a control system.

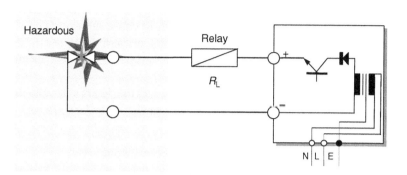

Figure 8.8
A typical application

The object of such a loop would be to transfer the status of the switch from one circuit to another. It would not be good engineering practice to try to switch high-power electrical circuits directly from field contacts. On conventional relay-logic control systems interposing relays are used between the field and the contactors. These handle the interfacing to higher-power devices such as motor starters in a more convenient manner. PLCs have now replaced many of these relay logic systems but the same concepts apply. A simple example will be used to show the thinking behind the electrical safety aspects.

The circuit in Figure 8.8 shows a constant voltage supply (normally 24 V DC) connected in series with a switch (in the field) and a relay located near the supply. The supply is derived from AC mains, rectified, smoothed and regulated. The regulator is shown in simple form for ease of explanation. The relay would be chosen according to its function. It must operate adequately with the voltage and current levels in the field circuit. It must provide the correct rating for contacts to drive other external circuits. There may be other low-power relays for logic and/or interlocking purposes or may be contactors for higher-power purposes.

The 'IS circuit', according to the definition, must consider all the electrical devices and connecting cables which are part of the instrument loop in the hazardous area, and the part or parts directly connected to these devices, even though they may be located and operated in the safe area. Devices that connect into circuits going into the hazardous area can also affect the safety.

The making and breaking of the contact will cause a spark. The voltage and current fed to the circuit will influence the size of the spark. It will also depend on the loop inductance of the circuit as a whole (ignoring the inductance of the relay coil at this stage). The resistance of the relay coil will limit the supply current value. Suppose that the supply was 24 V DC and the coil resistance was 100 Ω. The current would be 240 mA and the power in the coil would be 5.76 W. These are not unrealistic values if the relay was a contactor.

Comparing these values (24 V DC, 240 mA) with the permitted voltage and current values, they are below the IIC curve and therefore would not cause ignition. The resulting spark would not be incentive. Additional safety factors, discussed in Section 8.3.4, are applied to ensure this is the case.

The inductance of the relay would increase the size of the spark dramatically. When the field switch in Figure 8.8 goes to open circuit, the current falls to zero. The collapsing magnetic field of the coil cuts its own turns and produces a high reverse voltage called a 'back-emf'. The voltage is so high that it jumps the breaking contacts of the switch, producing a large spark. The stored energy held in the coil is therefore dissipated in that spark which could well be large enough to cause ignition. The value of inductance of any coil is complex in nature and depends primarily on the number of turns, the cross-sectional area covered and the magnetic properties of the medium within the area formed by the turns.

Other mechanisms that could influence the energy levels in any part of the circuit, which enters the hazardous area (in no particular order), are:

- The coil could become a short circuit, or shorted turns may decrease its resistance. In this case the current would rise.
- The supply could fail in two ways as seen in Figure 8.9. Fault 1 shows the collapse of the regulation circuit in a conventional series regulation system. Fault 2 is the worst-case failure mode, where the mains become connected to the output terminals of the supply by some internal catastrophic failure. This is more serious in that the instrument loop could have the mains supply directly across it. For the purposes of IS there is no distinction between these modes of failure.
- The insulation of the relay could break down such that voltages applied to the contacts may become connected to the coil circuit.
- The cable or connections forming the circuit in the hazardous area may be broken by external mechanical influences. A spark could occur at other places in the circuit other than at the switch.
- External energy could enter this system owing to mechanical faults in cabling, running in parallel with this cabling.

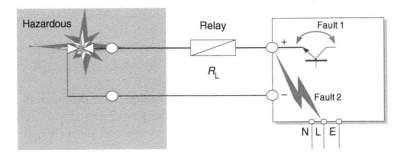

Figure 8.9
Mains faults in typical application

Of the greatest concern would be the fault mode where the live mains terminal was to become connected to the supply circuit (Fault 2). This would elevate the whole circuit with respect to earth. Under this single fault condition, the mains voltage superimposed on the low-voltage circuit may not be detected until a connection in the hazardous area is inadvertently made. This fault could be internal to the supply or external invasion causing the elevation. These faults, however unlikely they may seem, need to be considered if safety is to be assured at the highest level.

Taking the following additional precautions may protect against all the above situations:

- Limiting the voltage across the external circuit into the hazardous area
- Limiting the current in the external circuit
- Limiting the fault energy from supply failure systems
- Limiting the stored energy in the hazardous area circuit
- Limiting the likelihood of invasion from external sources.

These are the basic requirements of any IS circuit. The last point can only be controlled by care taken during installation. This aspect will be discussed in Chapter 12.

8.4.1   Energy-limiting 'system'

The arrangement shown in Figure 8.10 is a simple assembly of components that operate together to limit energy flow into the hazardous area. Firstly, it limits the voltage in a circuit by the use of the Zener diode and, secondly, it limits the current that can be sourced from that voltage by the inclusion of a CLR.

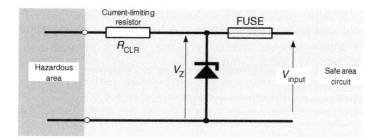

Figure 8.10
Voltage and current limiting arrangement

The Zener diode is connected across, in parallel or in shunt with the circuit to be protected. The operation of the Zener diode can best be explained graphically as shown in Figure 8.11.

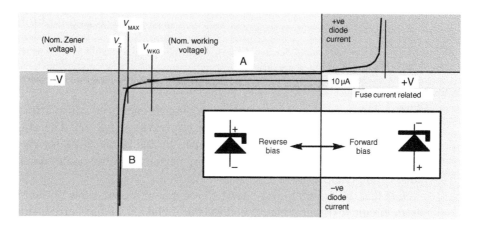

Figure 8.11
Zener diode characteristics

Used in the 'reverse bias' direction, the Zener diode does not conduct appreciably whilst the reverse voltage is well below the Zener voltage or Avalanche voltage as it is sometimes called. In the graph, Line A represents the leakage current of the order of Pico- or Nano-amperes at low voltages. This is considered insignificant until the working voltage ($V_{WORKING}$) is encountered. Line B (almost perpendicular to Line A) represents the constant voltage developed across the Zener diode when current flows through it. Where lines A and B meet is often referred to as the 'knee'.

The Zener diode junction 'senses' the voltage across it and reduces its resistance from infinity to a lower value. It tries to maintain a constant voltage across it by reducing its dynamic resistance. It may be thought of as acting like a voltage-sensitive resistor when an applied voltage exceeds the voltage $V_{WORKING}$. The supply is assumed to continue to provide current, which will dissipate power in the form of heat in the Zener diode. The highest voltage that can be developed across the Zener diode is known as the V_{MAX}.

Whilst the voltage across the Zener diode is 'clamped' at the maximum value, V_{MAX}, the highest current I_{MAX} that can be sourced from the circuit into the hazardous area is limited by the current limiting resistance (CLR) according to V_{MAX} divided by R_{CLR} as discussed in the theory above. This is said to be the 'safety description' of this network of components.

The safety description defines the open-circuit voltage and the short-circuit current that are available to the hazardous area. These values must remain below the values permitted by the standards when safety factors have been correctly applied. They may be expressed as a voltage, current and/or resistance. A typical example is 28 V, 300 Ω, 93 mA; 28 V divided by 93 mA gives 300 Ω.

This voltage and current will only persist until the safety fuse ruptures and clears the fault. Thus, the fuse has a most important role to play.

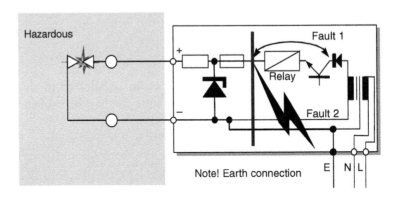

Figure 8.12
Energy-limiting arrangement in associated apparatus

This arrangement will not let through enough energy to cause a spark in the hazardous area during the normal operation of the circuit (Figure 8.12). The effect of the back-emf produced in the relay coil is also clamped by the diode in its forward-biased condition. The design of the safe area electrical circuit must accommodate the characteristics of the added components.

Note that the original relay specification may not be useable at this stage because of the extra circuit resistance introduced by the inclusion of the current-limiting resistor. This will be discussed more fully in the section on applications.

Clearly the Zener diode(s) will dissipate power from the source of supply during a fault where an overvoltage condition from the safe area supply exists. Given sufficient time under overload conditions, it would eventually overheat and fail if no other precautions were taken.

A suitably rated fuse is therefore interposed between the Zener diode and the source of supply such that when excess current is drawn by the Zener diode, the fuse will act to disconnect the whole circuit before any damage is done to the diodes. The breaking capacity of the fuse is specified in the standard at 1500 A for a maximum supply voltage of 250 V_{rms}. The fuse characteristics are carefully chosen such that it will operate reliably with the known parameters of the Zener diode combination.

The circuit must be referenced to earth. This is to ensure that the hazardous area circuit cannot float at mains potential.

The mechanical layout of the safe area arrangement must now be such that if Fault 1 or 2 occurs then there is adequate segregation within the design to ensure the limiting circuits cannot be bypassed. It is for this reason that the whole safe area device including the shunt diode network must be assessed for safety. This would become known as certified 'associated apparatus' (refer to Section 8.6).

8.5   The shunt diode safety barrier

The shunt diode safety barrier (hereafter referred to as a 'barrier') comprises the component arrangement discussed above but in a self-contained package. It may be interposed easily between safe and hazardous area circuits.

The arrangement shown in Figure 8.13 will provide energy limiting to the hazardous area under all supply failure conditions or sources of energy introduced into the safe area.

Note also that there is an extra connection to the barrier, which is referenced to the earth of the supply. It is this that provides the return path for fault current routing such that mains faults on the supply are detected and the fuse(s) act to clear the fault. With the introduction of the safety barrier, no failure modes of the power supply now need be further considered. Earthing is discussed in detail in Chapter 11.

The great benefit of using this type of external interface is that the safe area equipment does not need to be assessed for safety, provided that it meets certain simple criteria discussed in Chapter 12, on installation requirements.

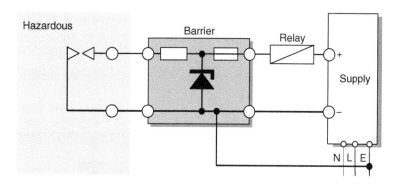

Figure 8.13
Barrier protected IS circuit

8.5.1   Component infallibility

The components that provide voltage and current limiting are therefore critical to the safety of any circuit to which they are applied. The consequences of their failure would render the circuit unsafe.

Understanding how they may fail will allow adequate precautions to be taken to ensure that certain failure modes do not compromise the explosion-protection integrity. If higher reliability devices are chosen, this reduces the initial risk. Such components are referred to as 'infallible components'. If these components are then backed up by duplication or even triplication, where necessary, then the likelihood of failure is reduced to such an acceptable level that the technique is considered to be the safest and may be used in a continuous hazard situation (Zone 0).

The standards therefore specify some conditions that must be applied to these components. These will be discussed now.

8.5.2   Zener diodes

Zener diodes are treated in IEC 79-11 under Section 13.7, which deals with the specific requirements of shunt-safety assemblies. Examining the failure modes of shunt Zener diode assemblies shows that they may fail to open circuit or short circuit. Zener diodes used in barriers are normally 5 W dissipation types of a special construction. The standard describes a temperature cycling test and a pulse testing technique necessary to ensure that the Zener voltage is consistent and will pick up any substandard Zener diodes.

8.5.3   Failure to short circuit

The voltage developed across the Zener diode would collapse to zero and so actually the more usual failure mode because the diode junction simply fuses together.

8.5.4   Failure to open circuit

The effect of the Zener diode blowing open circuit would remove its shunt path from the supply and allow the full fault voltage in the safe area through to hazardous area. The current would still remain limited by the circuit resistance but is likely to be higher than permitted due to the higher voltage. The open-circuit condition is considered unsafe.

It is for this latter condition that Zener diodes are duplicated or triplicated to achieve the category of safety previously described. Figure 8.14 shows a triplicated diode arrangement where the failure of any two diodes to open circuit still allows the third to perform the limiting. The circuit meets the criteria for safe in two countable faults of the infallible components: Ex 'ia'. The diodes used in this configuration are subjected to less stringent testing.

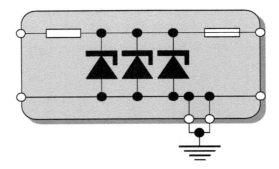

Figure 8.14
Barrier configurations

Alternatively, only two diodes, as shown in Figure 8.15, may be used in an Ex 'ia' barrier if a specific construction of the Zener is used and the Zener diodes are pulse-tested in accordance with specific requirements in the IEC 79-11 Standards. The representation shown in Figure 8.16 may well be used on the barrier and the user will not necessarily know which configuration is used inside.

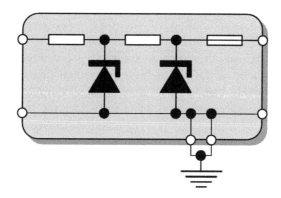

Figure 8.15
Duplicated zener barrier configurations

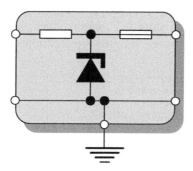

Figure 8.16
Barrier representation

The safety description voltage requires selection of the Zener diodes within bands of 0.1 V to achieve consistency and accuracy in the maintenance of the safety description. Zener diodes may be 'totem-poled' or stacked to give different safety description voltages. The connection of a Zener diode across the barrier is said to form a chain so it is the chain that must be duplicated or triplicated according to the type of barrier.

8.5.5   Resistor (current-limiting)

The resistor closest to the hazardous area terminals is required to be wire-wound or metal film. This is so that any failure of the resistance cannot reduce its value, which would increase the available current to the hazardous area circuit. Wire-wound resistors are preferred because their characteristics show that all failure mechanisms act to increase its resistance. This is the required condition. Some resistance elements such as carbon can fuse when stressed to give a lower-resistive value, which is clearly unacceptable. These types are specifically excluded from use.

8.5.6 Safety fuses

The fuse types are required to be high breaking capacity (HBC) and are generally ceramic powder-filled. This design of fuse does not allow the wire to vaporize and leaves a metal trace that could encourage tracking and arcing inside the fuse. There is no other failure mode to consider. The standard generally specifies the fuse design.

The resistance of the fuse is not insignificant. Its temperature coefficient is sometimes a problem in applications.

Taking the combination of the CLR, Zener diodes and safety fuse if any two components might fail then the circuit will remain safe. All these components are deemed infallible types and thus such an assembly is given the Category Ex 'ia'.

The likelihood of failure of these individual infallible components has been estimated as approximately 1 in 10^{16} per annum. This is accepted as extremely reliable.

Having chosen components to manufacture the barrier the construction itself must be reliable and of high integrity. The internal connections must not fail in any dangerous way and so specifications are included in the standard to which a barrier must conform. These include power dissipation and temperature rise under safe area fault conditions prior to the fuse rupture. The termination sizes and separation are detailed, governing but not specifying the arrangement of the barrier.

8.5.7 Barrier characteristics

There are two distinct sets of information that describe the characteristics of barriers. These are referred to as *Safety* and *Operational* characteristics. It is important that they are not confused. Figure 8.17 summarizes these differences. Illustrations of the use of these figures are discussed in the Applications section.

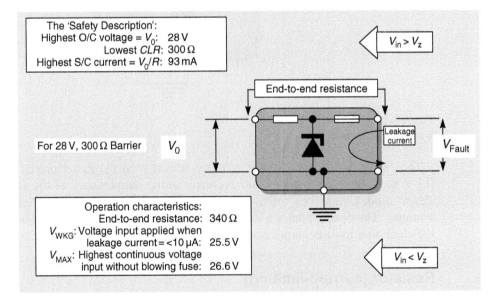

The 'Safety Description':
Highest O/C voltage = V_0: 28 V
Lowest CLR: 300 Ω
Highest S/C current = V_0/R: 93 mA

$V_{in} > V_z$

End-to-end resistance

For 28 V, 300 Ω Barrier V_0

Leakage current

V_{Fault}

Operation characteristics:
End-to-end resistance: 340 Ω
V_{WKG}: Voltage input applied when
leakage current = <10 µA: 25.5 V
V_{MAX}: Highest continuous voltage
input without blowing fuse: 26.6 V

$V_{in} < V_z$

Figure 8.17
Safety and operational characteristics

Safety

The safety characteristics provide the barrier's safety description, which is the highest output voltage under safe area overvoltage-fault conditions. These figures are necessary in order to safely connect the barrier into an IS circuit in the hazardous area. The safety aspects of barriers have been discussed but their use in the operation of an instrument loop must now be considered. This concept is introduced here. There are a number of common industry standard safety descriptions used. It is the safety description, which is of the utmost importance from the point of view of safety.

Operational

The operational characteristics are that necessary for the loop design in order for the instrument loop to work properly. The barrier must pass signals without distorting the signal's electrical properties. Careful design and selection of barriers must be used to ensure correct operation but the system as a whole must be safe and must work properly.

One important point is that the safety descriptions are deliberately used with particular fuse ratings. This is so that a hazardous area short circuit, which is safe from the point of view of explosion protection, will not blow the fuse if the state persists for a medium term. After a longer term short the fuse may well age and eventually blow.

Electronic overvolt protection techniques are built in to the design of some barriers. These are commonly referred to as semi-active barriers (as they are no longer the simple passive types described so far) and are more expensive than passive types. Their use is primarily for battery-backed systems where the float charge overvoltage may exceed the V_{MAX} and so the barrier fuse does not blow. They are only used on a few applications.

Other electronics are incorporated into the barrier in order to provide a floating supply for 4/20 mA transmitter applications.

8.5.8   Polarity of barriers

In Figure 8.18, a single barrier 'channel' is shown with different diode configurations to allow the channel to be operated +ve, −ve or AC (unpolarized) w.r.t. the earthy or return channel.

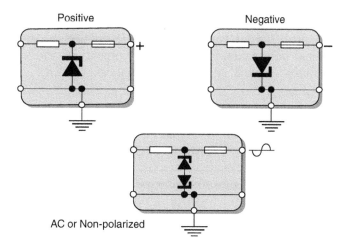

Figure 8.18
Barrier polarities

The need for polarization is best illustrated during discussions using applications to explain the theory. It is particularly important when using a number of barrier channels together. The primary reason that AC barriers are not used universally is that cable parameters values would be restrictive.

8.5.9   Multi-channel barriers

It is convenient to use two barrier channels together in an IS loop. The circuit in the hazardous area is therefore not directly connected but is still referenced to earth. This allows greater flexibility in the connection to the safe area circuit.

Note in Figure 8.19, that the circuit is referenced to earth via the chains of Zener diodes. It may take up any potential between 0 V (earth potential) and $\pm V_{MAX}$.

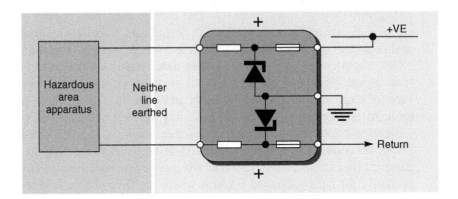

Figure 8.19
Two channel barriers

Two-channel barriers are usually of the same polarity but not always of the same safety description. They do provide for a simpler installation, reducing cost and space taken up. Two, three and four channel types are now common. Multi-channel types for specific applications such as strain gage bridges are also popular.

8.5.10   Typical barrier channel data

Some typical barrier types by safety description are listed in Table 8.3. Safety and operational information is provided for comparison and to assist in the understanding of the operation. Table 8.3 gives the essential data needed.

Safety Description			Max End-to-End Resistance (Ω)	V_{WKG} at 10(1) μA (V)	V_{MAX} (V)	Fuse Rating (mA)
V	Ω	mA				
1	100	10	120	0.3	2.0	250
3	300	10	318	(0.6)	3.6	250
10	50	200	85	6.0	6.9	50
15	100	150	155	12.0	13.0	100
22	150	147	185	19.0	20.2	50
28	300	93	340	25.5	26.6	50

Table 8.3
Common safety and operational barrier characteristics

8.5.11   Barrier characteristics descriptions

To understand the function of the values given in Table 8.3, definitions are given in Table 8.4.

Name	Description
1. Safety description	The safety description of a barrier, e.g. '10 V 50 Ω 200 mA', refers to the maximum voltage of the terminating Zener or forward diode while the fuse is blowing, the minimum value of the terminating resistor, and the corresponding maximum short-circuit current. It is an indication of the fault energy that can be developed in the hazardous area, and not of the working voltage or end-to-end resistance
2. Polarity	Barriers may be polarized + or −, or non-polarized ('AC'). Polarized barriers accept and/or deliver safe-area voltages of the specified polarity only. Non-polarized barriers support voltages of either polarity applied at either end
3. End-to-end resistance	The resistance between the two ends of a barrier channel at 20 °C, i.e. of the resistors and the fuse
4. Working voltage (V_{WKG})	The greatest steady voltage, of appropriate polarity, that can be applied between the safe-area terminal of a 'basic' barrier channel and earth at 20 °C for the specified leakage current, with the hazardous-area terminal open circuit
5. Maximum voltage (V_{MAX})	The greatest steady voltage, of appropriate polarity, that can be applied continuously between the safe-area terminal of any barrier channel and earth at 20 °C without blowing the fuse. For 'basic' barriers, it is specified with the hazardous-area terminal open circuit; if current is drawn in the hazardous area, the maximum voltage for these barriers is reduced
6. Fuse rating	The greatest current that can be passed continuously (for 1000 h at 35 °C) through the fuse

Table 8.4
Terminology definitions

8.5.12   Component arrangements

Physical barrier construction varies from one manufacturer to another but generally speaking all follow the same component arrangement shown in Figure 8.20. The infallible components are encapsulated such that they cannot be interfered with and the safety fuse cannot be replaced incorrectly. This is a requirement of the standards. If the safety fuse were to be replaced by a higher value, then the safety of the circuit may be compromised. Some manufacturers provide external fuses in the form of replaceable types that can be used to protect the internal safety fuse and/or can act as loop disconnect features.

The U_{MAX} is normally 250 V_{rms}. Barriers may be two- or three-diode construction types provided that they are Ex 'ia' rated for Zone 0 use. Connection to an IS earth is mandatory.

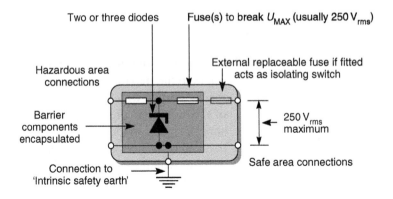

Figure 8.20
Barrier construction/connection

8.5.13   Combinations of barrier channels

Where more than one source of energy and power supplies the same IS circuit into a hazardous area, the combination of these safety descriptions must be assessed. In Figure 8.19, two independent channels, each with its own separate safety description, can provide increased energy and power into the circuit (under simultaneous fault conditions). The total current and voltage from all possible configurations of that arrangement must be examined. This is to ensure that power and energy levels from the combination are still within safe values for the given application and hazard.

8.5.14   Shunt diode safety barrier earthing

The requirement to provide an IS earth, as shown in Figure 8.21, is fundamental to the safe use of shunt diode safety barriers. The general principles of earthing and the reason for the IS will be discussed fully in Chapter 11.

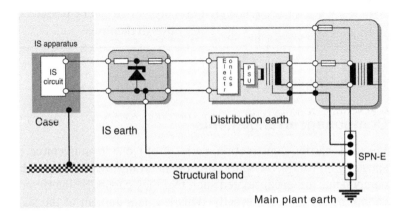

Figure 8.21
Barrier IS earth

The previous discussion on the principles of barriers showed a single barrier with its 'earth' connected to the mains earth of the safe area supply. This is acceptable only in order to demonstrate the principle. In practice, the earth connection on the barriers must be taken back to the plant earth associated with the instrumentation system's source of supply. This is normally defined as the point where the neutral of the distribution transformer for the instrument system is connected to the main earth point of the plant. Note that it may be single phase (as shown here for simplicity) or three phase. It is labeled as the IS earth and connects back to the star-point/neutral-earth ('SPNE').

8.5.15   Galvanic isolation

The inherent disadvantage of a shunt diode safety barrier is that its installation requires connection to an IS earth. This forces the safe area and hazardous area instrument loop circuits to be commoned at the same earth. The commoning effect is not itself detrimental, provided that the IS earth connection is correctly made and with adequate integrity.

If it were to become disconnected then the potential of the circuit in the hazardous area may become elevated w.r.t. the structural earth. A fault between the circuit and earth could cause incentive sparks or heat dissipation at the point of the fault. Where many barrier circuits are used together sharing a common earth, the effect of loosing the IS earth would be to increase the chance that stray paths would pose a danger. Clearly there are many possible fault scenarios, which could then occur.

The earthy connection of a barrier therefore may be considered as a direct route for current to flow from safe to hazardous area. No voltage or current limiting can be applied to this connection. It is a weakness in the arrangement.

The earth path, shared by the safe area circuit and the hazardous area circuit in a barrier, is broken by the electrically isolating arrangements in a 'galvanic isolator'. In this way, current cannot flow between the two parts of the circuit. A PD can occur which can be tolerated across the isolator, but this cannot jeopardize the safety, as a current cannot be injected into the hazardous area. The hazardous area circuit may be earthed at one point without causing currents to flow in that earth. It may be referenced to earth. The internal arrangement for a typical isolator is shown in block schematic format in Figure 8.22.

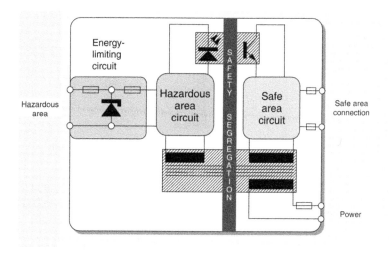

Figure 8.22
The galvanic isolator

Galvanic isolation achieves this by placing a method of signal isolation (compatible with the required application) between the safe and hazardous area circuits. Electronics in the safe area and the hazardous area circuits are necessary to communicate the signal over whatever isolating system is chosen. In the above case, optical isolation is used.

The system must also provide power to the hazardous area circuits and so a double-wound power transformer is used. It may be sourced from AC mains. 24 V DC-powered versions using invertors at higher frequency (reducing the size of the transformer and increasing the supply efficiency) are now more popular.

The design of the isolating components must be in accordance with the requirements of the standards. Manufacturers use specially constructed isolating devices. These have to be 'component approved' for use in the circuit. They are not 'infallible components', but are said to be ones 'on which safety depends'. Their design and construction provides guaranteed isolation to prescribed levels. Faults are considered but other mechanisms may be required to protect component-approved devices because they cannot be duplicated in a circuit. For example, thermal trips are embedded into mains transformers to protect against shorted turns causing overheating. Some 'component-approved' devices are designed and constructed by manufacturers in such a way as will allow their use in a range of isolating devices. Transformers, relays and opto-couplers are good examples of this.

The fault condition in the safe area side cannot permeate the insulating properties providing the isolation. The failure of the safe area supply becoming connected to the hazardous area terminals does not need to be considered.

The worst case scenario is the failure of the hazardous area circuit, where the secondary of the isolating transformer inadvertently becomes directly connected to the hazardous area terminals. The same energy-limiting network used on a barrier is necessary at the hazardous area terminals of an isolator to protect against this fault. No connection to earth is necessary. This is because no invasion can occur onto the hazardous area circuit from any other supply owing to the isolation provided. This can be observed in Figure 8.23. The energy in the hazardous area circuit is self-contained. The isolator will have a safety description in the same way that a barrier does.

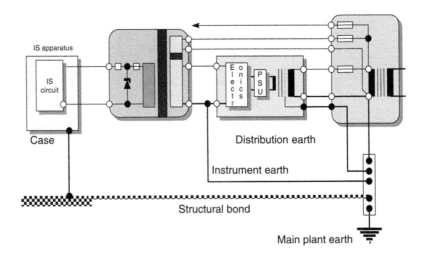

Figure 8.23
Effect of invasion on isolated circuits

8.5.16   Application of galvanic isolators

Galvanic isolators may be for analog, status or digital applications. The technique requires sophisticated electronics in both the safe and hazardous area circuits. The circuit must accept the incoming signal and suitably change its form to be communicated over some appropriate means of isolation. The other circuit must receive the signal from the isolating medium and convert it to a form that can be re-transmitted out.

In some cases the signal path must be bi-directional. In other cases the signal that goes from safe to hazardous are may not be the same as that which goes from hazardous to safe area. Two-wire systems using this format of powering and signaling to devices in the hazardous area will be discussed in the section on applications.

The safe area circuit can take up any potential completely independent of the hazardous area circuit. The hazardous area circuit may float or may be earthed as required depending on the circuit conditions. An example of this useful feature is the use of earth leakage detection, which can be applied in order to raise the integrity of the signal. This cannot be done with barriers.

The design of isolators is usually specific for their intended application. Other useful features are often designed into the isolator, as it is a convenient place to perform signal manipulation in the signal path.

The operational characteristics of isolators are different to their safety parameters. As with barriers, the safety description gives the highest possible values that will appear in the hazardous area under any fault condition. The operational characteristics are dependent on the application and will be declared by the manufacturer. The use of these figures will be discussed in the section on applications.

Originally isolators were expensive, but as demand has increased, the cost of manufacture has reduced. The applications are broadening to encompass custom design for special signal and power requirements.

8.5.17   Barriers vs isolators

There is no difference between the levels of safety achieved by these two techniques. The differences are in the application. Some advantages and disadvantages may best be drawn out at this stage.

8.6   Associated apparatus

The function of barriers and isolators, described above, limit or preclude excess energy from entering the hazardous area from the safe area. These devices are known as 'associated apparatus', and this term is defined below.

The IEC 79 Standard defines 'associated apparatus' as 'electrical apparatus which contains both IS circuits and non-IS circuits and is constructed so that the non-IS circuit cannot adversely affect the IS circuits'.

8.6.1   The certified interface

The block diagram in Figure 8.24 shows a typical system with certified apparatus in the hazardous area connected through an IS interface to uncertified safe area apparatus.

Such apparatus has the effect of electrically interfacing between the safe and hazardous areas. Barriers and isolators are known as 'IS interfaces' and are self-contained devices that interpose between the two circuits. Signals may pass in either direction and a limited amount of power is allowed to enter the hazardous area IS circuit. Certified interfaces usually pose industry standard safety descriptions and are easy to match with certified apparatus. Certified apparatus itself follows a standard form.

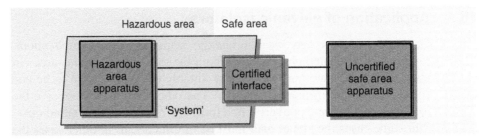

Figure 8.24
Modern IS system with associated apparatus (certified interface)

8.6.2   IS 'front-end' interfaces

The energy-limiting arrangement used for barriers and isolators can be built into certified safe area mounted equipment. A data logger or a DCS I/O card, for example, may have IS inputs directly onto its terminals and may not need an external interface. The complete arrangement containing the IS input and all associated equipment, i.e. the entire data logger or the whole I/O rack for the DCS, must be certified. Clearly this is convenient in many respects but is also very expensive to certify. The equipment is still associated apparatus by definition. It is said to have an IS 'front-end' or have 'internal interfacing'.

The electronic design of the front-end circuits will be the same as for barriers or isolators. Where galvanic isolation is employed, an IS earth will not be required. Where shunt diode barrier techniques are used then the IS earth is necessary. It may be integrated with other earthing arrangements and it may not be easy to understand how the earthing is actually arranged.

In the early days of IS (Figure 8.25 shows a typical arrangement) two pieces of apparatus were designed and built to work together. The electronic designer could choose operating voltages and currents in the circuit to suit the design, provided they did not exceed the maximum values permitted for safe use in IS circuits. The range of safety descriptions now available have been chosen to accommodate many standard applications thus avoiding the high cost of development and certification of this old arrangement and providing a greater choice of interface arrangements (barrier or isolator).

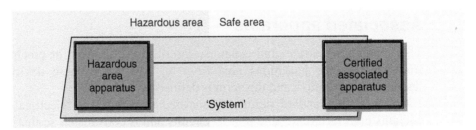

Figure 8.25
Older system concept

The use of internal IS interfacing is still encountered where there are particular supply/signal needs for specialist applications. However there are commercial implications in which a manufacturer will use a non-industrial standard safety description so that his equipment in the hazardous area may only be used with its interface or associated apparatus.

8.6.3   Electronic current and voltage limiting

Associated apparatus discussed so far has used Zener diodes and CLRs to perform the limiting functions. Most standard interfaces are built this way. Other methods of achieving this are by electronic means. Some equipments use elaborate supply systems to source power for a hazardous area apparatus where resistive limiting is operationally unacceptable.

This is achieved by the design of separate regulation circuits that are placed in series as shown in Figure 8.26.

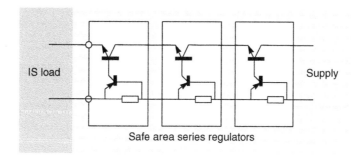

IS load   Supply

Safe area series regulators

Figure 8.26
Electronic limiting

If a semiconductor junction fails then it is more likely to fail to a short circuit. The junction fuses into a low resistance. In this case the low resistance would allow a higher voltage at greater current into the hazardous area. The preferred regulator design includes current-limiting or 're-entrant' characteristics for the overcurrent condition.

Three series regulators are required by the IEC 79 Standard but this still only permits certification to Ex 'ib' and are therefore restricted to use in Zone 1. In most cases the supply systems using this technique are for apparatus such as operator terminals that would never be mounted in Zone 0 and so this is an acceptable limitation.

8.7   Electrical apparatus in the hazardous area

Electrical equipment mounted and operated in the hazardous area must be designed such that it cannot cause ignition.

Electrical equipment surrounded by a potentially flammable atmosphere in a hazardous area must not be allowed to accumulate energy such that if released in an uncontrolled way, it could generate a spark or heat at sufficient levels to be incentive.

Some electrical devices such as a switch or contact in the hazardous area cannot store or generate any energy. A switch contains no element of inductance or capacitance. It cannot therefore store electrical energy. There is also no mechanism to generate electrical energy in a simple switch.

The switch-operated relay loop discussed previously, uses a switch mounted in the hazardous area connected to an interface. In this arrangement, the only possible source of power and energy into the loop will come from the safe area via the associated apparatus (interface, barrier or isolator).

The IS circuit can only receive excess energy from one other source, i.e. invasion from some external source. This is clearly undesirable and is a risk that must be reduced to acceptable levels. Such a risk is outside the scope of any individual assessment of safety and can only be controlled by rules governing the integrity of installation. These rules will be discussed in Chapter 12.

IS apparatus mounted in the hazardous area and connected to equipment in the safe area must be classified as one of the two types, as indicated in Figure 8.27.

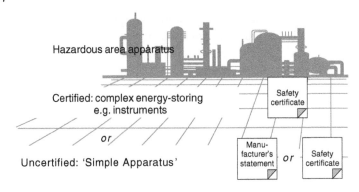

Figure 8.27
Apparatus permitted in hazardous areas

8.7.1   Simple apparatus

The simple apparatus concept is of the greatest importance to IS systems.

There are many instrument devices used in measurement and monitoring applications, which conform to this class of equipment. The simple apparatus class was conceived for this purpose. Clarifications over the original wording in the European Standard have been welcomed to provide even greater flexibility. The extract in Table 8.5 is taken from the IEC 79-11 Standard and its significance is explained subsequently.

'The following apparatus shall be considered to be simple apparatus:

Passive components e.g. switches, junction boxes, resistors and simple semi-conductor devices.

Sources of stored energy with well-defined parameters e.g. capacitors and inductors whose values shall be considered when determining the overall safety of the system.

Sources of generated energy e.g. thermocouples and photocells, which do not generate more than 1.5 volts, 100 mA and 25 mW. Any inductance or capacitance present in these sources of energy shall be considered as in b.

Simple Apparatus shall conform to all relevant requirements of this Standard but need not be certified and need not comply with clause 12 (marking). In particular, the following aspects shall always be considered:

Simple Apparatus shall not achieve safety by the inclusion of voltage and/or current limiting and/or suppression devices.

Simple Apparatus shall not contain any means of increasing the available voltage or current, e.g. circuits for the generation of ancillary power supplies.

Where it is necessary that the simple apparatus maintain the integrity of the isolation from "earth" of the intrinsically safe circuit, it shall be capable of withstanding the test voltage to earth ... of 500 V rms for 1 minute. Its terminals shall conform to ... stated requirements.

Non metallic enclosures and enclosures containing light metals when located in the hazardous area shall conform to ... IP 20.

When simple apparatus is located in the hazardous area it shall be temperature classified. When used in an intrinsically safe circuit within their normal rating and at a maximum ambient temperature of 40 °C, switches, plugs and sockets and terminals are allocated a T6 temperature classification for Group II applications and considered as having a maximum surface temperature of 85 °C for Group I applications. Other types of simple apparatus shall be temperature classified in accordance with ... this standard.'

Table 8.5
IEC 79-11 on simple apparatus

Examples of the first of the three categories above are switches, resistors, temperature-dependent resistors, LEDs, terminal blocks, connectors which are totally passive devices, as seen diagrammatically in Figure 8.28.

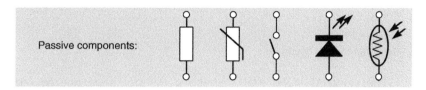

Figure 8.28
Passive simple apparatus components

In order to use the device for its intended purpose, power may need to be applied to simple apparatus from an IS source. This should not be confused with the device's ability to generate or store energy of its own volition. A switch requires an applied voltage to sense if it is conducting. These devices cannot in any way contribute to the energy in the circuit. There is therefore no limit to the number of simple apparatus devices included into any IS circuit. Terminals and connectors are also included here.

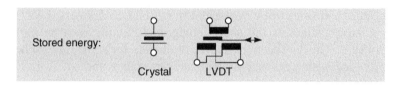

Figure 8.29(a)
Energy-storing simple apparatus

The second category of 'sources of stored energy within well-defined parameters' illustrated in Figure 8.29(a) covers any device containing inductive or capacitive elements where the total value can be examined under the L and C curves published in the standards. This clarification is the most useful but raises some questions. It implies that circuits which contain capacitance or inductance can be classed as simple apparatus if the total cumulative capacitance does not exceed the maximum permitted by the safety description of the feeding circuit (the associated apparatus). The values of L and/or C must be added to the actual cable parameters (which will be discussed in Chapter 12). Since the cable parameter concept is concerned with systems certification (see Systems), the effect of this is that even simple apparatus with significant values of L and C should be covered by a formal analysis of the safety parameters. This means that a systems 'certificate' or 'descriptive document' is necessary.

For example, the inductance of a magnetic pick-up coil may now be assessed to be within the limits of this simple apparatus clause, provided its total inductance together with the system cable matches the cable parameters of the interface.

In the third category, as shown in Figure 8.29(b) (refer Chapter 2) a thermocouple is an example of a device, which will generate energy. It is well below the maximum values stated and is therefore unquestionably a simple apparatus. Photovoltaic cells can generate in excess of 1.5 V in strong sunlight conditions and therefore may not always be considered simple apparatus unless use in low light levels is guaranteed. The separation of sources of stored energy from that of generated energy allows separate treatment of some apparatus to the advantage of the user. Piezo-electric devices used in ultrasonics and in vibration monitoring were previously precluded. There is no specific maximum energy limit in this international standard other than that represented by the curves. Older standards quoted 20 μJ (in IIC gases).

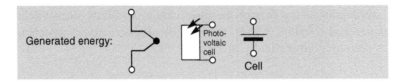

Figure 8.29(b)
Generating simple apparatus

The standard states that the above devices need not be 'certified' or marked in accordance with the standards. It is recommended that the user carefully documents simple apparatus with energy-storing capability.

This concept is not yet fully accepted in America and some other parts of the world.

Simple apparatus is considered as one of the categories of IS apparatus because failure need not be considered. In other words if simple apparatus devices fail then there is no risk that explosion-protection integrity will be compromised.

Some devices to be mounted in the hazardous area are unquestionably simple apparatus as in the above examples but they may still be certified and marked if it is in the manufacturers' best interest to do so. This will be for commercial reasons rather than technical ones. Examples of products dealt with in this way are resistive strain-gage bridges and some linear variable differential transformers (LVDTs).

Clause 2 clarifies the use of integrated circuits with an on-board ability to generate voltages. This has been the source of dispute for some time. Where an integrated circuit is provided with a 5 V supply and can generate + and −12 V rails, then this is specifically excluded from being classed as simple apparatus due to the complexity of analysis.

Simple apparatus will usually maintain a 500 V AC insulation (for 1 min) test to earth in line with other certified apparatus, unless it is designed to work with respect to earth (in which case it should be used with isolating interfaces).

The standard suggests that specialist advice be sought on sensors, which use catalytic reaction or other electro-chemical mechanisms.

8.7.2 Analysis of devices to assess compliance with IS

The analysis of field- or hazardous area-operated devices in circuits can be performed to determine the aspects of safety.

There are three main situations:

1. The conditions for simple apparatus are met. A device may be analyzed to ascertain if it may be connected and used as part of an IS circuit under the simple apparatus clause.
2. The conditions for simple apparatus are not met. If however stored or generated energy is above that permitted by simple apparatus, other clauses in standards may permit the use of such a device under either specific or limited conditions.
3. If the above criteria cannot be met, then the device or equipment must be designed to include specific safety components that limit the energy and power aspects. The assembly may then require formal apparatus certification.

8.7.3 Analysis application methods

Techniques for analyzing the energy and power dissipation of devices are relatively simple.

Generated mechanical energy can be determined by the formula:

$$E = \frac{1}{2}mv^2$$

Where
  E is energy in Joules
  m is the mass
  v is the velocity.

This is directly related to electrical properties where stored energy is determined by:

$$E = \frac{1}{2}CV^2 \quad \text{or} \quad \frac{1}{2}LI^2$$

Where
  C is capacitance in Farads
  V is the applied Voltage, or
  L is inductance in Henrys
  I is the current in Amperes.

A Piezo-electric device used for an accelerometer may generate a high voltage when subjected to shock. It would be necessary to calculate the capacity to store energy and to compare the values with those permitted in the standards.

In such cases some knowledge of practical values and other limitations will be necessary. Some assumptions and justifications will need to be stated in documentation of the analysis in order to prove whether the device complies with the requirements.

Circuits containing inductance and capacitance may be analyzed by simply adding together the total value of the like properties. This assumes the worst case failure condition.

8.7.4 Certified hazardous area apparatus

If the apparatus cannot be classified as 'simple' then it is normally assessed for conformance with a given standard. This enables a certificate to be issued (or 'approval' to be given) by a testing authority. The apparatus is termed 'certified apparatus' or 'certified hazardous area apparatus' and may be selected for use in a hazardous area. It must be installed and used in accordance with requirements discussed in Chapter 13.

Design of apparatus

The design of each type of IS apparatus is unique and depends upon its function and specification. The circuit of the apparatus can range from being uncomplicated, as in inductive sensors, to very sophisticated, for example, 4/20 mA process transmitters, multiplexers and display systems.

In each case, the designer must demonstrate that the voltages, currents and energy storage within the circuits and sub-circuits are securely controlled and adequately limited from causing ignition in the hazardous area. The requirements to be met are published in the standards. It is not the intention to discuss in detail all the possible techniques that can be used to design IS apparatus in this book. Some general and widely accepted techniques are shown for illustration.

Higher values of energy and power may be used in circuits, provided that adequate precautions are taken (to the satisfaction of relevant parts of the standards). Specific components will have been included in the design and are analyzed for failure in the same way as for a barrier circuit. Some common techniques to protect large reservoirs of energy are shown in Figure 8.30, to illustrate how energy storage and charge limiting is achieved. Resistors, diodes and Zener diodes are used in various ways, as shown.

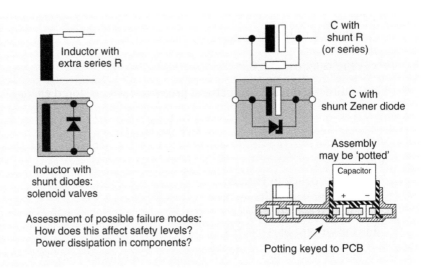

Figure 8.30
Treatment of energy-storing components in certified apparatus

The use of large value capacitors in a circuit design is generally avoided where possible. Where necessary, they are often combined with Zener diodes to limit the voltage applied under failure conditions to other parts of the circuit. Diodes or Zener diodes may be single, duplicated or triplicated, depending on the application.

Using a series resistor will limit the charge/discharge rate and so eliminate incentive sparks. A shunt resistance across a capacitor will allow the capacitor to discharge such that parts of the circuit become de-energized after a controlled time. The component with its safety component(s) may be encapsulated to raise the reliability and this will also help to dissipate heat, prevent the ingress of moisture and generally increase mechanical robustness from an operational point of view.

Where parts of circuits are isolated from each other, component-approved devices are used. Complex devices, such as process terminals, can be built up in separate sections in order to use more overall power.

Within IS apparatus, the areas of design that need special attention are:

- The selection of components, rating, size, characteristics
- The layout of the PCB and physical spacing of components
- Arrangement of terminations
- Heat dissipation
- Energy capture
- Integrity of joints and connection
- Documentation/drawings.

Segregation

The design and construction of the circuits must be of high integrity and so there is a great emphasis on the layout. Since each piece of apparatus is unique it is not possible to specify the detail of design in any document. Adequate separation distances are published in the standards and the minimum quality of PCBs and insulating materials are specified against the voltages used on the apparatus. Figure 8.31(a) illustrates how minimum separation requirements affect the layout of PCBs.

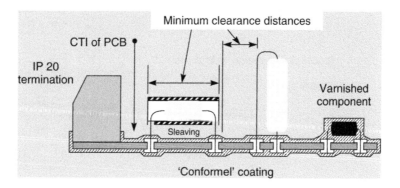

Figure 8.31(a)
Treatment of energy-storing components in certified apparatus

Creepage and clearance distances

Creepage and clearance distances are specified in the standards as shown in Table 8.6 and Figures 8.31(a) and (b). These figures apply both internally and externally to IS circuits and systems.

Peak Voltage	10	30	60	90	190	375	550	750	1000	1300	1575
	Distances in mm										
Creepage distance	1.5	2	3	4	8	10	15	18	25	36	40
Creepage distance under coating	0.5	0.7	1	1.3	2.6	3.3	5	6	8.3	12	13.3
Minimum CTI	90 90	90 90	90 90	90 90	175 175	175 175	175 175	175 175	175 175	175 175	300 175
Clearance	1.5	2	3	4	5	6	7	8	10	14	16
Distance through casting compound	0.5	0.7	1	1.3	1.7	2	2.4	2.7	3.3	4.6	5.3
Distance through insulation	0.5	0.5	0.5	0.7	0.8	1	1.2	1.4	1.7	2.3	2.7

Table 8.6
Creepage and clearance

Clearance is the shortest distance in air between two conductors. Creepage is the shortest distance over a solid surface.

In diagram A of Figure 8.31(b) these measurements would be the same. In diagram B a separating partition effectively increases the creepage value at higher voltages.

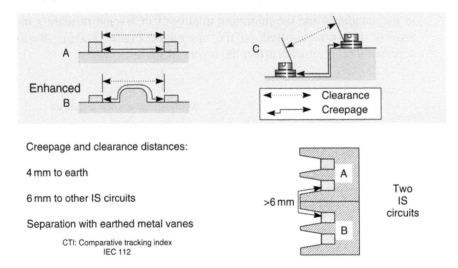

Figure 8.31(b)
Treatment of energy-storing components in certified apparatus

Where a circuit has two adjacent tracks, the voltage difference gives the appropriate figures of creepage and clearance. As an example, associated apparatus with a 240 V supply track adjacent to a 12 V track on a flat PCB would require 6 mm clearance and 10 mm creepage. In this case the 10 mm figure would prevail. Other internal apparatus requirements would be determined during the design of apparatus. There are many permutations in any individual design of apparatus. The requirements of different standards do vary slightly; this should be understood before apparatus design layout is finalized before certification.

Two independent IS circuit termination systems, for example, on interfaces or in junction boxes would require minimum creepage and clearance distances of 6 mm between circuits and 4 mm to earth because this is stated as a requirement in the IEC Standard.

Comparative tracking index (CTI)

The insulation quality of materials must also meet minimum requirements as stated in Table 8.1. The CTI is used to measure the effectiveness of a given material to resist surface breakdown and allow tracking from one conductor to another over the surface of the material. The test is detailed in IEC 112 and is used in the testing throughout the electrical equipment industry.

Two electrodes are positioned on the surface of an insulating material at a fixed distance apart. A voltage is applied and a salt-water solution is allowed to fall between the electrodes to encourage the track effect to start. The voltage and the number of drops used to cause breakdown are recorded as a measure of the index value.

This test is only of paramount importance to apparatus designers but not to those who select apparatus for use in hazardous areas, since the requirement will have been met at the apparatus design and manufacturing stage.

8.8   Enclosures

IS apparatus does not need any enclosure to be part of the method of protection. The standard requires that:

where intrinsic safety can be impaired by access to conducting parts, an enclosure of at least IP20 shall be provided as part of the apparatus under test.

The function of an enclosure is not therefore to provide mechanical strength, but to prevent inadvertent shorting or earthing of the current-carrying conductors of circuits. Occurrences of this on the same circuit will not cause concern but where two or more circuits share an enclosure, the combination of the circuits could pose a greater risk.

The IP 20 requirement is equivalent to allowing access of a terminal with a small screwdriver but not allowing the termination to be touched by the human finger. Thus termination must be deliberate.

Impact testing is not required as for some mechanical methods of protection. In these cases, degradation of protection could be caused by the ingress of solids or water if an enclosure suffered a damaged seal. Alternatively, if damage was too severe, short circuits could result in non-IS protection techniques. This would not be the case with IS apparatus and systems.

8.9 Temperature

The maximum temperature reached by components (safety or otherwise) within any piece of hazardous area apparatus requires detailed assessment.

During the operation of a circuit, heat will be developed from all components in which power is dissipated. The rise in temperature of the surface of the components, which may come into contact with a flammable mixture, will be measured to ensure that the T rating required (or to be achieved) is not exceeded.

This condition must be true under normal operation and under specified fault conditions for the apparatus under test. All components are normally examined. Heating from one component must not be permitted to affect another, particularly if it is a safety component. This assessment requires the judgment of skilled and experienced test personnel to interpret and apply the standards. Testing may reveal areas where apparatus needs to be redesigned or the layout changed in order to improve the acceptability of the design.

There are clearer guidelines for the definition of temperature ratings in the IEC 79-11 standard than there have been in previous or other standards. This was accepted from revisions to the EN 50 020: 1994. It clarifies a previously difficult area. Methods of assessing the ratings of conductors and copper track-work on PCBs used in certified apparatus are now provided. Tables give ratings that the designer must work to when aiming for T_4, T_5 or T_6 ratings.

There is a 'relaxation' of the rules for 'small components', which requires some explanation. It is recognized that where the surface area of a component is small, even though the surface area may reach the ignition temperature of a gas/air mixture, tests have shown that the heat transfer from the component to the gas is insufficient to allow ignition to take place. The standards give options under which high temperatures in small components are permitted.

For apparatus to be rated T_4, the total surface area of a component, excluding lead wires must be less than 20 mm^2 for a permitted temperature rise up to 275 °C. Above 20 mm^2 the power dissipation must be a maximum of 1.3 W. Greater than 20 mm^2 but less than 10 cm^2, the limit of 200 °C is given. This assumes a 40 °C ambient and further de-rating factors apply with higher ambient temperatures. T_5 and T_6 rating guidance are also given.

8.9.1 Methods for determining temperature rise

Methods for determining the temperature rise in components and conductors are detailed in all construction standards. The normal approach is to decide what T rating is required of a piece of apparatus and then to ensure that it complies with the requirements of the rating. This is achieved by either

1. Placing components in the circuit to ensure that currents are below specified limits, or
2. Demonstrating that the currents flowing under fault conditions are below those specified in the relevant construction standard.

The maximum permissible current I corresponding to the maximum wire temperature due to the effects of self heating ($t\,°C$) can be calculated by the following formula:

$$I = \left[\frac{(I_f^2 \times t(1+aT))}{T(1+at)} \right]^{\frac{1}{2}}$$

Where
a is the resistance-temperature coefficient of the material of the wire (0.004 265 K^{-1} for copper)
I is the maximum permissible current rms or DC in amperes
I_f is the current at which the wire melts
T is the temperature at which the wire melts (1083 °C for copper)
t is the maximum temperature of the wire due to self-heating.

Alternatively, tables given in the standards may be consulted which apply to printed circuit tracks, ribbon cables and ordinary conductors.

These calculations do not need to be performed for every individual conductor. The design and construction of apparatus should be assessed to determine whether compliance is general and demonstrate individual cases where the currents could reach higher levels. Signal paths would not normally require individual assessment but paths carrying earth fault currents would require calculations to be done.

8.9.2   Temperature rise calculations

Temperature rise in components can be calculated by data usually supplied by the manufacturer. The general formula applied is:

$$T_{FINAL} = T_{AMB} + (R_{TH} \times P_{DISSIPATED})\,°C$$

Where
T_{FINAL} is the expected final temperature of the component
T_{AMB} is the starting ambient temperature and is normally 40 °C
R_{TH} is the thermal resistance (in °C/Watt) of a component as stated by the manufacturer under specified conditions
$P_{DISSIPATED}$ is the power dissipated (I^2R) in the component.

8.9.3   Temperature assessment standards

Non-IEC 60079 compliant Standards specify different methods of assessing temperature. It is therefore important to understand and comply with the requirements for the particular standard for which certification or approval is being applied for.

Calculations may be provided to satisfy IEC/CENELEC certification requirements whereas the Canadian CSA rules, for example, emphasize testing under specified conditions.

8.10 The IS systems concept

An intrinsically safe electrical system is defined in European Normative EN 50 039 as:

> *An assembly of interconnected items of electrical apparatus described in a descriptive system document in which the circuits or parts of circuits intended to be used in a potentially explosive atmosphere are intrinsically safe circuits.*

This definition is taken from the European Standard EN 50 Series of which the majority of the IS Standard EN 50020 has been accepted into IEC 60079-11. The 'system' part, EN 50039, has to some extent been included in part in IEC 60079-14 (Installations) section 12.

8.11 An IS 'system'

A 'system' may comprise two or more pieces of apparatus connected together in some way as the part of an instrument loop in the hazardous area. Where two pieces of apparatus become connected together, then the combined capability of those individual pieces of apparatus to cause ignition must be assessed to ensure that the system complies with certain installation requirements.

This includes the interconnecting wiring, which has in itself the ability to store energy. This is considered in detail in Chapter 12.

An illustration of a system is shown in Figure 8.32 for the purposes of discussion.

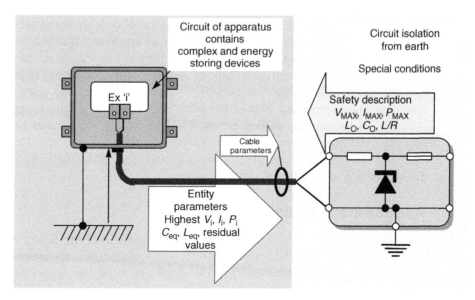

Figure 8.32
System concept

8.12 System documentation

The phrase in the standard, 'described in a descriptive system document', enforces the requirement that safety assessment of the interconnected system requires the designer to demonstrate how the system is safe when the apparatus is connected together in the way it has been designed.

8.12.1   System certification

Many industry standard 'systems' are covered by formal 'system certification'. This is where a manufacturer has his apparatus tested for safety compatibility with other specific pieces of certified apparatus (of his manufacture or of other manufacturers). A test house formally certifies the combination of each set of apparatus. The combination may be on one systems certificate or on individual certificates, depending on the requirements of the manufacturer.

The systems certificate will then state any special and/or installation conditions that are to be met for safety purposes.

This part is important because the complete 'system' also includes, for example, the interconnecting cable. This is not under the direct control of the testing authority, and the installation can only be considered safe if it has complied with all the requirements of installation.

The onus is therefore placed on the user to ensure that the installation conditions have been met and that adequate proof of this, including any calculations and measurements, necessary to demonstrate safety, are performed.

As a result the documentary proof that all is safe goes into plant safety documentation. The system certificate, apparatus certificate and a statement demonstrating those cable parameters have been adhered to form a 'system descriptive document'.

8.12.2   System descriptive document

The interconnection of apparatus to form a 'system' does not require a formal 'system certificate'.

The purpose of the 'system descriptive document' is to demonstrate how a system, in which a number of pieces of certified apparatus are interconnected, conforms to the IS apparatus and system requirements and is therefore considered to be safe.

This is a less formal approach than the system certification process and was conceived to offer greater flexibility for both the manufacturer and the user. The certified IS apparatus forms building blocks to solve applicational problems which can then be documented by suitably skilled personnel. The system is not used to its fullest extent because users often require formal systems certification.

In the following sections of this chapter, the processes of providing system documentation are discussed.

8.13   Assessment of safety

The assessment of safety is performed by a comparison of the inputs and outputs of the apparatus. The electrical connections to a piece of certified apparatus in the hazardous area may expect to receive energy. The certification of the apparatus will dictate how much energy can be received for the apparatus to remain safe.

8.13.1   Safety description

Associated apparatus, such as an interface, to which the certified apparatus is to be connected, may expect to give out electrical energy. The certification of this apparatus will dictate what the highest values will be under fault conditions. Since this device will provide the source of energy to the circuit, it will specify what the maximum energy storage parameters can be. These values are compared to the values seen at the power-receiving end added to the values, which are introduced into the circuit by the interconnecting cable.

If, according to the appropriate certification, the interface cannot provide higher output values than the apparatus in the hazardous area is permitted to receive, and the energy storage and cable parameters fall within the limits set by the associated apparatus, then the combination is considered safe.

The standard terms that describe the outputs and inputs parameters are as follows:
Output values from a 'power source' (associated apparatus): 'safety description'

Where U_O, I_O, P_O, is the maximum output of voltage, current and power, respectively.
Where C_O, L_O, is the maximum permitted values of capacitance and inductance respectively that may be safely connected to the output of the associated apparatus comprising other apparatus and cabling.

Note: U_{MAX} specifies the highest safe area supply voltage to the associated apparatus, normally 250 V AC rms.
Input values to a 'power receiver' (certified apparatus): 'entity parameters'

Where V_I, I_I, P_I, is the highest acceptable input value for V, I and P respectively.
Where C_i, L_i is the effective value of input C and L, respectively, that appears at the input terminals to this apparatus input.

This is sometimes written as C_{eq}, L_{eq}, where C_{eq}, L_{eq}, is the effective equivalent value of input C and L, respectively, that appears at the input terminals. It is equivalent in that it must be added to the cable parameters.

'Entity' parameters are sometimes referred to as 'modular' parameters.

8.14 Simple apparatus

Simple apparatus can appear in a system in one of the three forms. These are introduced here but discussed more fully in the context of applications, subsequently:

8.14.1 Passive simple apparatus

The passive form requires no further explanation because no energy contribution, whatever, can be made from this type. Most system certificates expressly permit the system to comprise simple apparatus without restriction.

8.14.2 Other (non-passive) simple apparatus

Where there can be stored energy in simple apparatus, IEC 60079-14-12 requires some assessment of the total cumulative effect. The treatment of this in a system depends on the type of device. Generally, if there is only one device in a system, then it is not considered significant.

8.14.3 Simple apparatus equivalence

Some apparatus is designed to present a 'non-energy storing' connection into an IS circuit. It is said to be 'like simple apparatus'. This useful facility is discussed in the section on Applications. It is sometimes referred to as having simple apparatus equivalent inputs or outputs. The apparatus is certified and the apparatus certificate will dictate what the maximum figures presented to the circuit will be. These are normally 1.5 V, 15 Ω, 100 mA with any L and C contribution stated energy storage. If a number of devices are used in the same circuit then the combination of devices will require assessment.

8.15   Safety parameters

It is possible that terminals on a piece of apparatus can both receive and source electrical values under different conditions. It is usual for these to be quoted; for example, a temperature transmitter connection to a thermocouple will have both input entity parameter and an output safety description. The use of these requires some expertise to correctly apply the figures.

In the scheme shown in Figure 8.33, and IS system is shown comprising, say, a 4/20 mA temperature transmitter and its associated equipment. Some detailed analysis is required to understand the safety and operational implications.

The IS concept generally only looks at the safety aspects of electrical apparatus and is not concerned with its operation. The user must satisfy him that the circuit will perform the task for which it is designed.

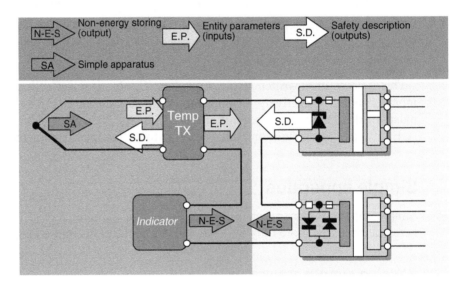

Figure 8.33
System analysis

8.16   Temperature classification of systems

With many intrinsically safe applications T_4 may be awarded automatically where the matched power dissipation is less than 1.3 W. Simple apparatus devices, which cannot dissipate power, are automatically awarded T_6.

The system takes up the worst case temperature classification of all the apparatus used in the hazardous area loop. The interface does not possess a T rating because it is safe area mounted but the system certificate or documentation must consider the power dissipation.

Normally, T_4 ratings for systems are adequate and almost invariably offered. There are very few vapors with ignition temperatures below 135 °C.

8.17   Systems concepts in other standards

The systems concept in CENELEC EN 50 039 is a specific document dealing with the relationship of apparatus when interconnected.

Although many other countries' standards do not publish separate 'systems' guidance, it is often, to some extent, included in the Apparatus Standard. A typical example of this is Australian Standard AS 2380.7 clause 2.9, which is entitled 'IS electrical systems'. It requires the 'electrical parameters and all characteristics of the interconnecting wiring ... to be specified in a descriptive system document or "Block diagram" '. It goes on to clearly specify that circuits in Zone 0 must use multicore cables with individual pair screens. Other minor differences are stated.

This, like many other standards, acknowledges the need to consider system phenomena but not with as much clarity as the new EN 50 039 that will eventually be brought into the IEC 79 arena.

8.17.1   Systems approach in North America

The approach in the US and Canada is somewhat less formal and more like the systems descriptive process in many ways.

The 'entity concept' provides the rules for the permitted interconnection of approved apparatus. Apparatus is approved with entity parameters, which must be matched in the same way as for the IEC requirements. However, the values for FM/UL/CSA-approved apparatus are slightly different from the CENELEC specification owing to the different interpretation of the *V/I* characteristics. The same graphs are used but reversed voltages are considered as additional to the main barrier channel voltage.

In other ways the same rules apply. All apparatus must have defined parameters even if they are non-energy storing. Simple apparatus concepts are thought of as too wide by these authorities.

8.18   Standards for Ex 'i'

The following standards govern this protection concept worldwide,

- Australia: AS 2380.7 governs this protection concept
- European Union: EN 50020 governs this protection concept
- IEC: IEC 60079.11 governs this protection concept
- United Kingdom: BS EN 50 020 governs this protection concept
- Germany: VDE 0170/0171 T.7 DIN EN 50 020 governs this protection concept
- France: NF EN 50 020 governs this protection concept.

9

Protection concept 'p'

9.1   General

This is one of most popular and widely used protection concept using 'the principle of separation' of three elements of the fire triangle in order to prevent ignition (Figure 9.1).

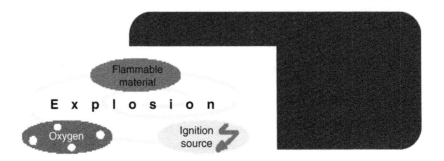

Figure 9.1
Explosion/fire triangle

This is a useful technique because an artificial 'safe area' having sufficient integrity can surround virtually any electrical equipment. The technique is flexible in that it can be adopted for many situations. The system is not used often due to,

- High cost and
- Inconvenience of equipment accessibility of this solution.

The system is expensive to operate and maintain because the clean air must be pumped and controlled by other equipments exposed to the hazardous area. The major benefit is that it can be used on very small enclosures up to complete control rooms. The problems associated with this tend to complicate area classification. A guaranteed gas-free air supply must be maintained so it must be piped in from a safe area. Disposal of used air, if it is likely to contain gas during the initial purge, must be handled such that it does not convert a safe area to a hazardous area.

Of the four techniques using the principle of separation, the Ex 'p' technique is very versatile. Any uncertified equipment may be placed in an enclosure where a inert gas or air is supplied and maintained at a slightly higher than atmospheric pressure level. Ordinary air is most commonly used although there may be cases where Nitrogen is preferred.

9.2 Definitions

Control of the atmosphere within a room or apparatus enclosure permits the safe use of electrical apparatus, which in the absence of the control would be unsuitable. The pressurizing or purging of the room or enclosure can achieve this. In some cases the two methods cannot be regarded as independent but for the purpose of this course the following definitions apply.

9.2.1 Pressurizing

It is a method of safeguarding, whereby air or inert gas in a room or enclosure is maintained at a pressure sufficient to prevent the ingress of the surrounding atmosphere, which might be flammable.

Where appropriate the pressure may be provided by a mechanical ventilation system.

A variant of this is also known as static pressurization. With static pressurization, the overpressure is created before the system is commissioned by charging the enclosure with protective gas and maintained solely by the sealing of the enclosure without any protective gas being supplied in the hazardous area. The protective gas must be inert. A maximum oxygen concentration of 1% by volume is permitted. Measuring equipment should be used to check this on every charging process.

9.2.2 Purging

This is a method of safeguarding whereby a flow of air or inert gas is maintained through a room or enclosure in sufficient quantity to reduce or prevent any *hazard*, which could arise in the absence of the purge. (To 'reduce' in this context means to reduce the risk of a *flammable atmosphere* occurring, thus permitting the use of electrical apparatus with a lower standard of safeguarding. Where the object is to 'prevent' a *hazard*, 'sufficient' shall take account of the highest likely rate of release of *flammable material* within or into the room or enclosure.)

Where appropriate the *purging* may be provided by mechanical ventilation of the forced or induced type.

9.2.3 Pressurizing/purging

This is a method of safeguarding employing both *pressurizing* and *purging*.

Pressurization with leakage compensation is characterized by the fact that the required overpressure is established in the interior of the enclosure after purging. With closed outlet, a supply of protective gas (instrument air or inert gas) is sufficient to compensate for the leakage flow from the pressurized enclosure and pipelines.

In the case of *pressurization with continuous flow* of protective gas, the overpressure is achieved by continuous flow of protective gas within the pressurized enclosure. Pressurization with leakage compensation and pressurization with continuous flow are based on a similar technical principle. However, the required protective gas flow rates differ greatly, leading to different designs. Due to its drawbacks in operation, static pressurization is not widely used. Continuous flow only offers advantages in the relatively rare case of internal release (analyzers).

For these reasons, the following sections describe the widely used method of pressurization with continuous leakage compensation.

9.3　Development of standards for Ex 'p'

The changes to the requirements in the European Standards on pressurization which arose during the transition to EC Directive 94/9/EC and the harmonization of regional standards at IEC level are somewhat more extensive than those for other types of protection. The technology of the pressurization usually results in equipment of Category 2 (for use in Zone 1) or 3 (for use in Zone 2). It cannot be used for Category 1 equipment (for use in Zone 0). For Zone 2, so far the VDE 0165: 1991 specify German national requirements for the simplified pressurized enclosure. On an international level, the IEC standard 60079-2 formulates requirements for the simplified pressurization with the defined type of protection 'p'.

There has been the European Standard EN 50016 applicable to Zone 1 equipment since 1995. This has the technical content largely equivalent to that in IEC 60079-2, type of protection 'p'. Only the special features of the equipment marking and a few other requirements in the EC Directive have been further incorporated. It has already been established that this standard meets the essential safety requirements in Directive 94/9/EC.

As already stated, the type of protection specified in EN 50016, pressurized apparatus 'p', in the main corresponds to design 'px'. The most important constructional requirements for pressurized systems remain the same. For example, the tightness, accessibility and mechanical strength of the pressurized enclosure are specified. The internal structure and the assignment of the equipment must permit trouble-free and complete purging of the enclosure during the purging phase. If apparatus which produces sparks during operation with an operating current of more than 10 A and a rated operational voltage of more than 275 V AC – or 60 V DC – are installed, spark and particle barriers must be provided for the outlets under precisely defined conditions. All actuators for control and signaling devices and inspection windows must, if they are fitted in the enclosure of the Ex 'p' cabinet, be subjected to thermal endurance test to cold and heat in accordance with EN 50014 followed by an impact test. Following that, the required IP degree of protection must be guaranteed.

Depending upon the type of protection pressurization 'px', 'py' or 'pz', graduated safety requirements are specified for the properties and functions of the equipment pressure and flow rate monitors used to monitor door status, etc. The manufacturer must provide the operator with all the information necessary for safe operation in the form of state diagrams, flow diagrams, etc. in a detailed instruction manual. The upper and lower limits and their tolerances for safety requirements should be specified by the manufacturer and noticed to the user. The functional sequences of the safety devices and the required response in case of malfunctions or faults should also be described.

9.4　Construction requirements

9.4.1　Ex 'p' pressurizing

The construction standards require the strength of the enclosure to be designed such that it will withstand certain overpressure limits. The clean gas supply must be assessed for its ability to maintain pressure without overpressure. Depending on the zone of use, interlocks are required that cater for warnings when pressure is lost in Zone 2 applications. In Zone 1, loss of pressure is required to disconnect electrical supplies and I/O in order to ensure that safety is preserved.

This sometimes causes difficulties because it may not be practical to suddenly shutdown part of a plant for operational safety reasons. Hence, the design and integration of Ex 'p' apparatus requires very thorough design considerations from all angles (Figure 9.2).

The initial purging routine must be carefully examined such that no pockets of gas are left in the enclosure. The volume and shape of the enclosure and the properties of the hazardous gas must be taken into consideration when determining where to site the inlet and outlets. Normally five volumes of air change are required.

The supply of clean air must be from a safe area. The discharge during the purge has to be carefully considered on each installation because it is likely that contaminated gas put into a safe area will create a hazard. Discharging into a hazardous area may require the fitting of a flame trap.

A pressure switch normally monitors the internal/external pressure differential. If this reduces beyond certain limits, action must be taken according to the zone of use. This means that the pressure switch itself will need to be Ex protected since it is normally mounted on the Ex 'p' enclosure and is part of the overall protection system. The monitoring system to which the pressure switch is connected may be mounted adjacent to the enclosure or may be situated in a safe area. If mounted in the hazardous area, it will need to be Ex protected in its own right. Most Ex 'p' arrangements therefore use a system of Ex-protected devices working together. This is not often realized.

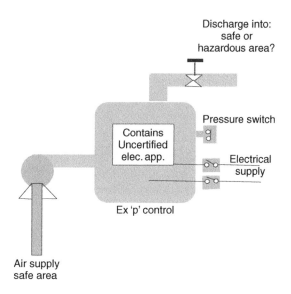

Figure 9.2
Ex 'p' pressurization

Each electrical supply or instrument circuit will require disconnection when pressure is lost and this is normally done in the safe area. Contactors/switches may be mounted in the hazardous area if suitably protected.

9.4.2 Examples of pressurizing

- A totally enclosed motor situated in a *hazardous area* is sealed as far as possible a pressurized with air or inert gas.
- An instrument cubicle (into which *flammable materials* are not introduced) situated in a *hazardous area* is sealed as far as possible and pressurized with air or inert gas.

- A control room (into which *flammable materials* are not introduced) situated in a *hazardous area* is sealed as far as possible and double doors form an air lock. The atmosphere is maintained at a pressure above that outside the room, the requisite pressure being provided by a mechanical ventilation system, which draws its air from a *non-hazardous area*.
- The interior of a motor of the force-ventilated type situated in a *hazardous area* is maintained at a pressure above that outside the motor, the pressure being provided by the separate fan supplying the motor cooling air which is drawn from a *non-hazardous area*.

9.4.3   Ex 'p' purging

Another variation of this technique is known as 'continuous purging' or 'continuous dilution'. It is based on the same principles but is adapted for use where the clean gas is used for cooling purposes or to dilute a potentially flammable mixture. A gas analyzer sampling tube may bring vapor into the enclosure. When released, the vapor must be adequately diluted to below its lower flammable limit before discharge.

In this implementation, a flow switch as shown in Figure 9.3 replaces the pressure switch. This is in order to maintain the designed flow rate through the enclosure.

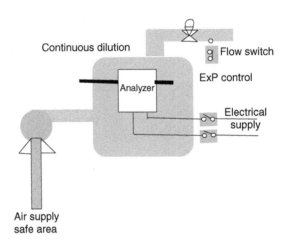

Figure 9.3
Ex 'p' purging or continuous dilution

There are many variations on this technique. Documentation should provide detailed information on the specific requirements for each application in order that maintenance can be effectively carried out without compromise to the safety limits designed and certified for the specific enclosure. Procedures for work on Ex 'p' equipment must bear in mind the design constraints and the application of equipment in the cabinet before being drawn up.

There are applications requiring the siting of unprotected electrical apparatus in a hazardous area where the Ex 'p' method is the only viable option. It can be easily adapted for an infinite range of enclosure shapes and sizes. Certification permits application-dependent variations, provided they conform to the general principles laid out in the standard.

The use of inert gases such as nitrogen may be employed but there are other safety implications in using this. The uncontrolled discharge of nitrogen can pose a risk to personnel in that if the nitrogen content of breathable air is raised then it becomes toxic.

9.4.4 Examples of purging

Extract ventilation is applied to a paint-spray booth to draw off all *flammable vapors* in order to reduce or prevent a *hazard* in the surrounding area. This is a technique similar to that applied to a fume cupboard.

A coating machine head applying a *flammable liquid* is fitted with an extract hood to draw off all *flammable vapors* in order to reduce or prevent a *hazard* in the surrounding area.

A hopper handling a *flammable dust* is equipped with an extract hood to draw off airborne dust in order to reduce its accumulation in the surrounding area.

Control room complexes may be protected using variations of this technique if adequate precautions with air conditioning for breathable air and fire protection and containment are taken.

This technique has been used in the past to place VDUs in hazardous areas. Access for maintenance may require equipment to be taken into a safe area, as the rules do not permit live working. Procedures must be written to define how each installation may be operated, serviced and maintained. The technique is expensive to install, operate and maintain. Air must be cleaned and pumped, and there is a cost associated with this.

9.4.5 Examples of pressurizing/purging

A room in a *hazardous area* containing instruments analyzing *flammable gas* is provided with a flow of air sufficient to prevent a *hazard* arising from leakage inside the room. The air pressure inside the room, being above that of the surrounding atmosphere, also prevents the ingress of a *flammable atmosphere* from outside.

An instrument analyzing *flammable gas* in a *hazardous area* is enclosed and provided with a flow of inert gas sufficient to prevent a *hazard* within the instrument arising from internal leakage. The inert gas pressure, being above that of the surrounding atmosphere, also prevents the ingress of a *flammable atmosphere* from outside the instrument.

9.5 Principles of application

The type of protection, pressurized apparatus 'p' encloses electrical equipments or systems representing potential ignition sources in a tight enclosure. Instrument air or inert gas is introduced into this enclosure until a defined overpressure in relation to the external atmosphere is achieved, which is then maintained during the operation of the system. This overpressure prevents penetration of flammable gas or combustible dust from outside into the enclosure and hence the coincidence of an explosive atmosphere and an ignition source. In principle, pressurization technology is also used for online analyzers, which in turn may be used to analyze flammable gases (or liquids). In such cases, flammable gases are fed via a pipeline to the analyzer in a pressurized enclosure. Any leakage in these pipelines or even in the analyzer may constitute an internal source of flammable gases inside the pressurized enclosure.

The area containing the flammable gas (i.e. the pipelines and analyzer) is described as containment system. Depending on the technical design of this containment system and gas feed system, it is described as infallible containment system (no release), a limited release system with a predictable maximum release rate or an unlimited release system. With unlimited release, overpressure must be created by an inert protective gas, which prevents oxygen from penetrating the enclosure. With limited release, a sufficiently large volume of air is used to dilute the combustible gas outside a small 'dilution area' so that an explosive atmosphere is unable to form.

Relative to the explosion triangle mentioned at the start, this means that there is no explosive mixture inside the enclosure (there is no flammable gas, or it is only present in amounts below the LEL, or there is no oxygen). In principle, any non-explosion-protected apparatus and system may be installed in the enclosure.

This type of protection is an interesting protection concept, in particular for operating high-power machines and systems in hazardous areas (such as motors, frequency converters) and where sparks could be created during normal operation.

However, it should be noted that this principle of protection, which at first glance seems so simple and plausible, requires a relatively high level of design and product engineering effort to achieve an overpressure in the enclosure with a sufficiently high safety level during the entire operating period.

9.5.1   Pressurizing and purging media

- Air is a satisfactory *pressurizing* medium and should be used in preference to inert gas because of the asphyxiation risk with the latter.
- Because a purged enclosure may have a *flammable material* released within it, *purging* with inert gas will usually provide a higher degree of safeguarding than *purging* with air. However, because of the risk of asphyxiation with inert gas, it is recommended that it should be used for small enclosures only, and that rooms and large enclosures should be purged with air at a rate sufficient to ensure dilution of any *flammable material*, which may be present to below its *lower flammable limit*.
- The source of air or inert gas shall at all times be free from all traces of flammable contaminant. Consideration should also be given to the need for drying or cleaning the air or inert gas.
- The source of supply of air or inert gas shall be reliable; if necessary standby system should be provided.
- In pressurized rooms, the pressure should be of the order of 6 mm water gage.
- It is to be noted that a pressure of 5 mm water gage is approximately equivalent to the effect of a 30 km/h wind on a vertical face.
- In pressurized enclosures the pressure should be above 5 mm water gage but in any case, under either normal or abnormal operation of the *pressurizing* system, should not exceed a value, which can distort or damage the enclosure.

9.5.2   Pressure and purge monitoring

- For pressurized rooms or enclosures the actions to be taken on pressure failure are given in the table below; they are dependent on the classification of the area in which the rooms or enclosures are situated and on the type of electrical apparatus installed in them.

Classification of the Area in which the Pressurized Room or Enclosure is Situated[a]	Action Required on Pressure Failure	
	'Non-sparking' electrical apparatus in the room or enclosure[b]	'Normally sparking' electrical apparatus in the room or enclosure[c]
Zone 1	Pressure-failure alarm	Pressure-failure interlock[d] and alarm
Zone 2	Nil (pressurization unnecessary)	Pressure-failure alarm

Notes:

a *Pressurizing* is not considered to provide a sufficient safeguard in *Zone 0*.

b 'Non-sparking' is used here in the general sense and includes *non-sparking* apparatus. '*Division 2 approved*' apparatus, and apparatus with *type of protection N*.

c 'Normally sparking' denotes apparatus, which in normal operation produces arcs, sparks or surface temperatures capable of igniting a *flammable atmosphere.*

d 'Pressure-failure interlock' means the automatic disconnection of the electrical supply to the apparatus within the room or enclosure on pressure failure (except in cases where this would create a dangerous condition in the process when other special precautionary measures shall be devised).

- For purged and pressurized/purged rooms or enclosures the action to be taken on purge failure shall be determined by taking account of the likelihood of release of *flammable material* within the room or enclosure, the type of electrical apparatus, and the classification of the area in which it is situated.

- The following methods are available for monitoring pressurized and purged systems:

  (a) A pressure-sensing device detecting the pressure in the room or enclosure.
  (b) A flow-sensing device (where appropriate) located in the outlet piping detecting the flow through the room or enclosure.
  (c) A pressure-sensing device detecting the pressure in the inlet piping.
  (d) A rotation-sensing device (where appropriate) on the pressurizing or extraction fan.
  (e) A fan contactor auxiliary switch (where appropriate).

Where a pressure-failure interlock and alarm are required method (a) or (b) should be used. Where a purge-failure interlock and alarm are required, method (b) should be used.

Where pressure or purge-failure alarms only are required any of the above methods may be used but it shall be noted that methods (c), (d) and (e) can give rise to false indications of state due to faults not associated with the sensing device (e.g. a blocked pipe downstream of a sensing device, a broken coupling or shear-pin, a welded-in contactor with fuses blown). These methods should therefore be used only where methods (a) and (b) are impracticable.

It may be noted that where more than one enclosure is pressurized or purged from a common header special care shall be taken in the selection and positioning of the monitoring devices.

- Before electrical supplies are re-established to apparatus contained within a pressurized or purged room or enclosure in which the pressurizing or purging has been interrupted, sufficient time shall be allowed for the complete scavenging of any flammable atmosphere, which may have accumulated therein. At least five changes of the enclosed volume are recommended; in many cases, it will be preferable to control this by an automatic timer.

  In the case of pressurized enclosures special outlet points, which can be temporarily opened for the purpose, may be required to ensure effective scavenging.

9.5.3   Special applications

Whilst many of the principles outlined above apply to apparatus such as instruments analyzing *flammable gas* and the rooms which contain them, further reference should be made to standards and to specialist codes of practice on the subject to arrive at the most economical way of protection in combination with other methods.

A typical example of this technique, as applied to a control panel is illustrated in Figure 9.4.

Figure 9.4
Switch rack being protected by Ex 'p'
(Source: R. Sthal)

9.6   Other design requirements

9.6.1   Room and enclosure design

- In the design of pressurized rooms the need for the following shall be considered:
    - Air-lock doors
    - Sealed windows
    - Sealed cable and other entries

– Controlled air outlets
– Adequate heating and ventilation.

- The temperature of the external parts of an enclosure to which a *flammable atmosphere* may have access shall be below the *ignition temperature* of the flammable *material* concerned.
- Where apparatus is placed in a pressurized or purged enclosure the design shall be such as to avoid abnormally high temperatures, which may lead to electrical faults.
- Pressurized or purged rooms and enclosures shall be suitably labeled to draw attention to their special nature.

9.6.2   Piping and ducting design

- It is preferable that the pressure inside any *pressurizing* or *purging* piping or ducting passing through a *hazardous area* should be above atmospheric. Where this is not practicable the piping or ducting shaft should be so constructed and maintained as to prevent the ingress of the surrounding atmosphere.
- For purged systems the outlet points shall be sited sufficiently remote from the inlet to prevent recycling of the purge and be such that the discharged material does not invalidate the classification of the area into which it is discharged.
- In siting of the outlet points of pressurized or purged systems consideration shall be given to the possibility that hot particles or flames produced as a result of electrical faults, may be discharged through the outlet points.
- In order to comply with the Highly Flammable Liquids and Liquefied Petroleum Gases Regulations 1972 electric motors used in purging systems relying on mechanical exhaust ventilation must not be situated in the exhaust vapor path. This precludes the locating of motors of any type, including those in flameproof enclosures, in the exhaust ducts. The alternative methods available are bifurcated duct fans, centrifugal fans and belt-driven fans, where in all cases the motor is mounted external to the duct.
- Similar considerations to those given above should be given to exhaust systems fans handling *flammable dust*.

9.6.3   Auxiliary apparatus

(i) In a room or enclosure, which is normally pressurized or purged, a *flammable atmosphere* may exist before the system is established or may accumulate should the system fail. All motors, contactors, relays, flow switches, pressure switches, etc. associated with the *pressurizing* or *purging* system and located in the pressurized or purged room or enclosure shall therefore be suitably safeguarded. In the case of pressurized rooms or enclosures, such apparatus shall be suitable for operating in the classification of the area in which the room or enclosure is situated. In the case of purged rooms or enclosures where the *flammable atmosphere* may be produced internally such apparatus shall be suitable for operating in the conditions which will obtain in the absence of the purge.

(ii) Where it is necessary to have a room or enclosure illuminated when the *pressurizing* or *purging* system is inoperative the lighting fittings, switches, etc. within the room or enclosure shall be of a type suitable for operating in the circumstances described in (i) above.

9.7   Operation of ...

a typical protection pressurized apparatus 'p' with continuous leakage compensation

Figure 9.5 shows a schematic diagram of a system having a pressurized enclosure with continuous leakage compensation. It also shows such a system in practical use. The protective gas is supplied through an air supply regulator consisting of pressure reducer and digital valve. The set values for protective gas flow rate during purging and the enclosure gas pressures are measured and maintained by an orifice plate, pressure switch and a sensor module. The core of such a system is the control, which processes all signals sent from the measuring points, controls the temporal sequence of the system functions and, if the specified minimum values for flow rate and operating pressure are not attained, initiates the necessary actions to guarantee safety.

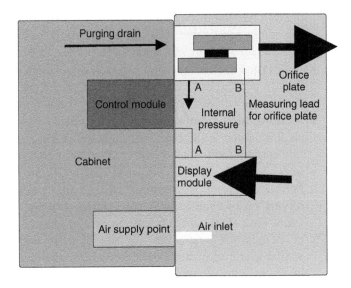

Figure 9.5
Switchrack type of protection Ex 'p' control with air outlets and orifice plate associated apparatus of intrinsically safe circuits explosion protected light fittings air supply regulator

The operation of a pressurized apparatus may be divided into three phases (Figure 9.6):

1. The preparatory phase
2. The purging phase and
3. The operational phase.

The preparatory phase starts when the control supply voltage is switched on. This is triggered by a signal from the control unit, the digital valve opens and protective gas flows into enclosure for the system to be protected. The overpressure, which is forming, lifts the lid of the orifice plate and flow starts. The pressure switches and sensor modules measure the actual values for the internal pressure and flow rate of the protective gas and compare them with the programmed set values. As soon as the minimum values are exceeded, a signal to this effect is sent to the control unit.

This starts the purging phase. The programmed purging time is counted down to zero in the control unit.

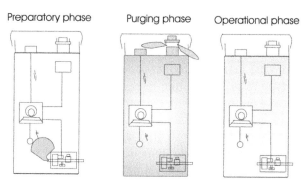

Preparatory phase   Purging phase   Operational phase

Figure 9.6
Typical course of the starting process for a control unit with type of protection pressurization 'p'

The system is now prepared for actual operation. The enclosure is purged with air or inert gas to remove any explosive mixture or to dilute it to a safe concentration before the ignition sources, in the form of the switching devices fitted, are switched on. The purging volume required depends upon the free volume in the enclosure and the prevailing flow conditions and is defined during the type testing of the system. Sensors and pressure-operating switches monitor the purging process continuously. If the minimum values for flow rate and pressure are not attained, the counter for the purging time in the control module is reset to the starting value and the purging process starts anew. If a maximum value for the internal pressure is exceeded, the purging process is completely terminated, and can only be restarted by actuating a reset command.

When the purging time is over, the operational phase starts automatically. The digital valve is closed for this. The leakage is compensated for by an integral needle valve, set to the required gas quantity. The internal pressure in the enclosure drops and the orifice plate closes. The seating thrust established at the orifice plate determines the working pressure. This must be maintained throughout the entire operation of the system to prevent penetration of flammable substances. A relay incorporated in the controls switches on the electrical equipment inside the pressurized enclosure. If the enclosure pressure falls below the set minimum values during the operational phase, all electrical equipments, which are not explosion-protected are switched off and a new purging phase is initiated.

As safe conditions are only achieved in the enclosure at the onset of the operational phase, the control module, the pressure operating switches and sensor modules, as well as the digital valve must obviously be explosion-protected. It is common to use type of protection flameproof enclosure 'd' together with the type of protection increased safety 'e' for the control module. All other components are intrinsically safe.

9.8   Testing

All the requirements of the type tests on which the manufacturer's declaration of conformity is based are as follows:

- Verification of maximum overpressure
- Tightness test
- Purging test
- Verification of minimum overpressure and
- Verification of internal pressure limitation
- Tests on the actuators and inspection windows to EN 50014 and
- The temperature test.

The routine tests to be performed by the manufacturers essentially comprise the functional and tightness tests.

However, with the type of protection pressurization 'p', it is frequently necessary to produce unique items or product series in very small numbers and hence the demarcation between type tests and routine tests is not as clear as for other types of protection. This means, for example, that determination of the minimum purging time requires the pressurized system manufacturer to operate equipment that is more complex. This is because of the requirement to simulate the dilution process with the aid of test gases and evaluated with measuring equipment. The required purging time is determined from this. However, for enclosures with a simple geometry, five volume changes have usually been found sufficient for the purging. In the national German foreword to the Standard EN 50016, reference is made to the fact that it may be possible to dispense with individual tests. The IEC draft Standard 60079-2 states this explicitly.

The co-existence of IEC Standards and EN Standards represents a special problem for manufacturers operating on a global scale. Therefore, IEC and CENELEC have in principle agreed on a simplified procedure for harmonizing the standards. It is not clear how far this will affect the standards on pressurization.

Here, we should explicitly point out again that standards *per se* do not have any legal significance. The Directive 94/9/EC, mentioned above, is legally binding in the European Community. This directive includes a description of essential safety requirements and test procedures for explosion-protected equipment, which must be performed for legal reasons. It may now be assumed that equipment built in accordance with an EN Standard in the series EN 50 … meets these requirements.

However, testing laboratories are not obliged to use a standard for testing equipment, they may verify conformity of equipment with the essential safety requirement of the directive in other ways. For use of Category 3 equipment in Zone 2, the manufacturer may issue the declaration of conformity. With an IEC Standard, it is necessary to establish in individual cases the degree to which the specifications satisfy the essential safety requirements in EC Directive 94/9/EC if they are to be applied in the EC for explosion-protected equipment. This also applies, if yet unharmonized IEC standards already have an EN number (usually in the series EN 60 …). However, this kind of adoption, as will also be performed for the pressurization (EN 6 007 9-2), indicates that, regardless of the legal formalities, the technical content is to a large extent identical to the that of the corresponding harmonized standards in the series EN 50 … For dust explosion protection, there are draft standards at IEC level. However, it is doubtful whether pressurization is in fact a suitable type of protection for dust. Dust layers may be removed by mechanical cleaning. Providing adequate dust tightness may prevent the penetration of dust into an enclosure. Moreover, purging and continuous flow tend to be counterproductive as they swirl up dust and hence can create an explosive dust atmosphere.

9.9   Summary

The type of protection-pressurized apparatus 'p' provides an internationally recognized method of explosion protection, which has proved its worth for special tasks over many years.

With the completion of international standards and the precisely defined requirements for Zone 2 applications – including, where appropriate, dust explosion protection – this type of protection will retain and increase its significance. The great advantage of this type of protection is that it places no (or only a few) additional requirements on the equipment installed inside the enclosure. It is therefore extremely versatile in use, including also special applications.

Lastly, it will be appropriate to state that,

Type of protection pressurized apparatus 'p' is a practicable solution for explosion protection for more complex electrical equipment.

9.10   Standards for Ex 'p'

The following standards govern this protection concept worldwide,

- European Community: EN 50016, October 1995 Electrical apparatus for potentially explosive atmospheres. Pressurized apparatus 'p'
- IEC: IEC 60079-2 Electrical apparatus for explosive gas atmospheres – Part 2: Pressurized enclosure 'p' 2/2001
- South Africa: SABS IEC 60079-2; 2001, Electrical apparatus for explosive gas atmospheres Part 2: Pressurized enclosures 'p'
- Australia: AS 2380.4 governs this protection concept
- Germany: DIN EN 50 016 or VDE 0170/0171 T.3 governs this protection concept
- France: NF EN 50 016 governs this protection concept
- United Kingdom: BS EN 50 016 governs this protection concept.

10

Other concepts

10.1 General

There are additional concepts that are not so widely used, but have been developed and find their use in specific applications. These, when used in conjunction with one or more of the popular methods of protection, can really lead to economically safe solutions.

These will be discussed further in the following pages.

10.2 Ex 'o': oil filling

This is one of the separation methods where the oil is used as separation medium. The Ex 'o' method shown in Figure 10.1 was originally conceived for high-power equipment. It provides explosion protection on a similar basis to Ex 'p'.

The definition of Ex 'o' protection concept in standards is,

A type of protection in which the electrical apparatus or parts of the electrical apparatus are immersed in a protective liquid in such a way that an explosive atmosphere which maybe above the liquid or outside the enclosure cannot be ignited.

The difference is that it uses a liquid medium (to separate a source of ignition from a hazardous gas/air mixture) rather than a non-combustible gas such as air. The presence of the oil permits moving parts such as the contacts of circuit breakers and is helpful in quenching the arc produced. It is more commonly used with power transformers where its secondary function is to provide cooling and enhance insulation properties. The choice of oils is critical because it must not give off toxic or hazardous fumes, particularly when subjected to heat and arcing conditions. Cooling, quenching and degradation characteristics must be matched for any given application. Unfortunately, popular oils have silicon-based additives which are highly detrimental to platinum gas detection elements. This technique is therefore to be avoided on sites handling hydrocarbons where gas detection is most common.

Its use in instrumentation is minimal with only one known application of the protection of circuit boards in a PLC.

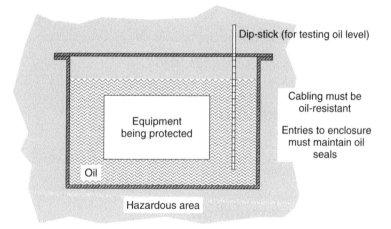

Figure 10.1
Ex 'o' oil immersion

10.2.1   Standards for Ex 'o'

This protection concept is governed worldwide as follows:

- European Union: EN 50 015
- Germany: DIN EN 50 015 and VDE 0170/0171 T.2
- France: NF EN 50 015
- United Kingdom: BS EN 50 015
- Australia: AS/NZS 60079.6
- IEC: IEC 60079.6.

10.3   Ex 'q': quartz/sand filling

This is another of the methods of protection under concept of separation. SABS / IEC 60079-5 describes the requirements for the protection of apparatus using sand or silica-based (quartz) beads to fill a container such that a flammable gas atmosphere is kept from coming into contact with a source of ignition. The standards define this concept as hereunder,

> *A type of protection in which the parts capable of igniting an explosive atmosphere are fixed in position and completely surrounded by filling material to prevent the ignition of an external explosive atmosphere.*

It is rarely used on its own and is mainly found in combination with the other construction techniques described here. It cannot be used in situations where movement is required, i.e., for the protection of relay contacts.

Modern instrumentation applications do not use this technique but old mining telephones were certified with components and terminals mounted in a compartment, which was filled with quartz beads. The chokes of Ex 'e' fluorescent luminaries are often protected like this.

The standard restricts the use of this technique to 16 A, 1000 VA and 1000 V.

This technique was limited to use in Zone 2. More recently, when used in conjunction with Ex 'e' components, for example, mounting in Zone 1 has been permitted.

10.3.1   Standards for Ex 'q'

The following standards are applied for this protection concept:

- European Union: EN 50 017
- Germany: DIN EN 50 017 and VDE 0170/0171 T.4
- France: NF EN 50 017
- United Kingdom: BS EN 50 017
- South Africa: SABS IEC 60079-5
- IEC: IEC 60079-5
- Australia: AS/NZS 60079.5.

10.4   Ex 'm': encapsulation

Encapsulation is used to prevent flammable gases from reaching a potential source of ignition in its own right. It may be thought of as the method of electrical protection using solids whereas Ex 'o' uses liquid and Ex 'p' uses gas.

The standards define it as follows:

Protection of electrical components by enclosure in a resin in such a way that an explosive atmosphere cannot be ignited during operation by either sparking or overheating which may occur within the encapsulation.

That is encapsulation is used to prevent flammable gases from reaching a potential source of ignition within the encapsulated apparatus.

Apparatus certified Ex 'm' was formerly treated as 'special' Ex 's', owing to the lack of a suitable standard. IEC 60079-18 permits the apparatus to be encapsulated in compounds of various types in order to prevent a flammable gas atmosphere coming into contact with excessive heat. Sparks cannot normally occur in such circumstances and some rules for prevention must be obeyed. The technique can be of great benefit in providing robustness and reliability. Encapsulating components provides for greater shock resistance, reduced environmental effects and better rejection of chemical attack.

The method is not often used on its own but combined with other methods to solve application problems. The technique is often used with IS where apparatus is certified Ex 'm', but a distinction should be made between potting of components to comply with IS requirements and certified encapsulation of apparatus in its own right. This is a subtle difference and requires detailed knowledge of a certification application to determine why it has been necessary to certify Ex 'm'. 'Potting' is often used as a way of raising the integrity of limiting components fitted across energy-storing components.

Repair of Ex 'm' apparatus is precluded since it is almost impossible to reclaim encapsulated components without damage.

10.4.1   Standards for Ex 'm'

The standards which are applicable for this type of protection concept all over the world are as given below:

- European Union: EN 50 028
- Germany: DIN VDE 0170/0171 T.9
- France: NF EN 50 028
- United Kingdom: BS 5501
- Australia: AS 2431

- IEC: IEC 60079.18
- South Africa: SABS IEC 60079-18.

10.5   Component certification

Component parts to be included in larger arrangements may be 'component certified' for some flexibility. In an Ex 'e' junction box for example, the enclosure will be impact-tested. The terminals to be used within will be component-approved.

The main uses of this technique are found in higher-power circuits such as induction motors, fluorescent lighting fittings, junction boxes and terminal housings. The German standards from which this came promote the use of toughened plastic cable sheaths on permanent installations, as opposed to the more expensive steel wire armored cable used elsewhere.

When applied to junction boxes, an Ex 'e' enclosure is given an 'enclosure factor' when certified. This represents the highest number of 'terminal-amps' permitted in the box. Terminals mounted in the box must be component-approved. The total of terminal-amps must be calculated and must be equal to or less than the enclosure factor.

10.6   Special type of protection 's' (IEC Concept Code Symbol Ex 's')

Where electrical apparatus can be shown to be safe in a given hazardous atmosphere but does not conform to any of the other recognized techniques of protection, Ex 's' permits the use of the equipment.

10.6.1   Definitions

The term '*approved*' is applied to apparatus, which has been approved by *HMFI* for use in specified *flammable atmospheres*. Such apparatus does not comply with any British Standard but is considered safe for use in *hazardous areas*.

The standards define this protection concept as hereunder,

> *A concept which has been adopted to permit the certification of those types of electrical apparatus which, by their nature, do not comply with the constructional or other requirements specified for apparatus with established types of protection, but which nevertheless can be shown, where necessary by test, to be suitable for use in prescribed Zones or hazardous areas.*

In UK the apparatus, which was approved by HMFI before 30 September 1969, is listed in HMSO Form F931 – Intrinsically safe and 'approved' electrical apparatus. Similar apparatus will in future be certified by *BASEEFA* as apparatus with *type of protection 's'* (special protection) although *HMFI* may still approve apparatus in certain special cases. *Type of protection 's'* is the term applied to apparatus which has been certified by *BASEEFA* as having been examined, type-tested where necessary, and found to comply with *BASEEFA* Certification Standard SFA 3009.

Examples of '*approved*' apparatus are:

- A battery hand lamp
- A combined compressed air-driven generator and lamp.

Examples of apparatus with *type of protection 's'* are:

- A factory-sealed fluorescent hand lamp with flexible cable
- A potted solenoid for valve operation complete with cable.

10.6.2   Conditions of use

'*Approved*' apparatus and apparatus with *type of protection 's'* are suitable for use in *Zones 0, 1* and *2*.

'*Approved*' apparatus is suitable for use only in the gases and vapors specified in the approval certificate.

Apparatus with *type of protection 's'* is normally suitable for use in all gases and vapors associated with Groups IIA, IIB and IIC except where the certificate states otherwise.

Apparatus with *type of protection 's'* is marked with a temperature class (T_1–T_6 in accordance with IEC 60079-4 or any other standard) and shall not be installed where *flammable materials* are used which have *ignition temperatures* below the maximum for that class.

'*Approved*' apparatus and apparatus with *type of protection 's'* shall not be used in any *flammable atmosphere* for which it has not been certified. No modification, addition or deletion to such apparatus shall be made without the written permission of the approving or certifying authority (such permission shall be obtained through the manufacturer of the apparatus) unless it can be verified that such change does not invalidate the certification or approval.

It shall be noted that care is taken to obtain the correct dry cells for torches as the use of high-power cells may invalidate the approval or certification.

When selecting apparatus special care shall be taken to ensure that the apparatus and its component parts are constructed to guard against electrical and mechanical failures in the intended condition of use.

Particular attention shall be given to the need for weatherproofing and protection against corrosion. An example of this is described in Figure 10.2. In this arrangement, a platinum coil is housed in a small volume enclosure, which is open at one end and covered with a sintered metal disk. The disk will permit a flammable gas mixture into the enclosure. The effect on the protected by a sintered metal disk is awarded to any apparatus that does not conform to the above but can be shown to be safe for the duty it is designed for.

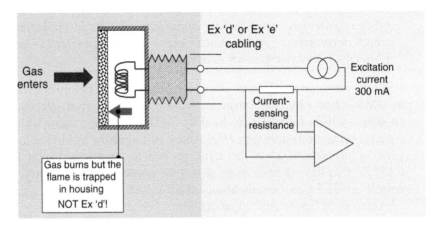

Figure 10.2
Ex 's' gas detection arrangement

SABS/IEC 79-14 installation recognizes Ex 's'.

The arrangement in Figure 10.2 shows heated sensors, which are housed in small volume enclosures. These are open at one end and covered with a porous metal disk on the other side. The disk permits a flammable gas mixture into the enclosure. As can be seen this arrangement does not conform to any of the known explosion-protection techniques. But we can prove that by appropriate tests that it is safe for the duty it is designed for, and, according to the definition of Ex 's', can be certified for use in a hazardous area.

10.6.3 Standards for Ex 's'

In Australia and New Zealand, AS 1826 presently governs this protection concept but is under review and likely to be replaced by AS/NZS 2380.8.

10.7 Multiple certification

From the discussion on the principles of the various types of protection, it will be realized that not all techniques are suitable in all circumstances. Indeed, not all techniques are *available* in all circumstances. It is becoming more commonplace to certify apparatus with more than one type where advantages are provided or where difficulties are overcome.

In Figure 10.3 the arrangement of a gas detector using three types of protection is shown to illustrate how the designer has employed these to allow the flexibility of construction, operation and ease of maintenance.

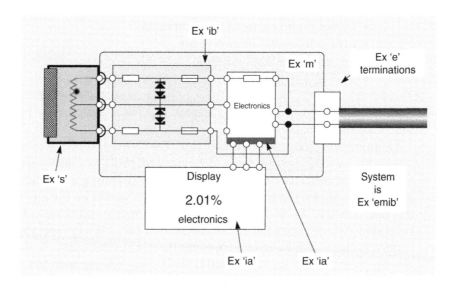

Figure 10.3
Gas detector with multiple certification

The gas detector is usually mounted in Zone 1 or Zone 2.

The detector head normally requires a high current of 300–500 mA. The supply voltage needed to the electronics at this current does not permit the system to be made completely IS.

A permanent Ex 'e' supply is acceptable for mounting in Zone 1 or 2. The Ex 'e' terminations are used to provide the supply to an Ex 'm'-encapsulated module containing the electronics for ranging and communication. The electronic module has an associated apparatus function to provide power and signaling to the indicator unit. This includes switches to function test the module. The indicator may be locally mounted as an integral part of the apparatus or may be remotely mounted from the detector head.

The high-current detector head has a barrier-like arrangement, specially designed for the current required (at a low voltage). The advantage of this configuration is that, using the live working capability of IS, the detector head (and the indicator) may be removed and replaced with minimum down-time and without risk of causing ignition, if a flammable atmosphere were to be present.

The most important type of protection for switchgear is 'flameproof enclosure', usually in conjunction with 'increased safety'. Switchgear does produce sources of ignition in normal use and therefore 'increased safety' alone is not applicable as type of protection for switchgear, since 'increased safety' is based on the principle to avoid sources of ignition by additional measures. However, 'increased safety', in conjunction with 'flameproof enclosure', is excellent for switchgear and control gear.

Modern, explosion-protected luminaries also use a combination of several types of protection to achieve the best results with regard to safety, function and economy.

10.8   Selection of certification method

The choice of the type(s) of protection available to a system designer when selecting apparatus to use in a hazardous area is based on a number of criteria. These are:

- Apparatus group required
- Zone of use
- Power requirements of the equipment
- Live working/calibration needs
- Environmental consideration
- Risk of damage.

More criteria are discussed in Chapter 12. The decision may be exemplified by the following situation:

An instrument loop may be protected by Ex 'i' or Ex 'd'.

A status switch application is to be located in a position where there is considered to be an increased risk of mechanical damage. (Loading arms and jibs are examples of possible situations.) In this case the preference would be for Ex 'i'. This is because if damage occurred to an Ex 'd' switch using mains voltage levels there would be an increased risk of explosion if the conductors became exposed and shorted. Personnel shock risk may also be increased under these circumstances. A standard contact could not be protected by Ex 'q' (Quartz) or Ex 'm' (encapsulation) because it could not then operate. However, a proximity switch uses the properties of ordinary encapsulation to enhance its mechanical performance whilst being compatible with IS techniques.

10.9   Apparatus for use in dust risks

Generally, we have gone in detail into protection of apparatus used in gas and vapor environments. We will briefly look into the methods of protection to limit the risks in a dusty environment.

10.9.1 Principles and standards

Some of the methods of safeguarding described earlier in this volume (namely *segregation*, *intrinsically safe systems*, *pressurizing* or *purging*) can be used to permit the safe use of electrical apparatus in dust-risk situations. However, the more widely used method of safeguarding is that of:

- Enclosing the apparatus so as to limit the amount of dust which may come into contact with sources of ignition such as arcs, sparks or hot surfaces within the apparatus and
- Ensuring that when the apparatus is operating at its rated capacity the surface temperature of the enclosure does not exceed the *ignition temperature* of the *flammable dust* in cloud or layer form.

Enclosures providing protection against dust entry are conveniently identified in terms of the nomenclature for degrees of protection of enclosures defined in SABS IEC 61241. This nomenclature is not yet being widely used but it is expected that it will be applied to electrical apparatus in the future.

Of the degrees of protection referred to above, two (IP 65 and IP 54) are considered suitable for use in dust risks. Until enclosures so designated are commonly available, the following guidance is given on types of enclosure offering equivalent protection.

10.9.2 IP 54

Enclosures that are constructed and tested in accordance with the requirements of IP protection (refer Appendix) afford this degree of protection.

In some cases standard *totally enclosed* weatherproof apparatus or apparatus with a *flameproof* enclosure of a weatherproof pattern meets these requirements, except that additional sealing may be necessary for spindles and shafts.

The tests required in BS 3807 may be carried out by any responsible organization but it is usual to use the facilities of the SABS Test house. In all cases, a certificate of test shall be obtained from the testing authority.

It is likely that in the future there will be a British Standard for dust-tight apparatus which also will afford protection equivalent to IP 65 and which will include requirements for surface temperature.

10.9.3 IP 65

In most cases standard totally enclosed weatherproof apparatus or apparatus with a flameproof enclosure of a weatherproof pattern is sealed sufficiently to afford this degree of protection.

In some cases, a standard flameproof enclosure also affords this degree of protection except that the sealing may need to be improved by the use of weatherproofing techniques.

In the case of the shaft of a squirrel-cage motor in a standard *totally enclosed* or *flameproof* enclosure the seal afforded by the grease in the bearing is adequate to provide protection to this standard.

10.9.4 Conditions of use

It is considered unlikely that dust can enter an IP 54 enclosure in sufficient quantity to produce a dust cloud or a dust layer capable of propagating combustion. However, where a *flammable dust* is liable to *train fire*, ignition by arcs or sparks of a small quantity of dust lying inside an enclosure may spread to the external atmosphere via deposits on the flanges and cause a fire or explosion in the surrounding area. In such cases, an enclosure to IP 65 may be required. Such an enclosure may also be required if the dust is electrically conducting.

While determining the methods of safeguarding, which shall be used in the various zones, particular attention shall be paid to the following:

- For *intrinsically safe apparatus*, where the *flammable dust* is sufficiently conducting effectively to short circuit any insulators on which it can settle, the apparatus should be contained in an enclosure to IP 65 when used in *Zone 20* and in an enclosure to IP 54 when used in *Zone 21* or *Zone 22*.
- In *Zone 21* or *Zone 22* standard industrial electrical apparatus, whether it is sparking or non-sparking in *normal* operation, may be used, provided it is contained within an enclosure conforming to the requirements of standards.
- For all types of apparatus the user shall ensure that the temperature, which the enclosure will attain with the internal apparatus operating at rated capacity, is below the *ignition temperature* of the particular *flammable dust* in cloud or layer form.
- When selecting enclosures special care shall be taken to ensure that exposed surfaces are reasonably flat and free from features which may collect dust, and that gaskets are in one piece and sufficient in thickness and width to ensure a permanent satisfactory joint. Special care shall also be taken to ensure that the apparatus and its component parts are constructed to guard against electrical or mechanical failure in the intended conditions of use. Particular attention shall be given to the need for protection against corrosion.

11

Earthing and bonding

11.1   Earthing

Correct 'earthing' is primarily required for the assurance of general electrical safety, reducing the risks to both human life and installations. The relevance to explosion protection is covered thoroughly from the explosion point of view with some initial general discussion to ensure that the intentions of the earthing philosophy are fully understood. The principles of electrical earthing are agreed upon internationally, though there are differences as to how these principles are best achieved in practice.

Electricity supply regulations may state specific requirements but these do not conflict with the intent of the standards and codes of practice for IS and other explosion-protected installations. This is a common misunderstanding of requirements.

There are many terms for 'earth' used in related electrical industries. The terms are imprecise and sometimes lead to confusion, thereby causing questionable levels of safety. In this discussion it is necessary to clarify the terms by defining the intention of the specific electrical connections used.

Electrical 'earthing' is required for five main purposes:

1. To reduce the risk of personnel shock
2. To operate electrical protective devices
3. To guard against lightning surges
4. To control electrostatic discharge
5. To minimize electrical interference.

These aspects will be discussed in detail in the following text.

11.2   Personnel safety

The effects of electricity on humans depend on the level of current and where it enters and leaves the body. Research shows that the limbs have a resistance of about 500 Ω. The central torso has a very low resistance value owing to the high water content. The effect of electricity penetrating the skin can be liked to the characteristics of a Zener diode with a reverse breakdown voltage of between 5 and 10 V. This depends on the individual's skin characteristics and the tendency to dry or greasy skin.

Hence the equivalent circuit of a human can be drawn as in Figure 11.1.

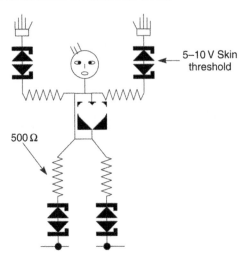

Figure 11.1
Human electric circuit

Under worst-case conditions, the threshold of sensation is at about 1 mA, according to health and safety experts. At about 5 mA the resulting shock is said to be disturbing, causing surprise in the recipient. This can cause the recipient to sustain injury in other ways.

Currents in the range 6–30 mA can cause temporary paralysis. This explains why humans cannot let go of conductors under shock conditions.

Between 1 and 5 A, ventricular fibrillation can occur. This means that the heart muscles may become out of synchronization and so the heart cannot pump blood around the body efficiently. A current of 10 A, through the heart, causes cardiac arrest.

The likelihood of high-current density through the heart mostly occurs when the left hand and the right foot become the connection points. Hand-to-hand connection can still be fatal.

Comparing these current values with the circuit for the body, and taking the worst case of a mild shock being permissible (with a current of 4 mA), then the maximum voltage desirable to be encountered by humans is of the order of

$$10\,V + 10\,V + \big((500 + 500) \times 4\ mA\big)V = 24\,V$$

This explains and reinforces the popularity of 24 V supplies for instrument systems. Under these conditions it is permissible to work live on such low-voltage equipment without fear of injury due to shock. In practice, most people do not feel anything below about 40 V. Instrument systems sometimes use 110 V AC configured as 55-0-55 with the secondary center tapped to earth. This practice is American in origin but has not been internationally followed due to the popularity of 220/240 V equipment.

However, there must always be the concern that heat generated from low-voltage systems at high currents can cause injury in the form of burns. So even low-voltage systems must be treated with due respect.

This means that under normal or fault conditions on a plant, the resistance of return paths must be sufficiently low so as not to produce voltages greater than, say, 25 V peak. Prospective fault currents may be calculated with a target earth loop impedance to prove personnel safety aspects are satisfied.

11.3 Hazardous area considerations

Structural or fault currents arising from electrical equipments operating in hazardous areas must not become a source of heat or sparks. Equipments must be adequately earthed to ensure that connections are of high integrity and low impedance. This must be adequately maintained throughout the life of the installation and equipment.

Fault paths are well defined and it is vital that correct fusing and overcurrent detection/protection is applied to power circuits such that any fault is cleared in an appropriately short time.

There is equally the concern that high-fault currents flowing, even during the brief time between the occurrence of the fault and the operation of the electrical protection, can cause elevated voltages to appear at other places in the hazardous area. Adequate isolation between circuits prevents this posing a risk. The isolation may need to be proved adequate periodically during the life of the plant.

From the curves and tabulations for gas ignition it will be realized that at less than 10 V, ignition by sparking becomes significantly less likely. If earth fault path impedance is kept below values, which will generate this voltage, then it follows that the power dissipated, and therefore the heating effect, must also be relatively low. Maintaining low values can be considered an acceptable risk. The quality of the earthing and bonding conductors on a plant must be a regular requirement for inspection.

Figure 11.2 shows a prospective short-circuit current of 4000 A through a 0.005 Ω resistance generating a 20 V elevation across conductors in the hazardous area. The power dissipated in the s/c resistance is 80 kW.

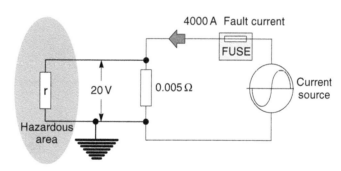

Power dissipated in $r = 20^2/r$ W
Time to clear the fault at 4000 A?
Can the connection at 'r' be intermittent?

Figure 11.2
Hazardous area fault, voltage elevation

If the conductors in the hazardous area were also terminated in 0.005 Ω, i.e. a short-circuit path perhaps through the earth structure of the plant, the dissipation would be 40 kW. The heating effect would be dramatic if the fuse did not operate.

The 10 V level is used as a rule of thumb to judge whether the quality of a loop bond is adequate. In practice very low values can be achieved with good cross bonding. This is however often taken to the extreme and an excess of copper is used in unnecessary places. This has the effect of increasing the time and cost of maintenance.

11.4   Earthing and bonding

There is a subtle but essential difference between earthing and bonding, which must be understood.

'**Earthing**' is where a low-impedance path is provided in order for return currents to operate electrical protection devices such as fuses and overcurrent trips in an appropriately short time. This is shown in Figure 11.3.

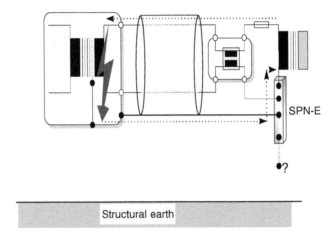

Figure 11.3
Earth path

'**Bonding**' is where voltage differences between electrical conducting parts are eliminated. This is shown in Figure 11.4.

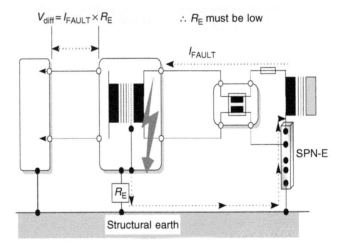

Figure 11.4
Bonding

The bonding of all the cases to the structure and the structure to the SPNE forms the bonding paths.

International Electrical Supply Regulations (ESRs) that cover fixed electrical equipment and installations require that there be an earth return that is backed up by a physical connection to 'terrestrial' earth. In this way there are two return paths acting in parallel, which enhances the integrity of an earthing system. One path is an earth path because its primary function is to conduct fault currents. The other connection is to ensure that significant voltage differences do not appear between devices. Refer to Figure 11.5.

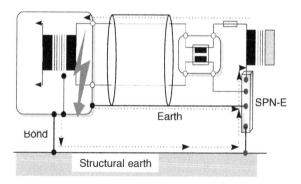

Figure 11.5
Earthing and bonding in parallel

Terrestrial earth, in this context, is taken as the art of providing a large surface area of a conducting material (usually copper) that is buried beneath the earth's surface and is in contact with soil such that the impedance between the conducting material and the soil is sufficiently low for the intended purpose of the connection. The structure of a plant is required to be bonded to the terrestrial earth. During the course of this discussion, these terms will retain these definitions.

The 'earth' and 'bond' conductors act in parallel. This is an advantage in that the two paths reduce the impedance. One path may be viewed as backing up the other, lest it should fail.

Consider a fault where the live conductor to the primary of the load transformer touches the metal (conducting) case, as shown in Figure 11.4. The resultant current in the live must be returned to the neutral of the supply transformer. This is in order that the circuit protection device works to clear the fault. The impedance of the distribution transformer neutral to structural earth is critical for safety. The connection from structural earth to terrestrial earth is not critical when considering these faults.

11.4.1 Power distribution

Transmission of electrical power from generator to consumer is normally done at relatively high-voltage levels in order to minimize losses. Typically, the supply to an industrial plant would be from a high-voltage substation owned by the distribution company. This would feed a local low-voltage transformer on site or adjacent to it. Here the line voltage would be reduced to 220/230 V AC for general services such as lighting, office equipment and instrument systems. High voltages three-phase supplies for high-power motors may be derived at the substation. 11 kV three-phase 'incomers' may be transformed down to 6.6 kV and 3.3 kV to 440 V, depending on the plant power requirements.

Since the ground's resistance through which earth fault currents may pass is uncontrollable and may be relatively high the fault conditions cannot be adequately protected by fuses alone, and 'out-of-balance' current detection is used.

Transformers used in distribution systems are inevitably 'double wound', which means that they have full electrical isolation between primary windings and secondary windings. The level of insulation will be tested to well above the working voltage of the transformer. With these high-voltage transformers, the design and testing must be such that the occurrence of breakdown of the primary (passing on the high voltage to the secondary) is negligible. In this way the fault protection for each section is contained and there is no need to consider the effects of high-voltage fault currents occurring in a lower-voltage section. See Figure 11.6.

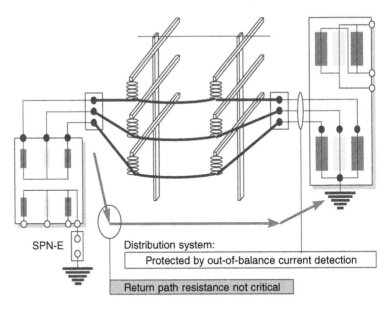

Figure 11.6
Distribution system supply

Electricity supply regulations (ESRs) require distribution transformers' secondary neutral (of a single-phase system) or the star-point neutral (of a three-phase system) to be bonded to a 'substantial' earth mat. The way this is done will be defined in the specification of the local ES authority applicable. Conventionally the neutral and the earth mat are connected together at a substantial un-insulated busbar in a main switch-room and is referred to as the star-point-neutral/earth busbar (SPN-E). This requirement dates back to the ESR in 1905, 1937 and 1944. It acts as a well-defined reference for the purposes of both earthing and bonding as discussed.

The accessibility and distance from the plant it is serving is sometimes a problem for which some solutions are offered later.

11.5   Static electricity

Static electrical charge is caused by the forced separation of molecules of non-conducting materials. Movement of the material or friction against the material can cause the charging effect. When wrenched apart, a surface charge builds up a high PD between the separated materials. The size of the charge depends on the violence with which this separation occurs. The non-conducting materials act as a dielectric, and allow the accumulation and storage of the charge as in a capacitor.

11.5.1   Generation of static

This poses a risk when handling certain materials such as hydrocarbons because they naturally have a low conductivity, and static charges are easily generated. Since the vapor is flammable the sudden discharge of static can produce ignition.

Explosions and fires in oil and gas tankers have occurred where static has been blamed as the cause. After a tank has been emptied, vapor may be present having been mixed with air that has been sucked into the tank as a result of emptying. This provides almost ideal conditions for static to cause ignition during the start of cleaning operations.

Drum filling is a good example to illustrate the principle (see Figure 11.7). Where the liquid enters the drum leaving the supply pipe nozzle, the flow profile deposits a charge on the nozzle and the opposite charge accumulates in the drum depositing itself on the metal of the drum. The charge appears between the nozzle and the filling cap rim. To equalize the charge it is necessary to bond the drum to the nozzle such that the charge is dissipated. As the nozzle is moved toward the drum at the start of the filling cycle and away at the end this is where any initial PD may cause a spark. Bonding must be in place before and after the filling event to ensure safety.

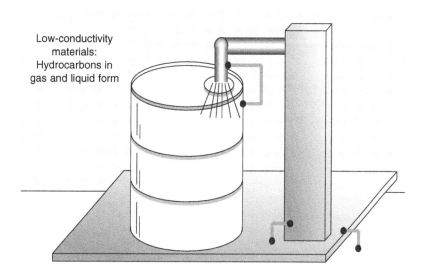

Low-conductivity
materials:
Hydrocarbons in
gas and liquid form

Figure 11.7
Drum filling static risk

The connection to terrestrial earth is not fundamental to safety, as is often misunderstood. It is important to identify where the charge will accumulate. It is preferable to bond the charge-collecting conductors as close to the accumulation point as possible. This is not always practical as in the case of road tankers. Specific information on this is published by road tanker legislative organizations.

Techniques to control static involve the use of chemical additives to raise the conductivity of liquids. Static combs are used to collect and dissipate charges. Locally induced ionization and artificially raised humidity can assist in some circumstances.

Effective bonding is essential, but reducing pumping rates will also reduce static build up.

11.5.2 Lightning

Lightning is generated by the movement of water vapor inside clouds. Up-draughts of warm air in contact with down-draughts of cooler air in large volumes and close proximity generate a vast static charge that build up to an extremely high voltage. It cannot be dissipated because of the low conductivity of pure water vapor. When the accumulated charge becomes so great that it breaks down the dielectric strength of the vapor, the result is seen as lightning. In circumstances where the charge is built up between the cloud and the ground, the discharge occurs when a lightning 'leader' ionizes the air between cloud and ground providing a lower resistance path for the majority of the charge to pass. The resultant current can be in the order of 20–100 kA peaks.

The current pulse shape has a very sharp leading edge and a much slower decay. Because the rate of rise of current is so high and it contains a high-frequency element, the impedance of conductors becomes significant.

The current from lightning rises from 10 to 90% of peak in typically 8 μs and decays to about 50% after 20 μs from the 10% rise point. The pulse shape is said to be an 8/20 wave and this is used as the model for designing lightning/surge protection systems and calculating the effect of the surges on circuits.

Tall constructions on any plant are susceptible to direct strikes. Lightning protection conductors are installed, not in order to handle the vast currents that flow but to conduct substantial but smaller currents that discharge clouds and prevent higher voltages from developing to the point of breakdown. In these cases the impedance connection to earth is critical. The current multiplied by the impedance will cause a rise in potential above ground for the duration of the strike. The lower the impedance the lower the potential. Clearly if other paths act in parallel with the structure then current will be shared and may travel through the plant to get to earth.

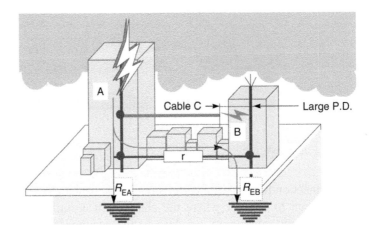

Figure 11.8
Static form of lightning

In Figure 11.8, showing a simplified plant diagram, a strike to building A will be conducted to earth but some current will be diverted through resistance R to the earth of building B. The majority of current will be encouraged to flow to earth via R_{EA} which should be maintained as low as possible. The remaining current will flow via the distributed resistance in the interconnecting bonding of the building. A potential gradient will exist across the structure that must be kept low by good bonding.

An oil refinery is a good example of where there is an ignition hazard combined with susceptibility to lightning strikes. Tall refraction towers act as charge collectors as they approach the clouds. The plant areas are spread out with a network of interconnecting steel pipes. A strategy of good bonding is essential to minimize the risks if lightening is commonplace in the plant's location.

There are two situations that cause concern for instrumentation systems:

1. Firstly, the effect of a lightning strike on a plant will undoubtedly induce surges in other metals in the vicinity of the building through inherent coupling capacitance and inductance. This includes cables. Instrument cables can be particularly susceptible to these surges. High potentials are liable to breakdown insulation and/or damage delicate instrumentation unless adequate precautions are taken.

2. Secondly, an instrument cable C can carry the full potential rise of the strike to building A across to building B and so the danger and the damage may not just be local but may be distributed.

11.6 Clean and dirty earthing

In any electrical system using AC supplies, current will flow in earthing and bonding paths. These are unavoidable and have to be coped with in the course of devising strategies.

There are two reasons for this.

11.6.1 Parasitic capacitance

The capacitance that exists between an electrical conductor operating at some voltage above earth and the surrounding conducting materials of structural metalwork is referred to as 'parasitic capacitance'. In the example of the electric motor, given in Figure 11.9, the capacitance between the field windings and the metal frame would allow the motor frame to attain an elevated potential if the earth returns back to the SPNE remained unconnected.

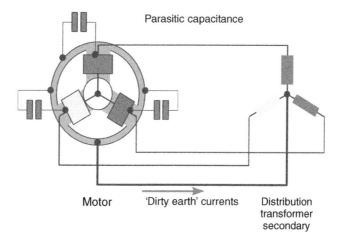

Parasitic capacitance

Motor   'Dirty earth' currents   Distribution transformer secondary

Figure 11.9
Stray currents from parasitic capacitance effects

Bonding the metalwork to the neutral of the distribution transformer eliminates any voltage elevation. However, it allows a small AC current to flow through the structure depending on the size of the capacitance, the applied voltage and the construction of the motor.

Any electrical equipment such as motors and transformers having this conduction path will exhibit these characteristics and will require bonding to earth. When the equipment is switched on the capacitance will charge up and an increased current will initially flow as a result. In steady state conditions the currents flowing in the structure of equipment may be substantial and they will become additive in bonding and earthing conductors.

11.6.2   Fault currents

Where a 'phase to frame' fault occurs, as in Figure 11.4, very high current will flow in the earth path in order to operate the feeder's protective device. A high current dumped to earth during such a fault will cause a dramatic elevation in the potential at the point where the current enters the earthing system. This situation must be cleared quickly by the fuse or breaker.

The concern for equipment operating in hazardous areas is that the currents flowing in earth returns are not interrupted such that sparks can occur or terminations allowed deteriorating such that heat can be generated at that point. Earthing must be substantial and proven.

11.7   Electrical interference

Electrical noise can be generated by the same mechanisms as discussed earlier. Capacitive and inductive coupling effects are responsible for noise that is induced into sensitive measurement and instrument circuits in many cases. This is where there is inadequate separation of signals from outside interference.

In Figure 11.10, the parasitic capacitance between the primary and secondary of the mains transformer couples the mains voltage to the secondary circuit. The resultant voltage appears on the input to the signal amplifier. When the field wires are connected, the capacitive coupling to other structural earths provides a current that develops a series mode voltage on the amplifier. This manifests itself as noise on the input.

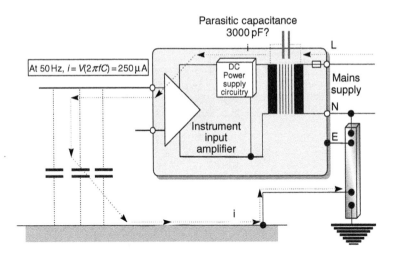

Figure 11.10
Internal parasitic capacitance effects

The cure for this is to take the following precautions, seen in Figure 11.11.

- Earthing the electrostatic screen of the transformer will reduce the coupling capacitance effect and conduct stray currents back to the source of supply.
- Referencing the 0 V rail of the input system to a common potential reference point such as the return of the source of supply will conduct any coupling currents through the power supply circuit back to that point.
- The use of screened cable should be used in the field to connect the sensor to the input amplifier. The screen must be earthed at the same reference point to where all the stray currents are connected, as they are all then returned to the source of supply.

- The screen of the signal cable must only be earthed at one point. If it is earthed at the field end, then currents derived from the parasitic capacitance in field equipment may try to use this as an additional return path.
- The screen will conduct other capacitively coupled currents away, thereby not permitting them to invade the sensitive signal circuit.

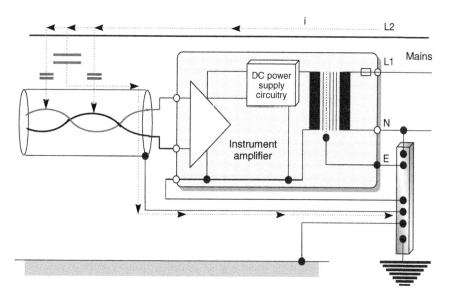

Figure 11.11
Correct reference strategy for instrumentation

The use of twin-twisted cable for balanced systems will also improve the rejection of interference. If interference becomes coupled to one of the conductors then it will also be coupled to the other conductor by the same amount. The signal appears in series mode but the interference appears as common mode. It is presented to an amplifier across an inverting and a non-inverting input, which rejects the common mode and passes the series mode signal almost perfectly (see Figure 11.11).

11.7.1   Interference rejection with barriers

The same criteria apply when preceding the input of a safe area-mounted device with a barrier. The barrier busbar conveniently becomes the focal point for all earths associated with the signal circuits. Care must be taken to ensure that only signal paths are connected into this arrangement. Adopting the strict philosophy of earthing screens at one point, that being the barrier busbar, ensures that coupling currents converge and are returned to the SPNE consistently (Figure 11.12).

It must be realized that there is still current flowing in the IS earth, albeit relatively small. Disconnection of the IS earth may not be indicated by noise appearing at the amplifier inputs but is likely to cause elevations in the common mode voltage seen in the circuit. This may be acceptable to the input amplifiers and the circuits generally, but on a large system it could become dangerous. This is because if the system became inadvertently earthed by a fault in the hazardous area, the o/c voltage and the sum of the s/c currents, due to leakage, may then be enough to cause heat or sparks.

Return currents that are allowed to invade signal paths may result in the signal becoming distorted or swamped by the noise. Clear segregation between earth arrangements and then connection at a common point will assist in de-coupling interaction.

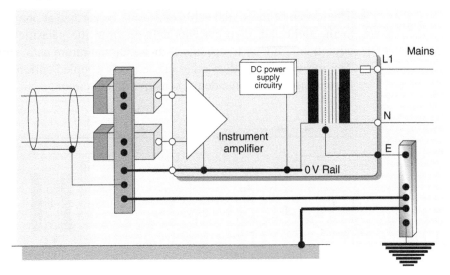

Figure 11.12
Correct earthing of safe area systems

11.8   Earthing terminology

Two common and more meaningful terms are of some help in distinguishing the different earth paths.

The term '*dirty earth*' is applied to the return path conductor that can carry the prospective fault currents from electrical equipment. As a result the conductor is likely to shift in potential by a greater amount (with respect to the neutral of the supply system which is the ultimate reference). Fault currents are inseparable from the parasitic capacitance currents, which will flow along the same route, although they will be at a much lower level.

The term '*clean earth*' can only apply to return-path conductors, which cannot carry fault currents from electrical equipment. This, by definition, is impossible to control or guarantee since any circuit to earth will conduct fault currents back to the return of the source of supply.

It is more correct and important to recognize that the name of the 'earth' path tends to be defined by what is connected to it. It is therefore more logical to describe the earth by the circuit types from which it is gathering 'stray' currents. Hence power, instrument, computer, structural and terrestrial earths are more meaningful names but their function requires clear definition in plant safety documentation so that their use is not misunderstood. There is little concern that the risk of ignition will be increased by incorrect earthing but the operational aspects of systems may be severely hampered.

11.8.1   Sneak paths

One of the problems encountered with carelessly referenced systems is the devious route that unsuspecting currents can take. These are difficult to find but essential to solve as they will cause interference and may pose a danger in the hazardous area. An example of this is where screens have been connected at both ends during plant commissioning. A new large motor was subsequently installed adjacent to the field-mounted instruments and severe interference is seen on the instrumentation. The problem is that structural earth currents have been introduced into the instrument and IS circuits. These situations manifest themselves in curious way.

Attempts at solving sneak paths must be done with extreme caution, as disconnecting unsuspecting earths may have unpredictable results. Ideally, the plant should be made safe before attempts are made. Otherwise a carefully thought-out strategy of approach is required. This means isolating and disconnecting each circuit in turn whilst noting any improvement seen until the problem is solved. Progressive improvement suggests there is a cumulative effect because a common point to that group of circuits is at the center of the problem. Such situations are difficult to generalize.

11.8.2   Power faults with IS apparatus

IS apparatus in the hazardous area normally requires to meet an isolation test requirement of 500 V AC applied to the internal circuits with respect to the case (which is normally connected to the structural earth). The insulation must not breakdown within 1 min. The purpose of this test is to ensure adequate isolation during power fault conditions whilst elevated voltages may appear across this insulation. Invasion of power must not occur on one IS circuit which could elevate voltages and pose an increased risk on other circuits.

This can happen in a number of circumstances discussed as follows.

In the situation shown in Figure 11.13, where a power fault touches the signal line in a simple installation, the fuse in the barrier is the weakest link and will undoubtedly blow before the distribution fuse. Assume a current of 1 A flows before the fuse ruptures. If the resistance of the IS earth is 1 Ω then the voltage elevation (potential difference, PD) seen at the instrument earthy terminal will be 10 V.

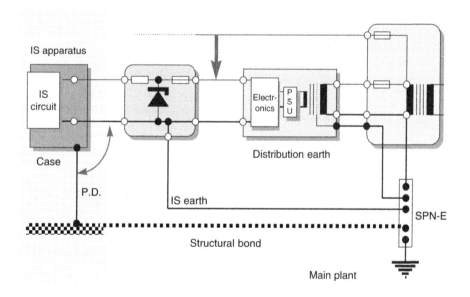

Figure 11.13
Power fault to safe area channel

The PD would exist only for the duration between the fault occurring and the fuse rupturing to clear it. In this example there would be no serious risk due to the margins built in to the safety characteristics.

If no isolation were found in the hazardous area instrument then the fault current would share through the structural return path. This would not be unsafe because the current is still limited by the CLR in the barrier.

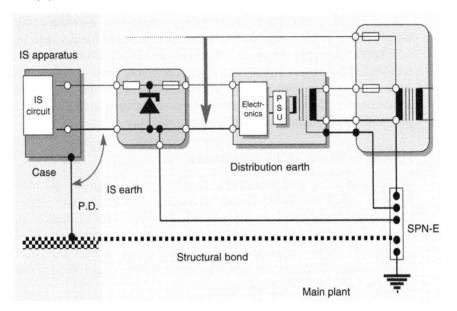

Figure 11.14
Power fault to earthy terminals

In Figure 11.14, the fault touches the earthy conductor of the safe area barrier circuit. The current is not limited by the barrier in any way and is dependent on the loop resistance it encounters. The distribution fuse may be 5 A and the peak current may be say 100 A before fuse rupture. The elevation on the busbar could peak at 100 V if the IS earth was 1 Ω. Clearly the PD generated could be higher or reaching close to the limit of the supply momentarily.

In a third situation, shown in Figure 11.15, an earth fault may occur in the hazardous area as shown.

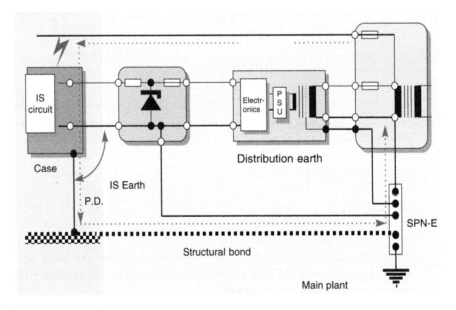

Figure 11.15
Hazardous area power fault

It could be argued that this failure is more likely to occur than the other two situations. This is because the mechanical construction and reliability of low-power instrumentation in cabinets is higher than plant-mounted high-voltage and power equipment such as motors which run at higher temperatures where breakdown of insulation is more likely to occur. The fault current flowing can be much higher in this situation and the structural earth resistance through which it must flow will determine the PD experienced across the earthy line and structural earth. 500 V is internationally accepted as being a realistic peak value needed to protect the integrity of the IS circuit under this condition.

11.9 Connection of earthing systems

The points discussed so far can be summarized into current earthing practice where it can be seen why the philosophy of earthing has evolved to that required in the standards and codes of practice.

11.9.1 Typical earthing arrangement

The diagram in Figure 11.16, illustrates a simple arrangement showing effectively five separate types of 'earth' connection, converging at the SPNE.

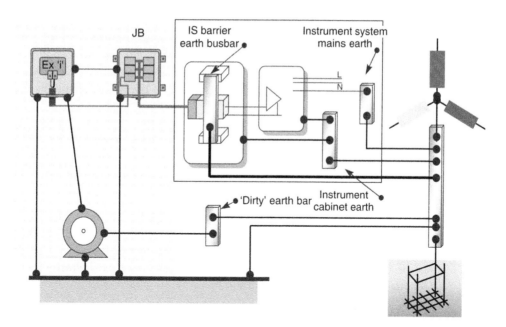

Figure 11.16
Earthing strategy

- The *structural earth* may not be a physical cable but nevertheless requires representation in this diagram, as it is a part of the bonding system.
- The *dirty earth* bar will collect all the earth returns from armored cables to power equipment and circuits in the hazardous area. The fan out to other equipment may use sub-busbars but all eventually converge on one dirty earth bar that is ultimately connected back to the SPNE.
- The chassis or cabinets of instrumentation equipment will inevitably be connected to structural earth via the steel-work of the plant construction.

Cross-bonding between equipment is necessary to ensure that harmful potentials cannot develop. It happens as a natural process when the equipment is mounted and bolted to frames that support other equipments. Where there is a risk that bonding may be inadequate, devices are strapped with bonding conductors but this is sometimes taken to unnecessary extremes. It is therefore quite natural that the 'dirty' and 'structural' earths appear very similar. This will be discussed more in detail later.

- The *IS earth* goes directly to the SPNE and must not be connected to any other earth system.
- The *instrument system mains earth* in this example is used to terminate all the supply earths, which would normally be commoned together as part of the interference control strategy. This bar should not be used for any signal earths.
- The *instrument cabinet earth* connects the chassis of the instrument cabinet(s) to the chassis and metalwork of all the equipments as part of the bonding required for personnel protection. This is distinct from the instrument system mains earth because some equipment manufacturers specifically used not to internally connect the chassis to the mains earth. This practice is unfortunately changing but its disadvantage is that it reduces the control over interference. As the quality of insulation and components used in modern industrial equipment rises then this does become less important and these two earths may be more conveniently treated as one.

Switch mode power supplies are becoming more common in safe area use but their great disadvantage is that they are considerably noisier than analog types. It is most important that they are correctly earthed to the mains supply earth for safety and noise control reasons as the filter capacitor network on the mains terminals must have a return path to earth.

Double-insulated equipment mounted in plastic cases is rarely used in safe area industrial applications because it cannot meet EMC requirements. Referencing of the supply systems is still important.

11.9.2 IS earth specification

The IEC 60079-14 in line with most codes of practice requires that the barrier busbar earth is connected back to the earth reference (usually the SPNE) busbar in not less than 4 mm^2 copper (or equivalent) cable with a resistance of less than 1 Ω. The IS earth must not be connected to any other earth system and must be identified as an IS earth (see Chapter 7 of this Book).

11.9.3 IS earth installation

It has become an almost industry standard approach to use a pair of 10 mm^2 cables for the IS earth connection. The resistance of this is extremely low at about 0.005 Ω for a 100 m run. The method of identification is not specified and no one way of doing it has emerged as preferred. One suggested way was to wrap the cable pair with turns of blue insulating tape every half meter. This is seen on some plants but whatever method is chosen should be clearly identified by the plant safety documentation (see Chapter 12).

11.9.4 IS earth testing

It is becoming common industrial practice and is highly recommended that earth circuits be connected using twin parallel conductors.

In Figure 11.17, the IS earth is linked to the SPNE by two conductors. In this way the loop integrity can be easily checked by breaking one conductor and measuring or monitoring the loop resistance. Separate connections for each of the two conductors onto the busbars at each end effectively duplicates the connection, thereby lowering the overall resistance and raising the integrity by including redundancy. This same strategy can be followed for all the busbar interconnections.

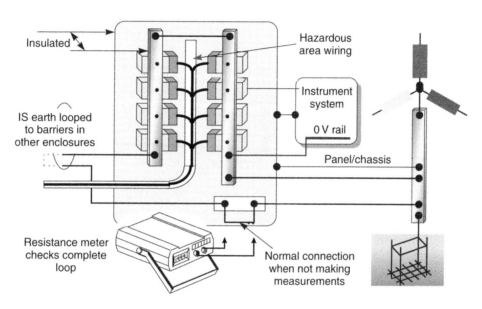

Figure 11.17
IS earth testing

IS earths from other cabinets may be connected by looping through. This is acceptable, provided that the integrity of the system as a whole can be proven. If connection using a tree structure is used then each interconnecting branch must be proved to the next one in the chain and the furthest branch must still meet the requirement of connection back to the SPNE with a resistance of less than 1 Ω.

11.9.5   500 V isolation test

The ability to withstand a 500 V rms test for an IS circuit conductor has become internationally accepted as the fundamental requirement for circuit isolation. The conductors of IS circuits in a cable must be separated from Earth *and* from any other IS circuit by the ability to withstand a 500 V rms test for 1 min. Where this requirement cannot be met, the installation conditions may require a special approach under certain countries' codes of practice. This is discussed more in the application section where examples are given of the accepted techniques for this approach.

11.10   Power supply systems

Where supply systems other than from national mains distribution are associated with IS then due regard must be given to the earthing integration. Locally generated supplies will be treated in a similar manner to AC mains distributed power and will be subject to the ESR discussed.

11.10.1   Offshore earthing

Offshore regulations permit the IS earth to be local to the barriers because the all-steel structure on which the whole installation is mounted is accepted as having such a predictably low earth resistance throughout the platform that no significant voltage rises will occur during fault conditions.

The regulations stipulate that an IS earth cannot be connected to the deck of the platform closer than 2 m from the point where the generator neutral-earth connection is made, as shown in Figure 11.18.

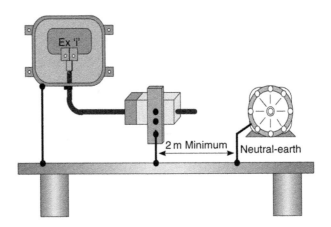

Figure 11.18
Earthing strategy offshore

11.10.2   Earthing skid packages

Self-contained skids built for installations must provide electrical systems that are compatible with the installation into which it is to be placed. If barriers are used then the IS earth should be made available for connection to existing IS earthing systems. In some cases the manufacturers may have earthed the barriers to the steel-work structure of the skid. If the application is for offshore then this may be acceptable but for onshore situations the IS earth must be kept separate as indeed should mains earths and signal screens so that they may be terminated in the proper place at the safe area end. See Figure 11.19.

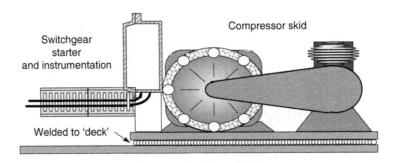

Figure 11.19
Earthing on skids

11.10.3 Earthing of battery systems

Battery power supplies of the various types are usually required to be fully floating. The prospective short-circuit current from batteries is extremely high and a great deal of damage can be done to batteries and the installation under unchecked short-circuit conditions. Including protection devices is desirable but reduces the supply integrity. Battery backing is normally specified for reliability and so fault mechanisms are considered and designed for in order to give the highest supply integrity.

By operating batteries without reference to earth (floating), a single fault to earth of one terminal would not cause disruption of the supply. Earthing the batteries on one side would mean that single fault of a second connection to earth (of the other polarity) would provide short-circuit conditions.

Operation of barriers requires the referencing to earth and this may not suit the supply system being battery backed. If the batteries supply is used via an invertor system that has full isolation then there is no problem. If the batteries are used directly then it is possible to use barriers but some compromises have to be made. Figure 11.20 shows some possible options.

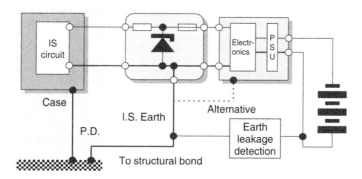

Figure 11.20
Battery-supplied systems

Galvanic isolators are therefore preferred for use with battery supplies for the obvious advantage that no earth referencing is required for safety purposes.

11.10.4 IS earthing on ships

The UK's IEE (Institution of Electrical Engineers) issues 'The Blue Book', currently the sixth edition, 1990, entitled, 'Regulations for electrical and electronic equipment of ships with recommended practice for their implementation'. Section 23 contains the regulations for power generation and distribution on tankers carrying flammable materials. They do not permit the hull to be used as a return path for fault currents.

The generation and distribution systems are fully floating and earth leakage monitoring is used on every section to assist in maintaining the integrity of supply. If, on one phase, a single fault to earth occurs, then fuses or circuit breakers do not disrupt the supply. Also, the fault can sometimes be cleared without disruption to supplies.

IS earthing is permitted, provided that it does not defeat the supply arrangements required. This is acceptable as adequate isolation is normally included upstream of barrier systems between the distribution secondary, the neutral of which is permitted to be 'referenced' to earth.

Lloyds Register of Shipping and other internationally accepted organizations produce documents, which think along the same lines. Such documents are not internationally agreed upon and in line with IEC thinking. This may come in time.

11.11 Portable equipment using batteries

The risk of mains invasion is completely removed from portable equipment that operates solely on internal batteries. Measuring and calibration equipment need only be treated as having safety parameters that require examination when connecting into a system. This is dealt with in the section on applications and systems.

11.12 Earthing arrangement standard solutions

Provided that the simple strategies explained above are used, problems encountered in designing and specifying IS earthing systems on installation can be overcome. Difficulties often arise when existing plants are extended. In the following section various philosophies are discussed to bring together some of the principles into practical use.

The simplest and most usual case, seen in Figure 11.21, on small installations is where the local transformer is relatively near to the control equipment and the cabinet earth can be conveniently taken back to the SPNE via the supply cable armoring. The IS earth goes back via two dedicated conductors.

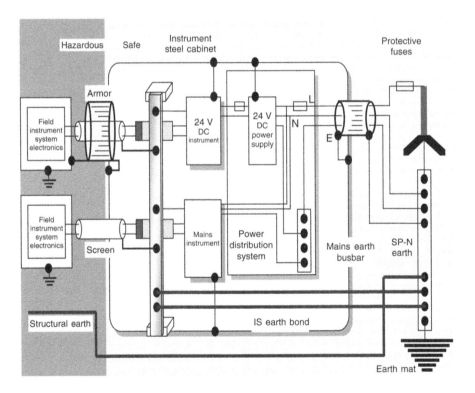

Figure 11.21
Typical system distribution showing earthing arrangements

The cabinet's internal distribution system has all the mains earths collected together onto a mains earth busbar. Note that in the hazardous area, the armor of the cable is connected, at both ends, to structural earths. The screens are only earthed at one point, being the barrier busbar.

In Figure 11.22, the same arrangement is used but the IS earth is direct to ground (referred to as a 'separate' or 'separated' earth). This demonstrates that soil resistance becomes part of the 1 Ω earth resistance argument. It is difficult to justify the acceptance of this because the resistance between the ground rods is no longer under the direct control of the user. This practice is still accepted in some countries and the IEC standards have not yet formally discouraged it.

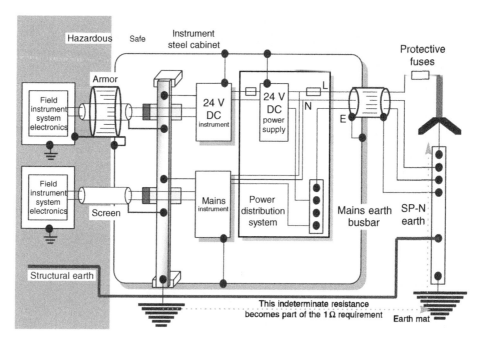

Figure 11.22
IS earth direct to local ground

The diagram shown in Figure 11.23 gives the arrangement used for the case where an intermediate transformer is used to supply the panel. If the transformer is merely to step down the voltage for the instrument system then it may not have the neutral earth connection made. In which case the upstream feeder transformer must still be the point to which the IS earth must be taken. This is normally only encountered on older installation, which has not been modified.

In Figure 11.24, the distance between the main feeder transformer and the panel is too long to tie to the IS earth. In another instance the supply may be delta and not star, perhaps because it is the only convenient feeder for the panel location.

The creation of a localized IS earth can be justified by the inclusion of an isolating transformer at which the secondary neutral is connected to an earth mat. Under these circumstances it is preferable to use out-of-balance current detection techniques to clear faults on the feeder because low-impedance fault paths to earth may not be seen by long-distance supplies fuse. The out-of-balance current detection would detect that the line current was not equal to the return neutral current and would therefore break the circuit.

This solution is recommended when ships such as tankers are to be fitted with IS barriers and equipment. It is also very useful in many other remote situations such as RTU outstations.

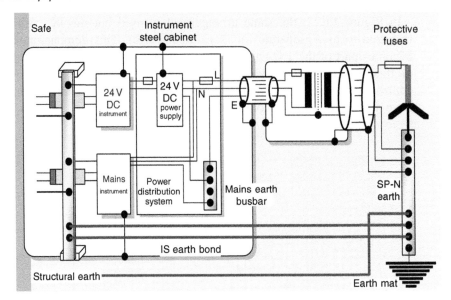

Figure 11.23
The intermediate transformer situation

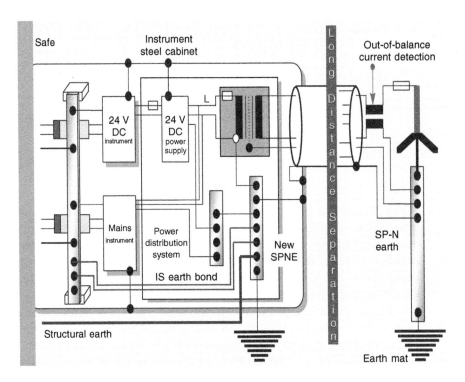

Figure 11.24
Local distribution transformer included with local IS earth

11.13   Earth loops

In Figure 11.15, showing a typical earthing arrangement, the connection between the structural earth and dirty earth busbar, which then meet at the SPNE bar, do form a closed loop. This loop will not introduce interference unless there were to be a strong magnetic field somewhere in the center of it.

The mechanism causing interference in a loop is one of induction from a changing magnetic field that falls within the area of the loop. Currents are induced into the loop by magnetic coupling in series mode. A loop formed around a device that possesses a varying magnetic field, such as a transformer or motor, will cause induction in that loop. Higher frequencies will couple to a loop more efficiently than lower ones.

It is unlikely that sufficient energy can be coupled into a looped circuit such that it can become a source of ignition. The coupling would need to be optimized to do this as in RF heating systems. It is more likely that interference will be injected into loops that form part of sensitive measuring circuits.

Provided that the loop area between the conductors is kept as small as possible then no ill effects should be observed.

Twin-twisted cables are an example of where the loops are kept as small as possible between the conductors. Thus induction to one conductor is most likely to induce equally on to the other conductor and the series mode induction is dealt with as common mode at the input to a differential or balanced-input amplifier. This is in the same way as capacitively coupled circuits can be arranged to reject noise.

Screening also helps to de-couple inductive pick up because eddy currents are induced into the screen and are conducted away. The eddy currents also oppose the magnetic effect that is inducing them, and this effect also helps to reduce the coupling to the signal on the wire.

11.14   Computer earthing

The electrical noise generated by the fast-switching digital electronics in computer systems must be kept away from sensitive analog circuits. The provision for earthing these systems in order to assist in the noise control is fundamental to their successful operation. Computer control system manufacturers often specify a computer earth as yet another dedicated earth path that requires integration into the strategy.

Reconciling the earthing arrangements for computers can become complicated, and a detailed knowledge of the internal supply and signal routing is necessary before the best arrangements can be made. If the computer system requires IS interfaces then the choice of devices may be influenced by the ease of accommodation of the earthing arrangements.

Generally there are two situations. Firstly if the computer I/O is fully and properly isolated then there should be no conflict in the earthing arrangements, and barriers may be used with ease. This is often the cheapest solution.

If the computer I/O is not isolated then IS isolators are preferable because no safety earth is required and the isolators provide some flexibility for extending the capability of the computer system.

PLCs are not greatly affected by this discussion since they are relatively small and self-contained. They are normally installed in close proximity to the barriers and other associated supplies and equipments. DCS is more spread out and requires more careful rationalization.

The connection XY, in Figure 11.25, may carry substantial noise and return currents from switch and Analog circuits to its power supplies. Its resistance must be kept as low as possible, preferably less than 0.1 Ω. The IS earth will not carry the instrumentation currents but is likely to have some larger noise currents flowing due to the presence of the computer. This is the preferred connection way from an IS point of view.

If the manufacturer specifies that the computer system must be connected directly to earth then the arrangement below is acceptable with known limitations.

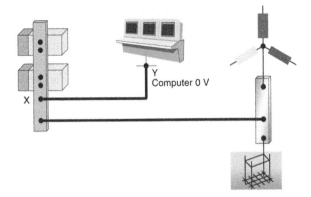

Figure 11.25
Computer referenced to IS system earth

In Figure 11.26 the computer earth becomes part of the protective earth route. It will probably carry substantial return currents (from X to Y) for the instrumentation. The resistance of the barrier busbar to SPNE link is therefore difficult to monitor because of these impressed currents on the measurement. It is an accommodated compromise from the recommended practice.

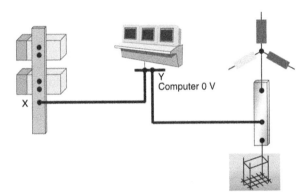

Figure 11.26
IS system earth referenced to computer system

Figure 11.27(a) shows the practical implementation of the requirements. The exact arrangement depends on the physical distances involved and the nature of the supply circuits to the computer for both main power and instrument loop power. If the computer has its own mains supply transformer that is also used to supply the I/O loops then this is the most desirable situation.

Figure 11.27(b) depicts direct connection to local earth mats for each of the systems. This is acceptable from an IS point of view, provided that the $1\,\Omega$ specification is maintained for the protective earth path between the busbar and the SPNE. This arrangement is ill-defined and liable to cause all sorts of problems.

Figure 11.28 is thought of as the solution to all the above situations. Its downfall is that because the loop formed can be over a wide area it is more susceptible to induced noise. Another problem is that power fault conditions in associated equipment may cause transients that can adversely affect the computer or instrument systems.

Other arrangements are seen on installations. Detailed analysis would be necessary before comment could be made on the acceptability and suitability from safety and operational viewpoints.

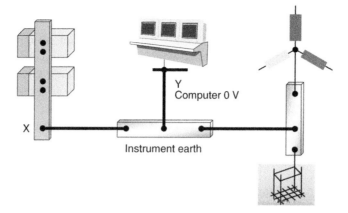

Figure 11.27(a)
DCS earthing, commoned

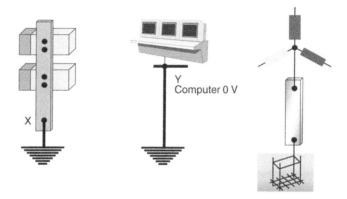

Figure 11.27(b)
DCS earthing (separate)

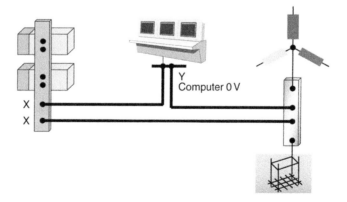

Figure 11.28
DCS earthing: cross-connection

   In the section on applications, examination of individual IS instrument loops, used as I/O for computer control systems, will describe how return currents impinge on the safety earth in more detail to allow a better understanding of the above.

11.15   Surge protection systems

Surge protection devices (SPDs) may be inserted into instrument cables in order to protect the instrument systems where there is a risk of invasion from induced secondary surges. An SPD works by limiting the energy passing through it into a protected circuit. In this way there are many similarities between an SPD and a shunt diode safety barrier. The main difference is in the magnitude of voltages and currents that devices must handle. Diverting large currents in a very short period of time require multistage protection techniques using a combination of Zener diodes, varistors and gas discharge tubes to clamp the applied voltage yet shift the current.

The excess energy is diverted to earth in the same way, as shown in Figure 11.29. The principal difference in the technique is the levels of energy that must be handled by SPDs. The quality of the connection to earth is at the heart of their success. Connection to terrestrial earth via the most direct and shortest route is essential for good performance. The local earth connection becomes the reference point for the system. The intention is that the electronics are all transiently shifted from one potential to another and back. If they all move together simultaneously then no PD can appear across the protected electronics to its detriment.

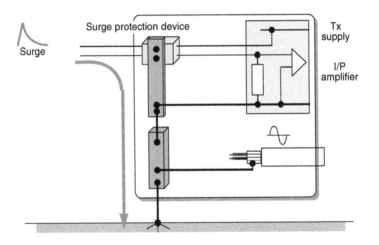

Figure 11.29
Surge protection

The use of SPDs with hazardous area circuits where shunt diode safety barriers are used requires careful planning of the earthing systems. The IS barrier and the SPD must inevitably be referenced to a common potential. On small installations it is common to move the IS earth arrangement toward the SPD earth as shown in Figure 11.30.

Bonding to the common busbar fulfills the criteria for both barriers and SPDs but requires the structural earth and mains earth to be commoned at this point also. The earthing of screens and armoring becomes more important in this situation and will be discussed later in this chapter. Clearly in the larger installations, this is not always possible to do in such an elegant way, and compromises have to be made.

This is where the use of galvanic isolating interfaces are preferred and there is no contention in the earthing requirements because no IS earth is required.

The isolator can absorb the sudden common mode change in potential with ease, providing that the potential does not appear across its input terminals (Figure 11.31).

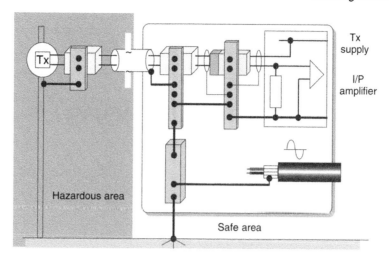

Figure 11.30
SPU with barriers

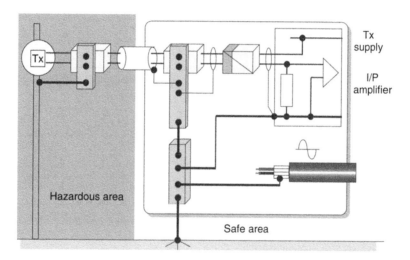

Figure 11.31
Isolators with SPUs

11.16   Standards and codes of practice

The standards and codes of practice that apply in various countries are not always specific in their requirements for earthing as applied to IS and related topics.

In the UK, BS 5345 states specific reference to the star-point neutral earthing system of the incoming supply. In Germany, VDE 0165 requires the creation of a reference potential to which plant and supply are connected. A Canadian Standard allows the designation of an earth reference but does not state to what else it is to be connected to other than the barrier earth.

The detail contained in this chapter has been based on practical experience of the interpretation of the more specific BS 5345 which many countries follow in the absence of their own standard.

12

Installations

12.1 Introduction to installation requirements

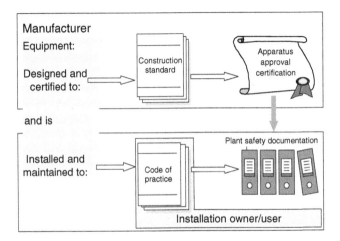

Standards and codes of practice

The manufacturer of electrical apparatus may produce and certify it to comply with a given standard. Once in the hands of the user, the manufacturer has no control over the safety aspects of the way in which it is selected, installed, commissioned, inspected and maintained. He cannot therefore be held responsible for the consequences.

The responsibility is placed on the owner of a plant, perhaps delegated to the user of the installation to ensure that it begins and remains as a safe installation. Hence there are separate requirements for the manufacturer and for the user of explosion-protected apparatus.

Each country may appoint, through its government, a technical body whose function will be to advise on the adoption of technical standards. The exact operation of this will vary from one country to another.

12.2 Installation requirements

In order to examine the installation requirements in this book, the first part of this section will look at the requirements of the International Standard IEC 60079-14: 1996: 'Electrical installations in hazardous areas (other than mines)'. The intention is to draw out the important requirements, comparing and contrasting them with other common codes of practice.

It should be realized that the IEC 60079 series was written to try to combine all the various national requirements for Ex-protected equipment and installations into one internationally accepted set of documents. It is generally recognized that there are shortcomings in the guidance that it gives. It tries to encompass the many different approaches that are accepted in different countries by becoming less specific about its requirements and leaving out valuable explanation of why it imposes conditions or alternative options. The current version of the standard does not always recognize that there are different approaches in place, but it is a first attempt to harmonize the installation in hazardous areas using an international platform in order to ease the difficulty in installing one piece of equipment certified in one country into an installation in another country.

Subsequently, in this book, the wider practical interpretation of the IEC and other codes that are still in place in many countries are discussed.

12.3   IEC 60079-14: standard contents

The contents of the IEC 60079-14 Standard are divided into sections. These are listed in Table 12.1 for the ease of reference.

Section	Content
1	Scope and object
2	Normative references
3	Definitions and terms
4	General
5	Selection of electrical apparatus (excluding cables and conduits)
6	Protection from incentive sparking
7	Electrical protection
8	Emergency switch-off and electrical isolation
9	Wiring systems
10	Additional requirements for type of protection 'd' – Flameproof enclosure
11	Additional requirements for type of protection 'e' – Increased safety
12	Additional requirements for type of protection 'i' – Intrinsic safety
13	Additional requirements for type of protection 'p' – Pressurized apparatus
14	Additional requirements for apparatus for use in Zone 2
Appendix A	Verification of intrinsically safe circuits with more than one associated apparatus with linear voltage/current characteristics
Appendix B	Methods of determining the maximum system voltages and currents in intrinsically safe circuits with more than one associated apparatus with linear voltage/current characteristics (as required in Appendix A)

Table 12.1
IEC 60079-14 section headings

Sections 1–3 give the conventional standard scope, cross-reference and definitions. Sections 4–9 give general requirements that pertain to all precautions taken. Sections 10–14 are specific to each type of protection recognized by the code. It is important to realize that not all types of protection accepted in various countries are covered by this document. There is an assumption of general compliance with the 79-14 Standard, even though more specific codes of practice may be in operation in particular countries or for particular contracts.

This discussion will consider the general requirements (Sections 4–9) and those specific to IS (Section 12) with the two annexes.

12.4   Other relevant installation standards and codes

In the appendix of this book, a list of countries and their generally accepted standards are given for initial reference. This list is not definitive or exhaustive and is included for guidance based on the best information at the time of writing. Changes were due in Europe but this has not happened yet; the old British Standards will eventually be made obsolescent. This may well have a worldwide impact because these are often used in countries that do not have local or national codes of their own.

The 'owner' (or 'user') organization of explosion-protected electrical apparatus should declare what standards are acceptable as part of its safety philosophy.

12.5   Safety documentation

It is an internationally accepted practice, if not actually construed in the law of countries that the owner of a plant or installation possesses documentation relating to safety issues. Such documents shall describe, and may be consulted on, all possible safety criteria affecting a plant. This is in order to ensure the safe operation and maintenance of a plant. This safety documentation has been introduced in earlier chapters in this book.

It will generally comprise the same information being amassed for the purposes of selection of equipment, design of the plant and operation of the process. The documentation and drawings produced by a design contractor, as part of the design, very often form the basis of this information.

Decisions to be made about the running of the plant, such as maintenance and periodic inspection (as required by IEC 60079-17 on inspection) can be made on the basis of information available from such safety documentation. The recording of philosophies and decisions should be maintained in the plant safety documentation. This forms a record of the management's safety attitudes and is a valuable procedure to satisfy the quality and insurance requirements placed upon plant operators.

The documentation can also be used to formulate the basis of training for employees, and contractors working on a site.

12.6   General requirements of the standard

The general requirements of IEC 60079-14 state that electrical installations in the hazardous area shall also comply with the appropriate requirements for installations in non-hazardous (safe) areas. The standard should be viewed as working alongside other documented requirements for electrical systems. There are no requirements in this standard that defeat or conflict with the object of other general electrical installation practices. It is the intention of this statement to reinforce the user's awareness of the need for personnel safety (shock and injury), fire prevention and overload protection.

In order to facilitate the selection of appropriate electrical apparatus and the design of suitable electrical installations, hazardous areas are divided into Zones 0, 1 and 2. This assumes that safe and all hazardous areas have been defined. It uses the phrase 'according to IEC 60079-10' which is the part dealing with 'Classification of hazardous areas'. However, a number of other industrial standards are used in practice that is industry-specific, and the standard recognizes this fact by declaring that other national standards are acceptable.

IEC 60079-14 goes on to say that electrical apparatus should, as far as reasonably practicable, be located in a non-hazardous area. Where it is not possible to do this, it should be located in the least hazardous area, thereby reducing the probability of contact with a hazard.

Apparatus must be installed according to its documentation. This will be discussed later in more detail.

On completion of an installation, an initial inspection (prior to the operation of the apparatus) is required to be carried out in accordance with IEC 60079-17: 'Inspection and maintenance of electrical installations in hazardous areas'. This will be discussed in the specific section on Inspection.

12.6.1 Documentation

IEC 60079-14 states that in order to install, or extend an existing installation, the following documentation is required, where applicable:

- Area classification documents
- Instructions for erection and connection
- Documents for electrical apparatus with special conditions
- Descriptive systems document for IS systems
- Manufacturer's/qualified person's declaration.

This statement implies that all information necessary for the correct and safe installation of apparatus must be available for the installer (and ultimately provided to the user as part of the plant safety documentation). The manufacturer must provide instruction manuals, certification documents and drawings that communicate sufficient detail of information to allow an assessment of the apparatus for the area in which it is to be operated. Other legislations require equipment to be accompanied by adequate documentation for its intended purpose of use.

Safe operation may require installation to the requirements of the standards to which the apparatus was certified and perhaps specific codes of practice.

For intrinsically safe systems, a system descriptive document is required which explains how the proposed equipment connected together to form a system is safe and compliant with the standard to which it was certified. This document will state the requirements for any special condition that must be applied.

This IEC standard document now clearly recognizes the responsibilities of the installer in placing equipment in a hazardous area. The declaration required may be taken on a number of levels. The person(s) undertaking the installation must be adequately trained and supervised. Supervisors must also be adequately trained. The declaration of correct installation in accordance with the information provided could only be made by a suitably trained individual or a direct supervisor who is suitably qualified. An individual must be given an appropriate level of supervision.

The standard cannot dictate the quality of documentation in the same way that it cannot specify what training and supervision is necessary. It infers that the owner of the plant carries the responsibility to decide and action these such that he can demonstrate that he has taken all reasonable practicable precautions to operate his plant in a safe manner.

12.6.2   Selection of electrical apparatus

All standards and codes of practice for installation require that a selection process has been performed (by the plant designer), prior to any erection of equipment, which ensures that the appropriate (type of) apparatus is chosen for a given installation. This selection process can only be done correctly when the hazards are known and defined.

Selection criteria depend on:

- Area classification (Zone)
- Temperature classification (T class), and
- Apparatus grouping (Group II, IIA, IIB or IIC).

These will be encountered by each piece of apparatus.

The definition of the hazard involves all of the above. Area classification only refers to the probability of presence of a gas or vapor. It has become common to refer to the 'hazard' which defines all these but has the wider meaning of including information on the properties of the materials that may influence the judgment of suitability. Chemical, environmental and mechanical influences of choice may come from this knowledge.

Figure 12.1 models a suggested routine for apparatus selection.

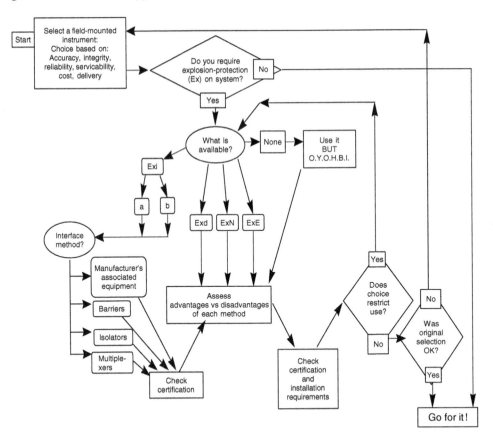

Figure 12.1
Selection process for apparatus

In the IEC 60079-14: Section 5, deals with the process of selecting apparatus according to the zone of use. It recognizes that different types of protection have different levels of integrity and their use must therefore be restricted to suitable zones as shown in Table 12.2. The risk must be balanced against the likelihood of the occurrence of a source of ignition.

Zone	Type	Letter	IEC 60079
0	Intrinsically safe	ia	–11
	Special (certified for Zone 0)	s	–0
1	Any type suitable for Zone 0		
	Flameproof enclosure	d	–1
	Pressurized apparatus	p	–2
	Powder filling	q	–5
	Oil immersion	o	–6
	Increased safety	e	–7
	Intrinsically safe	ib	–11
	Encapsulated	m	–18
2	See Section 12.6.3		

Table 12.2
IEC 60079-14 permitted zones of use

12.6.3   IEC 60079-14 requirements for Zone 2

IEC 60079-14 clause 5.2.3 qualifies four further acceptable conditions under which electrical apparatus may be used in Zone 2. This clause is reproduced as it appears in the standard and is then discussed:

IEC 60079-14 Requirements for Zone 2		
Clause	**Quote**	**Comment**
5.2.3	Apparatus for use in zone 2 The following electrical apparatus may be installed in zone 2: (a) Electrical apparatus suitable for zone 1 or zone 0 (b) Electrical apparatus designed specifically for zone 2 (for example type of protection n according to IEC 60079-15) (c) Electrical apparatus complying with the requirements of a recognized standard for industrial electrical apparatus which does not, in normal operation, have ignition capable hot surfaces, and	Equipment certified or approved to 'recognize' standards are permitted under this clause. North American Division 2 equipment is now acceptable.

(Continued)

Clause	Quote	Comment
	1. does not, in normal operation, produce arcs or spark, or 2. in normal operation produce arcs or sparks but the values, in normal operation of the electrical parameters (U, I, L & C) in the circuit including the cables do not exceed values stated in IEC 60079-11 with a safety factor of unity. The assessment shall be in accordance with the energy limited apparatus and circuits given in IEC 60079-15 (electrical apparatus type 'n'). Unless safety is demonstrated by test, a surface is presumed to be ignition-capable if its temperature exceeds the ignition temperature of the explosive atmosphere concerned. This electrical apparatus shall be in an enclosure with a degree of protection and mechanical strength at least suitable for non-hazardous areas with a similar environment. It requires no special marking, but shall be clearly identified, either on the apparatus or in the documentation, that it has been assessed by a person who shall: be familiar with the requirements of any relevant standards and codes of practice and their current interpretations; have access to all information necessary to carry out the assessment where necessary, utilize similar test apparatus and test procedures to those used by national authorities. apparatus in accordance with clause 5.2.4 (The rotating machine clause is not relevant to instrumentation applications).	Note that the word recognized is nowhere defined. This suggests the acceptance of the old 'self-certification concept'.
5.2.4	Selection of apparatus not available in accordance with IEC standards In order to ensure that the correct selection and installation of such apparatus (for example equipment marked 's' and identified with a zone of use in accordance with IEC 60079-0), reference should be made to the relevant national standard or code that deals with these aspects.	

These clauses are very useful in that they permit the installer some flexibility to use other techniques or perhaps variations of techniques that can be shown to be acceptable for Zone 2.

Type 'n' or 'N' and Ex 's' are formally recognized in this section although sections of the standard were not individually dedicated to their special needs. It permits the use of relevant standards operated by those who understand them. This is a further example of the standard throwing the onus onto the user, who must decide whether an organization is qualified to propose the use of a solution to his problem and can justify its safety.

Industrial grade apparatus not specifically designed for hazardous area use but which can be shown to be safe in normal operation under the type 'n' assessment constraints may be permitted in Zone 2 under this clause. Many countries and users have requested this approach in the past. The Dutch, heavily influenced by Shell, BP and the oil industry generally, where the hydrocarbon hazards are predominantly IIA and large areas are Zone 2, have developed a 'self certification' scheme that is acceptable under this clause.

Flexibility in the general marking of such equipment supports this argument. It is therefore particularly suitable for solutions to instrument problems where such problems can be confined to Zone 2. There is however a cause for concern that large areas will be incorrectly classified as Zone 2 to accommodate cheaper and lower-grade solutions that are not adequately defined and maintained, thus lowering the integrity of the protection.

Under clause 5.2.4, equipment certified to other standards of the non-incentive category may unquestionably be used in Zone 2. Old BS 6941 and 4683 instrumentation to the old-type 'N' standard equipment is therefore permitted.

12.6.4 Selection according to temperature classification and apparatus grouping

Clause 5.3 of this standard also clarifies how apparatus must be selected with a suitable temperature class and lists the classification's temperature ratings. For example, it states that the rating for the T_4 class of apparatus is 135 °C and that this is the maximum surface temperature of that apparatus. It then specifies that the ignition temperature of any gas or vapor must be greater than this temperature. This applies to all types of protection.

The ambient temperature range of unmarked apparatus is said by this clause to be –20 to +40 °C, unless, otherwise stated. This concurs with the previously made statement that the temperature classification applies at the maximum ambient temperature of 40 °C.

Clause 5.4 of this standard states that electrical apparatus of types 'e', 'm', 'o', 'p' and 'q' shall be of apparatus Group II. This is because these types of apparatus are not able to emit levels of fault energy by their very design or construction. Hence, sub-classification into energy-emission bands of IIA, IIB or IIC is not necessary.

However, types 'd' and 'i' do need to be sub-classified into the A, B and C sub-groups of Group II. This is because different apparatus can emit different energy levels. An enclosure may be certified Ex 'd' IIC or Ex 'd' IIB, dependent on its size and construction.

Clause 5.5 recognizes that other external influences that may affect the safe stating. An obvious influence is that of ambient temperature, while other considerations may be environmental such as vibration, weather protection and risk of damage from other appliances. The environment in which apparatus must operate may influence the decision on the suitability or not of different types of protection.

12.6.5 Supply systems

Sections 6–9 of IEC 60079-14 are primarily concerned with the electrical requirements of supply and cabling systems to explosion-protected installations. Much of this is not directly relevant to IS and is specifically excluded for IS circuits.

Section 6 in the standard declares that any contact with bare or exposed live parts, other than on IS circuits, shall be prevented. The standard does not say how it is to be prevented but clearly requires adequate insulation and shrouding of live parts. It recognizes that there is an adequately low risk associated with IS circuits and therefore specifically excludes Ex 'i' circuits in order to permit some applications where bare live conductors are accessible. Bursting discs and conductivity probes are examples of this situation that in any case does not occur frequently. This is taken as referring to instrument circuit terminations which are accessible whilst live and do not need shrouding such that it is easy to connect equipment such as test meters and/or hand-held controllers into.

Since there are different arrangements for electrical supply systems and the earthing provisions for these, the standard recognizes basic precautions to be taken. For example in type IT where the earth and neutral conductors are not connected (or only by a high impedance) it recommends the use of insulation monitoring devices to detect such a fault as a warning to failure.

This does not affect the discussion on earthing in this book. The use of 'supplementary equipotential bonding' to ensure that potentials between conducting parts (i.e. the metal casements of motors and transformers) to prevent dangerous voltages appearing between the frames of electrical equipment is implicit in any requirement.

12.6.6   SELV and PELV systems

The following relevant definitions categorize the level of voltage used in electrical systems associated with IS protection. This categorization is referred to in other IEC and national standards, and specifies precautions to be taken depending on the circumstances of use.

Extra-low voltage   Normally not exceeding 50 V AC or 120 V ripple-free DC whether between two conductors or to earth.

Low voltage   Normally exceeding extra-low voltage but not exceeding 1000 V AC or 1500 V DC between conductors or 600 V AC or 900 V DC between conductors and earth.

SELV (separated ELV)   An extra-low voltage system which is electrically separated from earth and from other systems in such a way that a single earth fault cannot give rise to the risk of electric shock.

PELV (protective ELV)   An extra-low voltage system which is not electrically separated from earth but which satisfies all the requirements for SELV.

The IEC 60079-14 standard recognizes the use of extra-low voltage systems as applicable to instrumentation by this reference. Circuits may or may not be floating with respect to earth, provided that they cannot give rise to the risk of shock. It therefore accepts the use of galvanic-isolating interfaces in IS circuits. Preferences are expressed toward a floating circuit being earthed at one point, or being monitored with respect to earth-by-earth leakage detection (ELD) systems.

An older Code of Practice, BS 5345, and some other well-known international documents have never been updated to include requirements for IS isolators.

The electrical protection section in the standard specifically excludes IS. With other techniques of protection, the intention is to ensure adequate protection or warning devices to operate without 'inadmissible heating' occurring. There is a danger that automatic disconnection of equipments could pose a greater safety risk than that of the risk of ignition alone. This emphasize the need for safety assessments covering all types of risk.

IS circuits, being energy-limited by such high levels of integrity arguably fall outside some national regulations for live working on electrical equipment.

12.6.7   Emergency switch-off and electrical isolation

IEC 60079-14: Section 8 deals with this subject by declaring that:

> *8.1*   *For emergency purposes, at a suitable point or points outside the hazardous area, there shall be a single or multiple means of switching off electrical supplies to the hazardous area.*
>
> *Electrical apparatus, which must continue to operate to prevent additional danger, shall not be included into the emergency switch off circuit; it shall be on a separate circuit.*
>
> *8.2*   *To allow work to be carried out safely, suitable means of isolation shall be provided for each circuit or group of circuits, to include all circuit conductors including the neutral.*
>
> *Labelling shall be provided immediately adjacent to each means of isolation to permit rapid identification of the circuit or group of circuits thereby controlled.*

Since this clause does not specifically exclude IS it is often misunderstood and instrument installations are sometimes taken to a costly extreme. Users have of each instrument loop. Some think that this is unreasonable. The provision for isolation of the 'supply' to equipment is inferred. To reduce the number of connections in an instrument loop must reduce the cost of design, purchasing, installation, maintenance, testing and inspection as well as enhancing the reliability. Sensible grouping of isolation can help in many circumstances.

12.6.8   Wiring systems

Section 9 of IEC 60079-14 gives the requirements for wiring systems. This includes cables, and the running and fixing of them.

Cables and cabling accessories including conduit should be installed in positions that prevent exposure to mechanical damage, chemical effects and heat. Where exposure is unavoidable, additional measures shall be taken or appropriate cables selected.

Where cables are installed inside the enclosures of apparatus, it is not considered subject to mechanical damage and heat. The certification of the apparatus will dictate the conditions under which the cable shall operate. Enclosures would be selected to withstand the ingress of specific chemicals.

However in the outside world of these enclosures, there is a risk that damage and exposure may occur and it is generally accepted that some form of mechanical protection is necessary, complying with the requirement of 'additional measures'. The form of that protection varies from one country to another, and from one user organization to another. It is mainly a matter of degree. The German Code of Practice allows the use of the very practical toughened outer sheath whilst other European and non-European countries demand 'steel-wire armoring' (swa). In the US, conduit has been, up to now, used almost exclusively.

The US tends to use standard cables and does not therefore have any problem in acquiring or running suitable types including the more specialized data communications types. These types of cable in swa versions are extremely expensive and somewhat rare.

Suitable cable glands must prevent flame propagation or vapor entrainment through conduit systems. The normally metal conduit becomes part of the equipotential bonding system and therefore measures must be taken to ensure good electrical continuity through it.

12.6.9 The 'Additional requirements for type of protection "i" – Intrinsic Safety'

IEC 60079-14 contains Section 12, which is entitled:

'Additional requirements for type of protection "I" – Intrinsic Safety'.

This bears a more detailed examination with some comment on the comparison with other codes of practice relating to IS. It is the aim of this discussion to draw out the acceptable and recommended ways of installing IS systems, demonstrating compliance with these rules. It is important to demonstrate that there is no conflict in the wording but this sometimes occurs in the interpretation.

Additional Requirements for Type of Protection 'i' – Intrinsic Safety		
Clause	**Quote**	**Comments**
12	Additional requirements for type of protection 'i' – Intrinsic Safety	
12.1	Introduction:	
	A fundamentally different installation philosophy has to be recognized in the installation of intrinsically safe circuits. In comparison with all other types of installations, where care is taken to confine electrical energy to the installed system as designed so that a hazardous environment cannot be ignited, the integrity of an intrinsically safe circuit has to be protected from the intrusion of energy from other electrical sources so that the safe energy limitation in the circuit is not exceeded, even when breaking, shorting or earthing of the circuit occurs.	The IEC 60079-14 standard recognizes that IS involves a different approach. Other Codes of Practice do not give this emphasis.
	As a consequence of this principle the aim of the installation rules for intrinsically safe circuits is to maintain separation from other circuits.	Encapsulates the essence of IS installation
12.2	Installations for Zones 1 and 2	Some additional rules apply in zone 0
12.2.1	Apparatus:	
	In installations with intrinsically safe circuits for zones 1 or 2, the intrinsically safe apparatus and the intrinsically safe parts of associated apparatus shall comply with IEC 79-11, at least to category 'ib'.	
	Simple apparatus need not be marked, but shall comply with the requirements of IEC 79-11 and IEC 79-0 in so far as intrinsic safety is dependent on them.	
	Associated apparatus should preferably be located outside the hazardous area or, if installed inside	5.2 covers selection of equipment according to zones

Clause	Quote	Comments
	hazardous area, shall be provided with another appropriate type of protection in accordance with 5.2.	i.e. Interfaces may be located in an Ex 'n' enclosure but can only be mounted in zone 2. See later in this manual
	Electrical apparatus connected to the non-intrinsically safe terminals of an associated apparatus shall not be fed with a voltage supply greater than U_m shown on the label of the associated apparatus. The prospective short-circuit current of the supply shall not be greater than 1500 A.	U_m is normally 250 Vac rms for most associated equipment. Older standards required a 4000 A breaking capacity but this is now seen as excessive.
2.2	Cables	
2.2.1	General	
	Only insulated cables whose conductor-earth, conductor-screen and screen-earth test voltages are at least 500 Vac shall be used in IS circuits	This is universally accepted as the required minimum grade of insulation or separation for circuits.
	If multi-stranded conductors are used in the hazardous area, the ends of the conductor shall be protected against separation of individual strands, for example by means of core-end sleeves.	Core-end sleeves are known as ferrules; many users fit ferrules as standard as it stops the cable identifications coming off.
	The diameter of individual conductors within the area subject to explosion hazards shall not be less than 0.1 mm. This applies also to the individual wires of a finely stranded conductor.	BS5345 required not less than 0.017 mm of conductor. In practice many organizations specify a site standard in excess of this at 0.5 mm^2
2.2.2	Electrical parameters of cables	
	The electrical parameters (C_C and L_C) or (C_C and L_C/R_C) should be known, or the worst case values specified by the manufacturer should be assumed for all cables used.	See section on cable parameters following.
2.2.3	Earthing of conducting screens	
	Where a screen is required, except as in a) through c) below, the screen shall be electrically connected to earth at one point only; normally at the non-hazardous area end of the circuit loop. This requirement is to avoid the possibility of the screen carrying a possibly incentive level of circulating current in the event that there are local differences in earth potential between one end of the circuit and the other.	See sections on cabling and Earthing where this area is discussed more fully.

(Continued)

Clause	Quote	Comments
	Special cases:	
	• If there are special reasons (for example when the screen has high resistance, or where the screening against inductive interference is additionally required) for the screen to have multiple electrical connections throughout its length, the arrangement of figure 2 may be used, provided that	
	1. the insulated earth conductor is of robust construction (normally at least 4 mm^2 but 16 mm^2 may be more appropriate for clamp type connections)	
	2. the arrangement of the insulated earth conductor plus the screen is insulated to withstand a 500 V insulation test from all other conductors in the cable and any cable armour.	
	3. the insulated earth conductor and the screen are only connected to earth at one point which shall be the same point for both the insulated earth conductor and the screen, and would normally be at the non-hazardous end of the cable;	
	4. the insulated earth conductor complies with 9.1.1;	
	5. the inductance/resistance ratio (L/R) of the cable installed together with the insulated earth conductor shall be established and shown to conform to the requirements of 12.2.5.	
	6. If the installation is effected and maintained in such a manner that there is a high level of assurance that potential equalization exists between each end of the circuit (that is between the hazardous area and the non-hazardous area), then, if desired, cable screens may be connected to earth at both ends of the cable and, if required, at any interposing points.	See Cable Parameters below
	• Multiple earthing through small capacitors (for example 1 nF, 1500 V ceramic) is acceptable provided that the total capacitance does not exceed 10 nF.	
2.2.4	Cable armour bonding	
	Armour should normally be bonded to the equipotential bonding system via the cable entry devices or equivalent, at each end of the cable run. Where there are interposing junction boxes or other apparatus, the armour will normally be similarly bonded to the equipotential bonding system at these points. In the event that armour is required not to be bonded to the equipotential bonding system at any interposing point, care should be taken to ensure that	The equipotential bonding system means the structural earth of the plant where the frames or chassis of equipment is generally cross-bonded to everything else. It is sometimes referred to as the 'equipotential plane'.

Clause	Quote	Comments
	the electrical continuity of the armour from end to end of the complete cable run is maintained.	
	Where bonding of the armor at a cable entry point is not practical, or where design requirements make this not permissible, care should be taken to avoid any potential difference which may arise between the armor and the equipotential bonding system giving rise to an incendive spark. In any event, there shall be at least one electrical bonding connection of the armor to the equipotential bonding system. The cable entry device for isolating the armor from earth shall be installed in the non-hazardous area or zone 2.	This refers to the use of plastic junction boxes that are increasingly popular.
2.2.5	Installation of cables	
	Installations with intrinsically safe circuits shall be erected in such a way that their intrinsic safety is not adversely affected by external electric or magnetic fields such as from nearby overhead power lines or heavy current-carrying single core cables. This can be achieved, for example by the use of screens and/or twisted cores or by maintaining an adequate distance from the source of the electric or magnetic field.	This area Specific requirements for IS cabling
	In addition to the cable requirements of 9.1.1, cables, in both the hazardous and non-hazardous area, shall meet one of the following requirements:	9.1.1 refers to general avoidance of damage to cables in the hazardous area.
	(a) intrinsically safe circuit cables are separated from all non-intrinsically safe circuit cables, or (b) intrinsically safe circuit cables are so placed as to protect against the risk of mechanical damage, or (c) intrinsically safe or non-intrinsically safe circuit cables are armoured, metal sheathed or screened.	See Multicore requirement.
	Conductors of intrinsically safe circuits and non-intrinsically safe circuits shall not be carried in the same cable.	See illustrations on segregation.
	Conductors of intrinsically safe circuits and non-intrinsically safe circuits in the same bundle or duct shall be separated by an intermediate layer of insulated material or by an earthed metal partition. No segregation is required if metal sheaths or screens are used for the intrinsically safe or non-intrinsically safe circuits.	Armouring not signal screens are meant here.

(Continued)

Clause	Quote	Comments
2.2.6	Marking of cables	
	Cables containing intrinsically safe circuits shall be marked. If sheaths or coverings are marked by a colour, the colour used shall be light blue. Cables marked in this way shall not be used for other purposes. If intrinsically safe or all non-intrinsically safe cables are armoured, metal sheathed or screened, then marking of intrinsically safe cables is not required.	Requirement comes from DIN VDE 0165 requiring light blue. ** note this important clarification. Relaxation of marking from older standards.
	Alternative marking measures shall be taken inside measuring and control cabinets, switchgear, distribution equipment, etc., where there is a risk of confusion between cables of intrinsically safe and non-intrinsically safe circuits, in the presence of a blue neutral conductor. Such measures include:	
	• combining the cores in a common light blue harness; • labelling; • clear arrangement and spatial separation.	
2.2.7	Multi-core cables containing more than one IS circuit	
	The requirements of this subclause are in addition to those of 12.2.2.1 to 12.2.2.6.	These clauses above cover general requirements and marking.
	The radial thickness of the conductor insulation shall be appropriate to the conductor diameter and the nature of the insulation. For insulating materials currently used, for example polyethylene, the minimum radial thickness shall be 0.2 mm.	This increased requirement is in line with other codes.
	The conductor insulation shall be such that it will be capable of withstanding an rms ac test voltage of twice the nominal voltage of the intrinsically safe circuit with a minimum of 500 V.	See Multicore cables section below.
	Multi-core cables shall be of a type capable of withstanding an rms AC dielectric test of at least:	
	• 500 V applied between any armouring and/or screen(s) joined together and all the cores joined together; • 1000 V applied between a bundle comprising one half of the cable cores joined together and a bundle comprising the other half of the cores joined together. This test is not applicable to and multi-core cables with conducting screens for individual circuits.	Older installations tended to use individually screened circuits but a trend is seen away from this to single overall screen use.

Clause	Quote	Comments
	The voltage tests shall be carried out by a method specified in an appropriate cable standard. Where no such method is available, the tests shall be carried out as follows: • The voltage shall be an ac voltage of substantially sinusoidal waveform at a frequency of between 48 Hz and 62 Hz; • The voltage shall be derived from a transformer of at least 500 VA output; • The voltage shall be increased steadily to the specified value in a period of not less than 10 s then maintained at this value for at least 60 s.	
2.2.8	Fault considerations in multi-core cables The faults, if any, which shall be taken into consideration in multi-core cables used in intrinsically safe electrical systems depend upon the type of cable used.	
	1. Type A Cable complying with the requirements of 12.2.2.7 and, in addition, with conducting screens providing individual protection for intrinsically safe circuits in order to prevent such circuits becoming connected to one another – the coverage of those screens shall be at least 60% of the surface area. No faults between circuits are taken into consideration.	Older type cable but still used. Type C & D specified by British and European Standards are not considered.
	2. Type B Cable which is fixed, effectively protected against damage, complying with the requirements of 12.2.2.7 and, in addition, no circuit contained within the cable has a maximum voltage U_O exceeding 60 V. No faults between circuits are taken into consideration.	
	3. Others For cables complying with the requirements of 12.2.2.7 but not the additional requirements of type A or type B, it is necessary to take into	

(Continued)

Clause	Quote	Comments
	consideration up to two short circuits between conductors and, simultaneously, up to four open circuits of conductors. In the case of identical circuits, failures shall not be taken into consideration provided that each circuit passing through the cable has a safety factor of four times that required for category 'ia' or 'ib'. For cables not complying with the requirements of 12.2.2.7, there is no limit to the number of short circuits between conductors and simultaneous open circuits of conductors which shall be taken into consideration.	
2.3	Termination of IS circuits In electrical installations with intrinsically safe circuits, for example in measuring and control cabinets, the terminals shall be reliably separated from the non-intrinsically safe circuits (for example by a separating panel or a gap of at least 50 mm). The terminals of the intrinsically safe circuits shall be marked as such. All terminals and plugs and sockets shall satisfy the requirements of 6.3.1 and 6.3.2 respectively of IEC 79-11. Where terminals are arranged to provide separation of circuits by spacing alone, care shall be taken in the layout of terminals and the wiring method used to prevent contact between the circuits should a wire become disconnected.	Requirements stated in this clause are to be applied to Junction Boxes where comments refer to hazardous area situations. 6 mm Creepage and clearance rules 4 mm to earth.
2.4	Earthing of IS circuits Intrinsically safe circuits may be either (a) isolated from earth, or (b) connected at one point to the equipotential bonding system if this exists over the whole area in which the intrinsically safe circuits are installed. The installation method shall be chosen with regard to the functional requirements of the circuits and in accordance with the manufacturer's instructions. More than one earth connection is permitted on a circuit provided that circuit is galvanically separated into subcircuits each of which has only one earth point.	Galvanic Isolators Shunt diode Safety Barriers or Galvanic Isolators These next two paragraphs are welcome additions to allow the use of multi-loop systems to solve application problems. Further welcome clarifications of the definition of Connection to Earth.

Clause	Quote	Comments
	In intrinsically safe circuits that are isolated from earth, attention shall be paid to the danger of electrostatic charging. A connection to earth across a resistance of between 0.2 MΩ and 1 MΩ, for example for the dissipation of electrostatic charges, is not deemed to be earthing.	More a concern in Zone 0
	Intrinsically safe circuits shall be earthed if this is necessary for safety reasons, for example in installations with safety barriers without galvanic isolation. They may be earthed if necessary for functional reasons, for example with welded thermocouples. If intrinsically safe apparatus does not withstand the electrical strength test with at least 500 V to earth according to IEC 79-11, a connection to earth at the apparatus is to be assumed.	See discussion on Earthing of IS installations to IEC Standards, below
	In intrinsically safe circuits, the earthing terminals of safety barriers without galvanic isolation (for example Zener barriers) shall be	
	1. connected to the equipotential bonding system by the shortest practicable route, or 2. for TN-S systems only, connected to a high-integrity earth point in such a way as to ensure that the impedance from the point of connection to the main power system earth point is less than 1Ω. This may be achieved by connection to a switch room earth bar or by the use of separate earth rods. The conductor used shall be insulated to prevent invasion of the earth by fault currents which might flow in metallic parts with which the conductor could come into contact (for example control panel frames). It shall also be given mechanical protection in places where the risk of damage is high.	
	The cross-section of the earth connection shall be either:	
	• at least two separate conductors each rated to carry the maximum possible current, which can continuously flow, each with a minimum of 1.5 mm^2 copper, or • at least one conductor with a minimum of 4 mm^2 copper.	
	NOTE – The provision of two earthing conductors should be considered to facilitate testing.	

(Continued)

Clause	Quote	Comments
	If the prospective short-circuit current of the supply system connected to the barrier input terminals is such that the earth connection is not capable of carrying such current, then the cross-sectional area shall be increased accordingly or additional conductors used.	
2.5	Verification of IS circuits	
	Unless a system certificate is available defining the parameters for the complete intrinsically safe circuit, then the whole of 12.2.5 (and its subclauses) shall be complied with.	Systems certification is formalized by the addition of this clause.
	When installing intrinsically safe circuits, including cables, the maximum permissible inductance, capacitance or L/R ratio and surface temperature shall not be exceeded. The permissible values shall be taken from the associated apparatus documentation or the marking plate.	
2.5.1	IS circuits with only one associated apparatus	
	The sum of the maximum effective internal capacitance C_1 of each item of intrinsically safe apparatus and the cable capacitance (cables generally being considered as concentrated capacitance equal to the maximum capacitance between two adjacent cores) shall not exceed the maximum value C_o marked on the associated apparatus.	This formalizes the discussion on the matching of cable parameters in the application section of this manual.

See below for an explanation of L/R ration |
	The sum of the maximum effective internal inductance L_1 of each item of intrinsically safe apparatus and the cable inductance (cables generally being considered as concentrated inductance equal to the maximum inductance between the two cores in the cable having the maximum separation) shall not exceed the maximum value L_o marked on the associated apparatus.	
	Where the intrinsically safe apparatus contains no effective inductance and the associated apparatus is marked with an inductance/resistance L/R value, if the L/R value of the cable, measured between the two cores in the cable having maximum separation, is less than this figure, it is not necessary to satisfy the L_o requirement.	
	The values of permissible input voltage U_i, input current I_i and input power P_i of each intrinsically	

Clause	Quote	Comments
	safe apparatus shall be greater than or equal to the values U_o, I_o and P_o respectively of the associated apparatus.	Determination of temperature
	For simple apparatus the maximum temperature can be determined from the values of P_o of the associated apparatus to obtain the temperature class. The temperature class can be determined by:	
	(a) reference to table 4, or (b) calculation using the formula: $$T = P_o R_{th} + T_{amb}$$	
	where T is the surface temperature; P_o is the power marked on the associated apparatus; R_{th} is the thermal resistance (K/W) (as specified by the component manufacturer for the applicable mounting conditions); T_{amb} is the ambient temperature (normally 40 °C).	
	And reference to table 1.	
	In addition, components with a smaller surface area than 10 cm^2 (excluding lead wires) may be classified as T5 if their surface temperature does not exceed 150 °C.	
	The apparatus group of the intrinsically safe circuit is the same as the most restrictive grouping of any of the items of electrical apparatus forming that circuit (for example a circuit with 11B and 11C apparatus will have a circuit grouping of 11B)	
2.5.2	IS circuits with more than one associated apparatus	
	If two or more intrinsically safe circuits are interconnected, the intrinsic safety of the whole system shall be checked by means of theoretical calculations or a spark ignition test in accordance with clause 10 of IEC 79-11. The apparatus group, temperature class and the category shall be determined.	
	Account shall be taken of the risk of feeding-back voltages and currents into associated apparatus from the rest of the circuit. The rating of voltage and current-limiting elements within each associated	

Clause	Quote	Comments
	apparatus shall not be exceeded by the appropriate combination of U_o and I_o of the other associated apparatus.	
	NOTE – In the case of associated apparatus with linear current/voltage characteristics the basis of calculation is given in annex A. In the case of associated apparatus with non-linear current/voltage characteristics, expert guidance should be sought.	
	A descriptive system document shall be prepared by the system designer in which the items of electrical apparatus, the electrical parameters of the system including those of inter-connecting wiring are specified.	
3	Installations for zone 0	
	Intrinsically safe circuits shall be installed in accordance with 12.2, except where modified by the following special requirements.	
	In installations with intrinsically safe circuits for zone 0 the intrinsically safe apparatus and the associated apparatus shall comply with IEC 79-11, category 'ia'. Associated apparatus with galvanic isolation between the intrinsically safe and non-intrinsically safe circuits is preferred. Since only one fault in the equipotential bonding system in some cases could cause an ignition hazard, associated apparatus without galvanic isolation may be used only if the earthing arrangements are in accordance with item 2) of 12.2.4 and any mains-powered apparatus connected to the safe area terminals are isolated from the mains by a double-wound transformer, the primary winding of which is protected by an appropriately rated fuse of adequate breaking capacity. The circuit (including all simple components, simple electrical apparatus, intrinsically safe apparatus, associated apparatus and the maximum allowable electrical parameters of inter-connecting cables) shall be of category 'ia'.	This preference is in line with the German Standard that does not permit non-isolated interfaces. This requirement is from BS5345.
	Simple apparatus installed outside the zone 0 shall be referred to in the system documentation and shall comply with the requirements of IEC 79-11, category 'ia'.	Surge suppression requirements are a welcome addition to this part of the standard

Clause	Quote	Comments
	If earthing of the circuit is required for functional reasons, the earth connection shall be made outside the zone 0, but as close as is reasonably practicable to the zone 0 apparatus.	This is considered to be quite explicit.
	If part of an intrinsically safe circuit is installed in zone 0 such that the apparatus and the associated equipment are at risk of developing hazardous potential differences within the zone 0, for example through the presence of atmospheric electricity, a surge protection device shall be installed between each non-earth bonded core of the cable and the local structure as near as is reasonably practicable, preferably within 1 m, to the entrance to the zone 0. Examples of such locations are flammable liquid storage tanks, effluent treatment plant and distillation columns in petrochemical works. A high risk of potential difference generation is generally associated with a distributed plant and/or exposed apparatus location, and the risk is not alleviated simply by using underground cables or tank installation.	
	The surge protection device shall be capable of diverting a minimum peak discharge current of 10 kA (8/20 μs impulse according to IEC 60-1, 10 operations). The connection between the protection device and the local structure shall have a minimum cross-sectional area equivalent to 4 mm^2 copper.	
	The spark-over voltage of the surge protection device shall be determined by the user and an expert for the specific installation.	
	NOTE – The use of a surge protection device with spark-over voltage below 500 V ac 50 Hz may require the intrinsically safe circuit to be regarded as being earthed.	
	The cable between the intrinsically safe apparatus in zone 0 and the surge protection device shall be installed such that it is protected from lightning.	

12.6.10 Summary of IEC installation rules

Some clarifications and additions to the original European Standards on which this was predominantly based are seen as useful moves toward harmonization. Earthing remains as one of the major differences and bears further discussion.

The US has developed the use of entity parameters to allow easier compatibility assessment of equipment and this is now being drawn into the standard.

12.7   Practical aspects of IS installations

For the purpose of seeing how the standards and codes of practice are implemented on an installation, a typical system may be thought of as comprising five main parts, as shown in Figure 12.2.

1. Safe area apparatus
2. Associated apparatus mounted in the safe area
3. Hazardous area cabling
4. Hazardous area junction boxes
5. Hazardous area mounted apparatus.

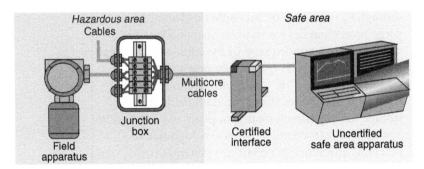

Figure 12.2
The five main parts of an instrument installation

Each part will be examined to study the IEC and other common requirements placed upon installers.

12.7.1   Safe area equipment

This applies to all electrical devices connected immediately upstream of the safe area terminals of any associated apparatus (IS interface device) (Figure 12.3).

The requirements for any safe area uncertified equipment is that they do not exceed a supply of U_M which is normally 250 V AC rms. Some associated apparatus may be certified with a different value of U_M if it is designed for a particular function. This is not common.

BS 5345: Part 4 additionally specifies the use of a 'double-wound' and 'suitably fused' mains transformer that is not fed from greater than 250 V AC rms.

The U_M maximum of 250 V is derived from a test originally in the European Standard where a tolerance of 5% is specified. This permits use with systems where 254 V AC is encountered.

When interfaces are connected to equipment containing switch mode power supplies, they generate higher voltages at a higher frequency in order to use smaller transformers. It is argued that the quality of insulation and isolation on the miniaturized transformers is usually of an adequately high specification and therefore the likelihood of failure is acceptably small.

Similarly, CRTs used in the displays of some control systems use 30 kV EHT levels and therefore cannot conform to this requirement. It is argued here that the EHT circuits are sufficiently remote from the interfaces that the likelihood of the simultaneous failure of all the interposing electrical systems involving CRTs is too small to pose a risk. In any case, the EHT source impedance is relatively high, reducing its ability to breakthrough and cause damage on downstream equipment.

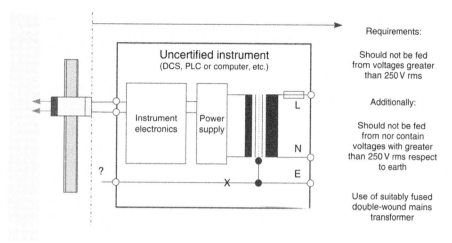

Figure 12.3
Safe area apparatus requirements

Most equipment conforming to normal industrial standards and grades of quality will conform to these basic requirements. The intent is to ensure that risk of breakthrough from supply to signal terminals are adequately low.

Installation of interfaces

The rules for the installation of the interfaces are not as detailed in the IEC standard as in the BS or EN documents. There are very few specific requirements for the use of interfaces. These are inferred from other statements. However, the onus is placed on the manufacturer to supply adequate information; the requirement to install in accordance with the manufacturer's instructions is clearly stated. Historically, this is because the original codes of practice only gave specific instructions on the use of shunt diode safety barriers as this was the only technology available. It relies on the earth connection for integrity and so had to be specified as a user requirement separately to the design standard. The development and acceptance of galvanic isolation techniques being used in different modes has effectively been included in the standards by requiring the designer/manufacturer to provide the necessary safety information. This explains many of the generalized forms of requirements in the IEC standards. The requirements are written such that they can be applied to all forms of apparatus used in the hazardous area whether associated or hazardous area-mounted.

The US has always placed a duty of providing adequate information on the manufacturer/supplier of equipment and so providing detailed information on installation is the responsibility of the manufacturer.

BS 5345 gives somewhat more guidance in specific areas. These are extended by 'common industrial practice' to associated apparatus of all types. The installation must comply with system documentation and the manufacturer's recommendations. In addition:

- The location of interfaces should be permanently marked to show the correct type of replacement barrier in each position. (This is owing to different safety descriptions available.)
- Barriers (interfaces) are normally mounted in the safe area at the nearest convenient point to the hazardous/safe area boundary.

- Maintain IP 20 and protect from unauthorized interference by an enclosure.
- Hazardous area mounting is permissible if the appropriate type of protection is provided for barriers and cabling:

  – Zone 1: Flameproof (Ex 'd') enclosure
  – Zone 2: Type 'n' enclosure.

Note: Barriers mounted in Type 'n' enclosures must have Type 'n' (BS 4683 or equivalent) 'component approval'. Isolators were never eligible for this type of approval because safe area circuits may not comply. In the same way neither barriers nor isolators comply with Ex 'e' approvals and cannot be mounted in Ex 'e' enclosures. Some apparatus is dual-certified (Ex 'eia') and may be mounted in the hazardous area. Ex 'e' installation codes have not been updated to reflect this modernization in technology and practice.

12.7.2 Cabling

Cable is not 'certifiable' in the same way that apparatus may be tested. There is, then, no such thing as IS cable. It is merely cable that conforms to the minimum specifications of the IS requirements as stated in either the standards, codes of practice for installation or in systems descriptive documents. The latter is more likely if there are special requirements for the cable.

Cabling in this case refers to the conductors and their insulation arrangements that are required from the terminations of one piece of electrical apparatus to those of another (the interconnecting cables of IS apparatus). It does not cover the internal wiring of any apparatus. This is, of course, a consideration of the apparatus certification after its design.

The minimum standards have changed very little over the years. The IEC 60079-14 Standard requires that the minimum size of cables is 0.1 mm^2 conductor with 0.1 mm thickness of PVC or equivalent insulation material around each conductor. This normally considerably exceeds the user's specification of cable for modern installations. As the quality of insulation has risen with developments in technology, the thickness has reduced.

The code of practice, BS 5345, requires 0.2 mm thickness and a minimum size of 0.017 mm^2 but this was modified in IEC 60079-14 to come into line with the German VDE 0165 Standard.

Other specific requirements are that the cable shall be identified as being part of an IS circuit. There are many misunderstandings about this. The requirement does not force the use of a particular marking regime but merely states that the user must decide on how the cable is to be identified as carrying IS circuits.

Identification of cables by color coding is popular and sensible. For example, it is common to use color-coded outer sheaths on various types of instrumentation systems:

- *Blue*: Process control systems
- *Red*: Fire and gas detection systems
- *Yellow*: Emergency shutdown systems.

Some of those circuits may be IS and others may be other types of protection. It is then quite common to run the IS circuits on separately installed cable trays where the tray itself denotes the use of IS. The safety documentation should therefore declare that all cabling run on, say, blue trunking shall be IS.

Some standards require that if a color is to be used for identification of IS circuits then that color shall be 'light blue'. If a user chooses a discernible shade of blue, then no other cable on the installation should be of that particular shade. This is sometimes difficult to control.

Crimping

It is accepted that crimps or ferrules are used on multi-stranded conductors whereas solid conductors can be terminated directly into terminal blocks on the apparatus. Some users require installations to use crimps on all terminations. There is no reason why this should not be done technically. The extra expense of crimps and labor is perhaps the only regulator.

Segregation of cables

The requirement to segregate cables may be interpreted in several ways. It is usual to separate power-carrying cables from instrument or signal cables to reduce the risk of noise induction. To obviate the risk of invasion then the standards require 'adequate' segregation. 50 mm is normally taken as acceptable where no mechanical protection is required. If the cables are adequately protected by armoring, then no segregation is required. Partitioning or back-to-back on cable trays are all acceptable from an IS standpoint.

Figure 12.4 gives some common alternatives.

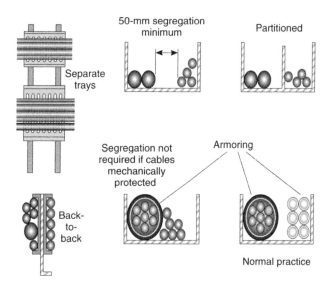

Figure 12.4
Requirements for IS and non-IS cables

Care should be taken when bending armored cable and the manufacturer's minimum radius of curvature figures should be heeded.

Cable conduits

The use of cable conduits is popular in the US and Canada. The National Electrical Code is written around the requirement to protect cables for all methods of protection and does not single out IS for special treatment at present. Conduit systems do have both advantages and disadvantages, which must be considered before use on an installation. They are costly to install, difficult to modify or expand once in place, and can entrain vapors internally unless properly installed and maintained. They are high reliability and durable.

Cable armoring and screening

The screening and armoring of cabling must be connected as in clauses 12.2.2.4 and 12.2.2.5 (respectively) of IEC 60079-14 (see above). This is for reasons stated in Chapter 6 on earthing practice. The continuity of these paths must be maintained through cable joints, junction boxes and marshaling cabinets.

In Figure 12.5, the single pair cable with screen in Case 1 shows the screen connected to the busbar earth.

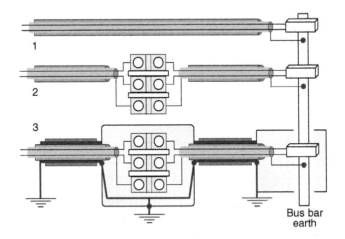

Figure 12.5
Armoring and screening continuity 1

Where the cable runs through a junction box, separate insulated terminals should be provided for cores and the screen as in Case 2.

The provision for armor is made on the JB framework as shown in Case 3. It does not matter that the armor is connected to earth at three points because it is considered part of the equipotential bonding system of the entire plant.

In Figure 12.6 the most common situation is seen where the multicore has individual screens and an overall screen. All the screens are single-point earthed at the busbar. The screen of the pairs is maintained through to the outgoing pair by the provision of its dedicated terminal, as in Case 2 in Figure 12.5. In the junction box, a separate terminal is provided for the overall screen, but it must not be connected or 'earthed'. The accepted way of installing this would be to label the terminal in a distinctive way and declare that it is only for single unearthed screens. This should be noted in any plant safety documentation.

In the arrangement shown in Figure 12.7, the multicore has no pair screens. The overall screen is cross-connected to each outgoing pair screen in a suitable way for the type of terminations used. This raises the question that it is not recommended that more than one conductor is terminated, unless both are crimped together.

Cable parameters

The values of inductance, capacitance and resistance in any IS circuit will include that contributed by the interconnecting cables between of pieces of apparatus. The distributed nature of these properties give any pair of conductors its characteristic impedance.

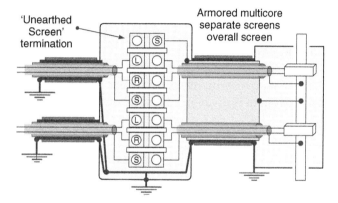

Figure 12.6
Armors and screens 2

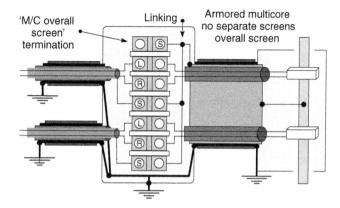

Figure 12.7
Armors and screens 3

It was realized in the 1960s, that when an IS interface is connected to a pair of conductors, a significant amount of energy could accumulate in the cable. A sufficient quantity of stored energy, if released, could give rise to incentive conditions. Limits were placed upon the amount of energy permitted from then on.

The energy stored in the cable is a function of the cable length and/or its physical construction. Cable capacitance is dependent upon the dielectric constant of the insulation material used. It is not easy to calculate this value. Measurements are often performed by the cable manufacturer as a check of cable quality and figures are normally stated.

Inductance is only dependent on the physical size properties of the conductors. A typical formula to calculate the inductance is given by:

$$L = 0.1 + 0.92 \, \text{Log}_{10} \left[\frac{(c-a)}{a} \right] \, \mu\text{H/m}$$

Where (in meters)
   a = radius of the conductor
   c = distance between centers of the conductors.

The limits on the stored energy permitted in the IS circuit are dictated by the IS apparatus which is connected to the circuit and capable of injecting voltage and/or current under fault conditions. The combined safety description of the apparatus gives this.

The safety parameters determined from this description specify the highest values of capacitance and inductance which may be connected. It is normally given for use in apparatus IIC Group.

It is the safety parameters that dictate the conditions for the cable (Figure 12.8). Calculations and measurements of these values follow. The documentation of these values for a system is given in Chapter 9 on Certification.

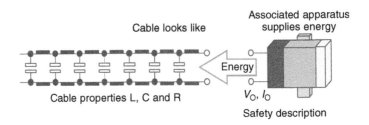

Capacitance: Limit determined by voltage: Energy $= \frac{1}{2}CV^2$

Inductance: Limit determined by current: Energy $= \frac{1}{2}LI^2$

Resistance is determined by conductor properties and size
and does not affect Safety... Only operational aspects

Figure 12.8
Cable parameters: L, C and R

Application of cable parameters

Cable parameters are often thought confusing and are incorrectly applied. IEC 60079-14 clause 12.2.2.2 does not state how to use the figures whereas other codes are more helpful.

The following discussion should clarify the process for considering the parameters.

Capacitance

The parameters quoted by the manufacturer in Figure 12.9 are typical. Note that the core to core was measured at 34.5 pF/m that is less than that stated by the manufacturer at 41 pF/m. It is the quoted figure that is normally used for parameter calculations.

Generally, the worst-case value is taken for whatever configuration of cable used. If all cores were to become connected in parallel and the safety description voltage becomes applied between the cores and earth, then, according to the values in Figure 12.9, 442.9 pF/m of capacitance is seen. Using a 28 V description, 0.13 µF are permitted in IIC vapors. This permits a length of cable of:

$$\frac{0.13\,\mu\text{F}}{442.9\ \text{pF/m}} = 293\,\text{m}$$

This length may well be long enough. There is no doubt that under any cable failure conditions where the cable becomes connected to the source in the worst way possible that this arrangement would be safe. If IIB or IIA vapors are encountered the margins for error are considerably higher.

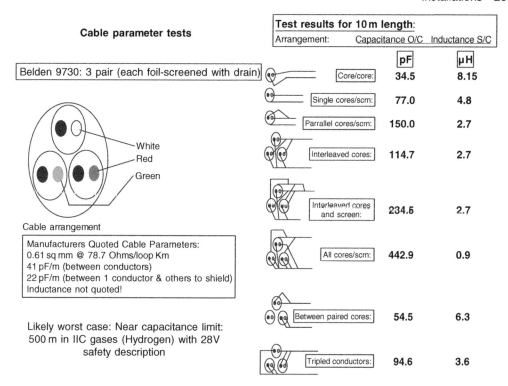

Figure 12.9
Cable parameter tests

The multiplying factors used to convert IIC into IIB and IIA figures are generally taken as 3 and 8, respectively.

Where the calculations show that the worst case was unacceptable, the arrangement would be reconsidered by saying that if the cable installed will conform to the multicore requirements then only the core to core/core to screen figures can apply, as the risk of cable failure to the worst case conditions is acceptably small.

Under these circumstances the permitted cable length would be a worst case of:

$$\frac{0.13\,\mu\text{F}}{150.0\,\text{pF/m}} = 86.6\,\text{m}$$

Inductance

The figure for inductance was not provided by the manufacturer. This is quite common. It was measured as 8 µH/m worst case. A figure of 4.2 mH is acceptable for IIC from a 28 V 93 mA interface. This gives a useable distance of:

$$\frac{4.2\,\text{mH}}{8\,\mu\text{H/m}} = 525\,\text{m}$$

It is often the case that the cable inductance cannot be met, and so an alternative approach is to consider the inductance/resistance ratio (L/R) permitted by the supply.

L/R ratio

This is explained and applied in IEC 60079-11. It is based on the premise that as the length of the cable increases, the resistance also increases, limiting the current that can flow through the inductance of the cable. This is true because the properties of inductance and resistance seen in the cable are effective in series. The energy stored by the cable inductance must be zero when there is no inductance. It must rise to some maximum value when the resistance is at its lowest and the inductance is at its highest proportion. It must fall again when the resistance approaches infinity because no current can flow. Therefore the maximum energy storage value can be calculated and expressed as a ratio of inductance to resistance (Figure 12.10).

The formula for determining the maximum *L/R* ratio for a resistive source is given in the standard as:

$$\frac{L_O}{R_O} = \frac{\left(8eR_I + \left(64e^2R_I^2 - 72U_O^2 eL_1\right)^{1/2}\right)}{4.5(U_O)} \text{ in H}/\Omega$$

Where

e is the minimum spark test apparatus ignition energy in joules and takes on the following values:

- Group I: 525
- Group IIA: 320
- Group IIB: 160
- Group IIC: 40.

R_I is the output resistance of the power supply in ohms
U_O is the maximum open-circuit voltage
L_I is the maximum inductance present at the power source terminals in Henries.

This includes a factor of safety of 1.5.

If $L_I = 0$, which is the case with a resistive supply then L_O/R_O is simplified to = $32eR_I/9U_O$.

This formula may be further simplified by applying the matched-power concept. The maximum energy is transferred when the resistance of the cable is equal to the resistance of the source. Further therefore the applied form becomes:

$$\frac{L}{R} = \frac{4L_P}{r}$$

Where

L_P is the permitted maximum value of the cable inductance for the maximum current which can flow.
r is the infallible resistance of the source of the supply.

Thus a 28 V 300 Ω 93 mA barrier, having an L_{MAX} of 4.2 mH would have an expected permissible *L/R* of:

$$\frac{4 \times 0.0042}{300} = 55\,\mu\text{H}/\Omega$$

- A cable of zero length has no inductance
- A cable of infinite length has infinite resistance, so no current flows

- At some finite length, the stored energy reaches a maximum value
- Finding and using this maximum value removes the need to consider the cable length.

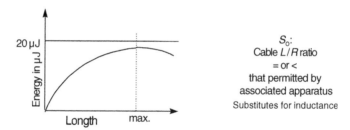

Figure 12.10
Cable parameters: L/R *ratio*

Provided that the chosen cable's value stated is not greater than that permitted by the interface, there is no restriction on the length of that cable due to inductance. Only capacitance need be considered.

This concept is not accepted in some countries.

Although discussed at some length, in reality, cable parameters are rarely a problem. Voltage and capacitance provide the most common limitation on two-wire instrument but only when lengths of 500 m are encountered in IIC atmospheres with 28 V safety descriptions. This aspect of IS should not be over-exaggerated.

Multicore cables

Additional requirements for multicore cables are included because the risk of sufficient energy being available to cause ignition is increased when several independent IS circuits are contained within one overall sheath.

These requirements are not stringent and are in line with good industrial practice.

- IS and non-IS circuits must not be contained in the same multicore.
- They should be run where risk of mechanical damage is slight.
- They should be fixed throughout their length.
- Each IS circuit is to occupy adjacent cores.
- No voltage should exceed a 60 V peak.

Identification of IS cables is required by labeling or the use of color coding as discussed.

Some standards stipulate the use of individual pairs with screens in multi-cores (Type A) where the circuits enter Zone 0.

12.7.3 Junction boxes

The reasons for mounting junction boxes in the hazardous area are essentially:

- To allow for the extension of cables to locations
- To allow round-routing of circuits to more than one piece of apparatus
- To allow the marshaling of separate cables from the field into multi-cores
- To house other compatible apparatus and facilities.

IEC 60079-11 states that all components carrying IS currents are required to conform to aspects of the standard but does not give any additional requirements.

The important points that emerge from this are summarized as follows:

- *Segregation*:   IS and non-IS circuits must be spaced at least 50 mm apart.
- *Terminations*:   These are 'simple apparatus' but must conform to minimum creepage and clearance distances. These shall be rated IP 20.
- *Enclosure*:   Mechanical protection of the enclosure is not necessary for safety purposes. Identification of the enclosure is necessary.

The wording of older standards (BS 5345 Part 4) states four major guiding rules for the approach to junction boxes.

- The need to avoid inadvertent grounding or shorting of conductors.
- Terminations should be reliable, properly sized, and maintain creepage and clearance distances – 6 mm to other circuits and 4 mm to earth ('simple apparatus').
- Glanding should maintain weatherproof box rating selected and should adequately support the incoming cables.
- The enclosure should be clearly labeled as carrying IS circuits.

Some codes of practice require specific additional information to be included such as circuit identification. This is usually done by loop identification systems such as the ISA system which is widely adopted. Other standards wisely suggest that adhesive metallic labels should not be used inside enclosures where there is a risk that the label may become detached and cause short circuits.

A common terminal arrangement is shown in Figure 12.11. Bottom entry is often used to reduce the risk that moisture may enter through cable gland systems.

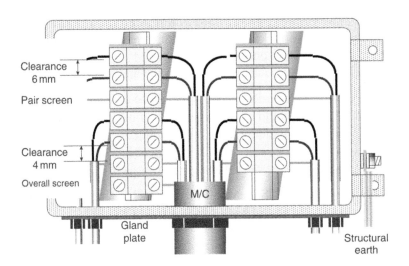

Figure 12.11
IS junction box arrangements

There is no restriction on the layout chosen or the number of terminals appearing in a box, unlike Ex 'e' protection. Any number of terminals may be used in a system because they are classified as 'simple apparatus'. The color of the terminals may be specified in some codes of practice. Germany and those who follow the VDE 0165 Code will use light or 'bright' blue.

In Figure 12.12 the need for extra caution is owed to the presence of higher-power Ex 'e' circuits in the same enclosure.

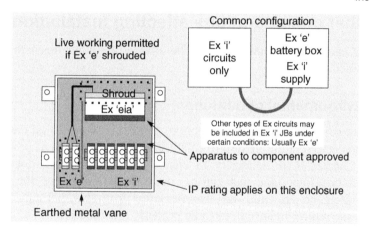

Figure 12.12
Ex 'i' and 'e' junction box arrangements

Ex 'e' supply systems are suitable for location in Zone 1. It is quite common and convenient for these to be located with associated IS circuits in instrumentation systems. This arrangement is becoming more popular where communicating devices need more local power. This solution of Ex 'e' power to Ex 'i' circuits will increase in use over the next few years.

However, the arrangement is not covered adequately by any code of practice. Rules of thumb have emerged which state a preference for shrouding the Ex 'e' terminals such that live work may be permitted on the Ex 'i' apparatus. Otherwise the cabinet would not be accessible other than by powering down or by 'gas clearance procedures'.

Another approach used is where battery supplies are localized with IS apparatus. The Ex 'e' cabinet would not normally be opened and so separating the major IS content into another enclosure simplifies maintenance procedures.

12.7.4 Field-mounted apparatus

Field-mounted apparatus must be installed according to the conditions of certification for safety purposes.

The apparatus certificate and the systems documentation will give the relevant electrical characteristics for installation. The manufacturer will also provide some instructions on safety and operational issues.

In general, apparatus in the hazardous area must be installed:

- Undamaged
- According to any special conditions of the certification
- Meeting any insulation tests (500 V AC 1 min) or comply with alternative arrangements
- With full and correct identification and labeling
- With correct earthing of the casing or enclosure
- With correct and working fail-safety duty, if applicable
- With correct cabling and safe parameters
- With appropriate environmental protection
- As fit for purpose for which it has been designed and selected.

12.8 Other considerations affecting installation

The integrity of installation may be affected by other mechanisms over which caution is advised.

12.8.1 Environmental conditions

With some techniques of protection, general environmental conditions are more likely to cause deterioration in system safety over time. Subsequent periodic inspection, discussed in Chapter 8, is necessary to ensure that deterioration does not reduce safety. It is therefore important that during the installation phase the correct and documented requirements are followed. Any deviation from this, perhaps owing to on-site situations not catered for in design, must be properly recorded.

Severe deterioration may cause installations to be redesigned. It is generally recognized that this will be one of the major threats to safety systems and it is imperative that good design and timely correction of problems will reduce risk of failure.

If deviation from a manufacturer or designer's requirements for installation are found these should have been thoroughly explained in suitable documentation, particularly where safety issues are affected.

12.8.2 Multiple protection methods

There are many pieces of apparatus that are certified with more than one technique of protection. No specific guidance on the method of approach for this situation is included in any standard or codes of practice to date.

Certification and manufacturer's instructions normally provide the caution that multiple techniques are used on apparatus. The current convention seems to be that the manufacturer will detail procedures for the installation and commissioning of these more complex protection systems. Gas detection systems have used this arrangement for some considerable time, being amongst the first systems affected.

With the increase in communications systems entering hazardous locations the use of multiple protection techniques will become more prevalent. The standards and codes of practice will inevitably lag behind technology on this issue.

12.9 Other installation issues

IS loop design criteria does not normally vary from one type of installation to another. Variations in the duty and perhaps integrity of applications are affected by the industry and the environment encountered.

12.9.1 Instrument model variants

In fact, there may be many mechanical variants of an instrument to suit different applications and environments but the common element in the considerations for IS is the electronics producing the 4/20 mA signal. The well-known Rosemount 1151 transmitter series is a good example of where one basic design covers almost the entire range of housings for the termination and electronic transmission systems. Note, for example, the power supply and output specifications in this book.

You will also find specific conditions for use in hazardous areas under a range of standards in this document. It would appear that the same electronics module will comply with the certification requirements with a number of different authorities but this is quite normal. Manufacturers will often produce one version of an instrument for non-hazardous locations and certified Ex 'ia', Ex 'd' or Ex 'n'. The only difference between instruments may well be the label.

Beware that some casings of instrument housings bare the words 'Do not open in a hazardous area' (or similar) even though the circuit is IS. This is a requirement of the Ex 'd' certification but does not necessarily apply to IS or Ex 'n'.

These comments reinforce the need for reliable identification of instruments.

12.9.2 Commissioning

There are no specific references to the function of commissioning other than those discussed in the maintenance and inspection section of this book.

Recent legislation affecting many countries does not permit 'live working' on ANY electrical equipment unless safety (from shock and damage) can be assured. This is said to include IS circuits, unless adequate practices and procedures are in place to allow work to be done.

Clearly one of the principal advantages of IS installations is that the risk of causing ignition is so low that live working from an explosion protection point of view is quite acceptable. Assessment of practices and procedures may be necessary to comply with other legislations.

13

Inspection and maintenance

13.1   Inspection and maintenance

IEC 60079-17: 1996 is entitled 'Recommendations for inspection and maintenance of electrical installations in hazardous areas (other than mines)'. It was first introduced in 1990.

It is introduced in the standard as being 'supplementary' to IEC 60079-14. It states:

Electrical installations in Hazardous areas poses features specially designed to render them suitable for operation in such atmospheres. It is essential for reasons of safety in those areas, that, throughout the life of such installations, the integrity of those special features is preserved; They therefore require INITIAL inspection and either:

- *Regular PERIODIC INSPECTIONS thereafter or*
- *Continuous supervision by skilled personnel and when necessary, maintenance.*

By this statement, the standard recognizes that just as selection and installation rules are provided in this book, the continued use of the apparatus in a safe way requires those rules to operate throughout the life of the installation. There is, therefore, a need to maintain the plant in a safe condition by ensuring that explosion-protected apparatus remains in the same or similar physical state to when it was first installed and commissioned. This can only be done by monitoring the factors which may affect the installation. If deterioration is apparent then some sort of corrective action must be taken to maintain safety (Figure 13.1).

Such factors include:

- *Deterioration in condition of equipment*:   Wear and tear through a process of aging may force the deterioration of seals, embrittlement and cracking of enclosures. Metal can 'relax' from stress and fixtures become loose if incorrectly designed and specified. Weathering, moisture and ambient temperature cycling can cause or accelerate imperfections in equipment. Constant vibration and mechanical shock are other causes.
- *Environmental attack*:   Chemicals, vapors and accumulation of condensates may cause primary or additional progressive failure of equipment. A combination of chemicals and other factors mentioned above may have unforeseeable dramatic corrosive results.

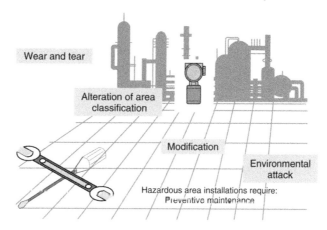

Figure 13.1
Preventative maintenance

- *Modifications*:   The installation may have undergone modifications during its life or during commissioning in order to improve response to control, quality or efficiency. Such changes may have been done hurriedly and without full control of circumstances. Plant additions and extensions are known to suffer from an increased risk of changing philosophies affecting overall safety. Such changes may demand new area classification.
- *Alteration to hazard*:   The introduction of new chemicals into a manufacturing process may require the modification of equipment to accommodate a higher-risk situation.

13.2   Integrity 'preserved'

The two approaches, given in the introduction as acceptable, are, in the first case, a more formal requirement for inspection. This comes predominantly from some of the older British and European Standards, and codes of practice. The second approach suggests that the requirement for formal inspection is more relaxed and places the onus in the domain of the 'maintenance' departments of user organizations. The installation should operate under 'continuous supervision' by people who know the plant and can perform maintenance at all levels. Some engineers and users are unsure of this approach.

The latter approach however is more in line with the operation of IS apparatus where safety is built in and cannot therefore be 'viewed' in the same way as potential danger may be spotted when looking at a flameproof enclosure.

13.3   Scope of IEC 60079-17

The scope of this document is stated as:

The report is intended as a guide for users and covers factors directly related to the inspection and maintenance of electrical installations within hazardous areas only. It does not include conventional requirements for electrical installations nor the testing and certification of electrical apparatus. It does not cover Group 1 (applications for mines susceptible to firedamp) apparatus. The report supplements the requirements laid down in IEC Publication 364-6-61.

Note: IEC 364: 1986: Electrical installations of buildings. Part 6: Verification. Chapter 61: Initial Verification.

Mining applications require a somewhat different approach because the construction of equipment to suit mining environments is peculiar to that industry.

The following definitions are taken from IEC 60079-17:

Maintenance

A combination of any action carried out to retain an item in, or restore it to, conditions in which it is able to meet the requirements of the relevant specification and perform its required functions.

Inspection

An action comprising careful scrutiny of an item carried out either without dismantling, or with the addition of partial dismantling as required, supplemented by means such as measurement, in order to arrive at a reliable conclusion as to the condition of an item.

There are three grades of inspection described which operate at progressively deepening levels of study.

Visual inspection

An inspection that identifies, without the use of access equipment or tools, those defects, e.g. missing bolts, which will be apparent to the eye.

Close inspection

An inspection which encompasses those aspects covered by a visual inspection and, in addition, identifies those defects, e.g. loose bolts, which will be apparent only by the use of access equipment, e.g. steps (where necessary), and tools. Close inspections do not normally require the enclosure to be opened, or the equipment to be de-energized.

Detailed inspection

An inspection which encompasses those aspects covered by a close inspection and, in addition, identifies those defects, e.g. loose terminations, which will only be apparent by opening-up the enclosure, and/or using, where necessary, tools and test equipment.

13.4   General requirements

The objective of this standard may be interpreted as providing guidance on the identification of defective apparatus such that it may be maintained to acceptable levels of safety.

The general requirements suggest that adequate, up-to-date documentation be available, specifically including previously recorded information. 'Sufficient' information should be available in order to enable maintenance to be carried out on hazardous area apparatus 'in accordance with its type of protection'. Knowledge of classification of the plant is also required.

Additionally, and once again, emphasis is placed on the suitability of personnel undertaking inspection and maintenance tasks to be suitably qualified and experienced, having undergone training and appropriate refresher training on a regular basis.

13.5   Inspections

The discussion that follows assumes that decisions are made by the plant owner about how safety is to be ensured in a given installation. This may vary from one location on a plant to another. The standard recognizes that each particular installation is unique and it does not attempt to dictate how, when, where and what to inspect and maintain, simply because of the differences encountered in each situation. The responsibility for safety, of which the inspection and maintenance is deemed to play a major part, is always placed upon the owner of the plant. The owner may, perhaps rightly, delegate this detailed decision-making to the plant operating organization that has a detailed knowledge of the process.

There are three 'types' of inspection. The type refers to the point-in-time or period of inspection and is laid out in Figure 13.2.

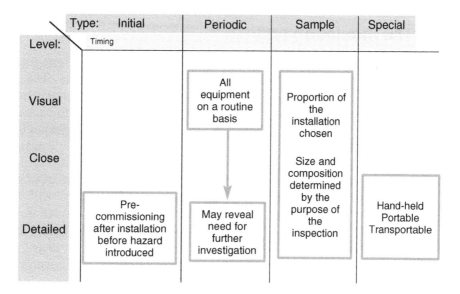

Figure 13.2
Inspection requirements: grades

After installation and prior to any hazard being introduced onto a plant, there is a specific requirement to perform a detailed inspection. Verification that the installation fully agrees with the documentation must be sought as a starting point in the life of the plant. The results of this as with all inspections shall be recorded.

Determination by the experience of operating industrial plants will enable decisions to be taken on periods between types and grades of inspection that are required to follow this initial inspection. The results of any inspection must be reviewed. Decisions are necessary on whether there needs to be any adjustment in the inspection or maintenance procedures, based on the interpretation of these results. Justification of these changes may be necessary by further safety audits. There are obviously cost implications in any change to be made, and appropriate levels of management should be involved in these reviews (Figure 13.3).

Where safety issues cannot be verified by a visual or simple physical inspection process, then it may be necessary to measure parameters. An example of this is the connection between the barrier busbar and the SPNE bar. The requirement is for less than 1 Ω. Clearly this cannot be verified by any other method than measurement.

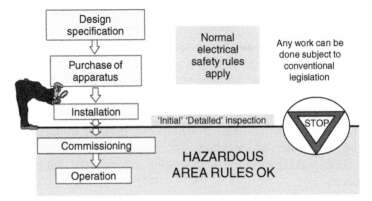

Figure 13.3
Inspection requirements

13.6   The insulation test

The requirement for cables to meet a 500 V AC insulation test for 1 min expresses the expected grade of cable required to preserve safety integrity. The safety of IS circuits, however, does not depend on a minimum finite insulation resistance level. Faults from core to core of a cable in the same IS circuit are taken into account and cannot cause ignition by design. Faults from cores to 'earth' are also designed to be safe by imposing a regime that minimizes risk. It is conceivable that a fault to earth at the same time as other faults occuring on a loop could pose a danger but these risks are minimized.

There is no specification for leakage resistance in any of the standards because this is an operational consideration. Poor insulation on RTD circuits will affect loop accuracy whereas switch-status loops are incredibly tolerant. The insulation resistance of a cable may degrade over time owing to aging and environmental attack. It is therefore important to interpret the requirement for inspection as detecting the collapse of cable integrity by the precursor of monitoring the long-term trend of insulation, even though it is of operational interest (Figure 13.4).

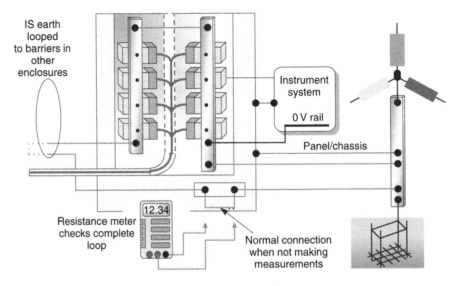

Figure 13.4
Inspection of IS earth integrity

The rigor with which insulation testing is performed depends on the environmental conditions of the plant. It is usual to apply 'sample' inspection testing on cables thought most likely to be affected. It is not recommended that all cables are periodically tested.

The testing of all loops and cables must be performed after a new installation is completed. This is easy to do whilst the plant is not hazardous (during 'closed' commissioning); however if the new plant is an extension to an existing installation that is already hazardous, care must be taken with measurement procedures to ensure safety.

13.7   Maintenance

There has been a significant change in the second issue of IEC 60079-17: 'maintenance recommendations' in the 1990 issue. This has become 'maintenance requirements' in the second issue in 1996. It has therefore become more detailed and more stringent in response to earlier criticism with respect to maintenance clauses.

13.7.1   General requirements

The general requirements for all types of protection are summarized as follows:

- Remedial action is to be carried out where necessary. Caution is expressed in that safety must be preserved prior to and during remedial work.
- Flexible cables are particularly prone to damage and may require special consideration in that more frequent inspection may be necessary.
- If any equipment is taken out of service then disconnected cables must be dealt with by correct termination in an appropriate enclosure, being correctly isolated and insulated or being correctly isolated and earthed.
- Where special bolts and other fastening or special tools are required, these items shall be available and shall be used.
- Environmental conditions including ambient temperature and chemical attack effects are required to be examined and verified for suitability.
- All parts of installations shall be kept clean, particularly where accumulations of substances can cause heat build up.
- Equipment generally should be undamaged. Weather proofing applied to equipment should be maintained effectively.
- Care shall be taken in respect of anti-condensation devices operating, vibration not loosening fixings and the generation of static electricity.

This standard expressly permits that live working may be carried out on IS apparatus and systems.

The introduction to this document supports the view that quality and not quantity of detective work in maintaining a plant is necessary.

13.7.2   Additional requirements for type of protection 'i' – intrinsic safety

The following extract is taken from the IEC 60079-17 standard and relates specifically to IS installations.

Additional Requirements for Type of Protection 'i' – Intrinsic Safety		
Clause	**Quote**	**Comments**
Section 12	Additional requirements for type of protection 'i' – Intrinsic Safety	
8	Type of protection 'i' – Intrinsic safety (see table 2 and IEC 79-11)	
8.1	Documentation	
8.1.1	The documentation referred to in table 2 shall, as a minimum, include details of: (a) circuit safety documents, where appropriate; (b) manufacturer, apparatus type and certificate numbers, category, apparatus group and temperature class; (c) where appropriate, electrical parameters such as capacitance and inductance, length, type and route of cables; (d) special requirements of apparatus certificate, and detailed methods by which such requirements are met in the particular installation; (e) physical location of each item in the plant.	Documents used are normally those produced by the plant designer and are those to which the construction and commissioning are undertaken
8.1.2	Labelling Labels shall be inspected to ensure that they are legible and comply with the requirements laid down in the appropriate documentation to ensure that the apparatus actually fitted is that specified	Compliance with original design requirements.
8.1.3	Unauthorised modifications The requirement to check that there are 'no unauthorised modifications' can present some problems, in that it is difficult to detect alteration to, for example, a printed circuit board. It may be possible to utilise the fact that the soldering associated with most repairs/alterations is not of the same type or quality as the original Photographs of the original boards, supported by listings of the key components upon which the safety of the circuit depends, may be useful.	See notes in section 8.7 below
8.1.4	Interface devices between intrinsically safe and non-intrinsically safe circuits Diode safety barrier installations shall be inspected to ensure that the correct types of barriers have been used, and that all such devices are firmly fixed to the barrier earth bar in a way, which gives good earth continuity.	

Clause	Quote	Comments
	Installations shall be inspected to ensure that relays which act as safety barriers between circuits and other devices with moving parts have not become damaged by repeated operation or vibration in a way that reduces the segregation afforded.	Lifted from BS5345 part 4 but not updated to cover other galvanically isolating associated apparatus
8.1.5	Cables	
	Installations shall be inspected to ensure that the cables used comply with the documentation Particular care shall be given when utilising spare cores in multicore cables containing more than one intrinsically safe circuit, and to the protection afforded where cables containing intrinsically safe systems and other cables run in the same pipe, duct or cable tray.	
8.1.6	Cable screens	
	Installations shall be inspected to ensure that cable screens are earthed in accordance with the appropriate documentation. Particular attention shall be paid to installations utilising multicore cables that contain more than one intrinsically safe system.	Plant installation policy should cover this requirement
8.1.7	Point-to-point connections	
	This check is only required at the initial inspection.	It is also required at the commissioning stage of any extensions to the plant.
8.1.8	Earth continuity of non-galvanically isolated circuits	
	The resistance of the earth connection between intrinsically safe circuits and the earth point shall be measured on initial inspection.	See Section 8.8 below.
	The measurement shall be made using a tester specifically designed for use on intrinsically safe circuits.	
	A representative sample of connections, selected by the responsible person, shall be measured periodically to confirm the continuing integrity of the connections.	
8.1.9	Earth connections to maintain the integrity of intrinsic safety	
	The resistance of the earth connections necessary to maintain the integrity of the intrinsically safe system (such as transformer screen earth, barrier relay frame earth) shall be measured as in 5.3.8.	*See below.*

(Continued)

Clause	Quote	Comments
	There is no requirement to measure the earth loop impedance of mains powered apparatus associated with intrinsically safe circuits other than that required for normal control room instrumentation to protect against electric shock. Since, in some equipment, the intrinsic safety earthing is internally connected to the equipment frame, any impedance measurements (such as between the earth pin of the plug and the equipment frame, or the equipment frame and the control panel) shall be made using a tester specifically designed for use of intrinsically safe circuits.	
8.1.10	Intrinsically safe circuit earthing and/or insulation	
	The insulation testing of intrinsically safe circuits is necessary to confirm that they are earthed or insulated from earth throughout, whichever of these conditions is required by the original design.	Procedures for testing are discussed in section 10.
	Insulation testing of intrinsically safe systems or circuits shall only be carried out using a test device specifically approved for connection to such circuits.	
	Where, in order to carry out these tests, the common earth connection to a group of barriers is disconnected, the tests can only be made if either the plant is free from hazard, or if power is removed completely from all the circuits which depend upon that common earth connection. This test is only required on a sample basis.	
8.1.11	Separation between intrinsically safe and non-intrinsically safe circuits.	
	Junction boxes and boxes containing safety barriers shall be inspected to ensure that they contain no wiring not specified in the documentation appropriate to any system passing through them. See also subclauses 12.2 and 12.3 of IEC 79-14.	(Requirements for Zones 1 & 2 and Zone 0 respectively in these clauses)

13.8   Testing

The testing of circuits referred to in the standards, primarily involves insulation testing. This should be undertaken with normal high-voltage test equipment at the pre-commissioning ('detailed') inspection stage. Immediately afterward it is recommended that low-voltage tests are carried out using certified equipment and some comparison is made. The results shall be recorded as required.

Thereafter periodic testing may be performed using low-voltage (certified apparatus) test equipment whilst operating to safe procedures even though the plant is hazardous.

The testing of other parameters is discussed in Chapter 10 of this book.

13.8.1 Trends

The standard correctly infers that the trend of performance of insulation is of the greatest interest. If the sample of measurements show a trend of maintaining adequate insulation properties then there is no cause for concern. Where deterioration is seen, the mechanism of that deterioration should be located and prevented. Alternatively, more suitable apparatus may have to be installed.

13.8.2 Useful data for collection

Other useful phenomena to record are the oscilloscope waveforms seen between the various earths. Interference patterns can be traced by observing new waveforms and disconnecting equipment power supplies and earths until a previous situation's waveforms are seen. In that case the last earthing system attached will somehow be the cause of the problem and further investigation may start from that point. This technique has been successfully used in the past but it requires a thorough and diligent approach to the investigation with due regard to safety. The use of equipment in the safe area earthing arrangements does not require any assessment. Disconnection of earths for tracing purposes will require careful evaluation.

13.9 Unauthorized modification

The standard states that inspection should look for signs of unauthorized modification but accepts that it is difficult. This form of inspection would need to be done by a person adequately trained on the specific apparatus type. This may need to be the manufacturer or his representative. The likelihood of this occurrence may be assessed and if adequate systems are in place to ensure that apparatus is repaired by the manufacturer then this level of inspection is not required.

13.10 Earthing integrity verification

Clause 5.3.8 and subsequent of this standard requires some clarification. Summarizing the requirements: where a barrier is used to interface a hazardous area circuit, it requires initial verification that the barrier is in fact connected to 'earth'. In addition, any part of the circuit in the hazardous area, or on associated apparatus in the safe area, where parts are earthed, and the connection must be verified. IS-approved test devices must, according to the requirements, be used for these purposes.

It goes on to say that if disconnection of the protective earth system is necessary to do this then the hazardous area must be made safe or the power must be removed from the instrument section under test. This shall be done on a sample basis.

As has been discussed, this requirement is obviated by the use of circuits that operate in a self-revealing fault mode. The use of intelligently assigned and set up alarms will assist in providing failure data. This is the current trend in system design.

Note that the standard is unspecific on values which are acceptable or unacceptable. This is one example of where the document has not succeeded in galvanizing the requirements of many different countries. There are divisions even within Europe about the methodology of the earthing requirements. This standard is less specific than existing national standards and therefore has not been adopted.

13.11   BS 5345 inspection requirements

In this still widely accepted document, it separates the requirements of 'initial' or pre-commissioning in Sections 18 and 19, and 'routine' or periodic inspections in Sections 18 and 20. The recommendation for the period is two years or more frequently for installations in 'particularly arduous environmental conditions'.

The problem with the document is that it has never been updated to include galvanically isolating interfaces and associated apparatus. It talks at length about the requirements of shunt diode safety barriers in Section 17, and is relatively specific about the requirements. As such it explains to some extent why it makes recommendations, which is why it has been so widely adopted and accepted.

The requirements of this code are often stipulated in addition to those in IEC 60079-17 because they are familiar to users. It is hoped that future revisions will become more definitive as has already occurred with the first revision in 1996. BS 5345 will not be updated further in preference to the adoption of 79-14 and -17.

Points worthy of note are mentioned here and clarify some of the requirements of IEC 60079-17.

- Use of 'properly trained personnel' is mentioned as the prerequisite for carrying out inspections. It also suggests that persons other than the installer should undertake the inspection.
- Initial inspection is charged with verification that the installation meets the requirements of the area classification, apparatus grouping and temperature classification as defined by the plant owners.
- The installation must comply with the (design) documentation and take into account any special conditions of installation.
- Labeling must be correct and apparatus properly identified.
- Apparatus must be undamaged after installation.
- Replaceable components such as lamps, fuses etc. should be checked to ensure the correct type has been fitted.
- Earthing shall be in accordance with clause 16 (see Section 13.10.1).
- Cable screens are to be in accordance with relevant documentation and clause 16.
- Interfaces should be properly fitted (meaning according to the manufacturer's instructions).
- Interface and junction boxes are to be inspected to check they contain no non-IS circuits.
- Cabling is to be inspected to ensure that only IS cables are used in multicore cables and protection is afforded where necessary.

13.11.1   BS 5345 clause 16

Further guidance is given on the 1 Ω connection between SPNE and the IS earth bar for barriers in clause 16.1. These requirements were discussed in Chapter 11 of this book on Earthing.

IS earth connections must be separate from plant earth and other electrical earths, except at one point – the main electrical system earth point.

13.11.2   Periodic inspection

Where apparatus has been operated in service, the above points are mentioned as forming the basis of the periodic inspection without the need for the initial classification audit. In addition, and with the benefit of experience, company's written procedures often include specific checks for contravention of creepage and clearance distances on termination system during inspections that could occur as a result of mechanical damage.

The recording of results is required by an effective and verifiable means.

Clause 21 begins by saying that, 'Where compliance with the appropriate documentation cannot be demonstrated by physical inspection, electrical testing should be carried out in accordance with …' other requirements in Section 21. Little of this advice has been brought into IEC 60079-17. It has however been interpreted in Chapter 10 of this book.

'Where compliance with the appropriate documentation cannot be demonstrated by visual inspection alone...'

Insulation testing (hazardous area)

When a cable fault is suspected:

Resistances of core to core and core to screen, typically

Pass > 10 MΩ
Fail < 1 MΩ

Intrinsically safe high-voltage insulation testers, certified to old standards (BS 1259) are available but are not recommended due to cable parameter limitations

Follow established procedures for testing
Be trained on the use of test equipment

Figure 13.5
Inspection requirements

Figure 13.5 suggests acceptability bounds for insulation testing that are not in the standards but have evolved as experienced based 'rules of thumb'.

14

Safe working practices

14.1　General

We have to remind ourselves that one of the real benefits of this discussion can be realized if we obtain a greater degree of loss prevention during the operation of the production facility. Having designed and installed as per the regulations it is of utmost importance to give adequate attention to the adoption of safe working methods and practices that will ensure:

MISHAP PREVENTION AND LOSS PREVENTION OF MEN AND MATERIAL

Most of us working with electricity take risks. Usually we get our jobs done without any harmful results. Mishaps or injuries usually result from an unclear understanding of a risk or danger.

The first part of this chapter is designed to help you eliminate or minimize mishaps. It also provides you with a good review of what to do in case of an accident with burns due to fire.

In this chapter we shall be looking at some of the more obvious practices that any electrical technician is supposed to follow in terms of:

- Safety
- Looking out for danger signals
- Inspection and maintenance requirements
- Maintenance and safe practices
- Safe methods of insulation testing and requirements for hazardous area
- Safe methods of earth integrity checking and the specific requirements for a hazardous area
- Safe practices after fire and electrical shock.

14.2　Safety observations

Working safely is the most important thing you can do. Because of their importance, several precautions are included as the first subject in this chapter. Of course, there are more precautions, but these are some you should definitely think about. The keyword here is *think*.

- Never work alone.
- Never receive an intentional shock (e.g. by testing).

- Only work on, operate or adjust equipment if you are authorized.
- Do not work on energized equipment unless absolutely necessary.
- Keep loose tools, metal parts and liquids from the electrical equipment. Never use steel wool or emery cloth on electric and electronic circuits.
- Never attempt to repair energized circuits except in an emergency.
- As a general rule, never measure voltages in excess of 300 V while holding the meter wire or probe.
- Use only one hand when operating circuit breakers or switches.
- Use proper tag-out procedures for regular and preventive maintenance.
- Be cautious when working in voids or unvented spaces.
- Beware of the dangers of working aloft. Never attempt to stop a rotating antenna manually.
- Keep protective closures, fuse panels and circuit breaker boxes closed unless you are actually working on them.
- Never bypass an interlock unless you are authorized to do so by the authorized persons, and then properly tag the bypass.

Some safety-related practices which need to be more strictly followed when operating machinery in hazardous area are listed below.

Warning signs

They have been placed for your protection. To disregard them is to invite personal injury as well as possible damage to equipment. Switches and receptacles with a temporary warning tag, indicating work is being performed, are not to be touched.

Working near electrical equipment

When work must be performed in the immediate vicinity of electrical equipment, check with the technician responsible for the maintenance of the equipment so you can avoid any potential hazard of which you may not be immediately aware.

Authorized personnel only

Because of the danger of fire, damage to equipment and injury to personnel, only authorized persons shall do maintenance work on electrical equipment. Keep your hands off all equipment, which you have not been specifically authorized to handle. Particularly stay clear of electrical equipment opened for inspection, testing or servicing.

Circuit breakers and fuses

Covers for all fuse boxes, junction boxes, switch boxes and wiring accessories should be kept closed. Any cover, which is not closed or is missing, should be reported to the technician responsible for its maintenance. Failure to do so may result in injury to personnel or damage to equipment in the event accidental contact is made with exposed live circuits.

14.3 Danger signals of electrical malfunctioning

Personnel should constantly be on the alert for any signs, which might indicate a malfunction of electric equipment, especially in a hazardous area. Besides the more obvious visual signs, the reaction of other senses, such as:

- Hearing
- Smell and
- Touch.

should also make one aware of possible electrical malfunctions.

Examples of signs which one must be alert for are:

- Fire
- Smoke
- Sparks
- Arcing
- An unusual sound from an electric motor
- Frayed and damaged cords or plugs, receptacles, plugs and cords which feel warm to the touch
- Slight shocks felt when handling electrical equipment
- Unusually hot running electric motors and other electrical equipments
- An odor of burning or overheated insulation
- Electrical equipment which either fails to operate or operates irregularly
- Electrical equipment which produces excessive vibrations are also indications of malfunctions.

When any of the above signs are noted, they are to be reported immediately to a qualified technician.

DO NOT DELAY.

Do not operate faulty equipment. Above all, do not attempt to make any repairs yourself if you are not qualified to do so. Stand clear of any suspected hazard and instruct others to do likewise.

14.4   Need for inspection and maintenance

Having stated that 'loss prevention' is one of the main objectives of any ongoing enterprise, it is needless to emphasize that this can be only achieved if all the apparatus, equipment and systems for hazardous area are designed, selected and installed with painstaking care. The equipment must also be maintained as close as possible to the state it was when it was new.

Electrical installations, which comply with standards and regulations, or with other statutory documents dealing with the same subject, possess features specially designed to render them suitable for operation in hazardous areas. It is essential for reasons of safety in those areas that throughout the life of such installations the integrity of those special features is preserved. They therefore require systematic inspection and if necessary, maintenance both initially and at regular periodic intervals thereafter.

It should be noted that the correct functional operation of hazardous area installations does not mean, and should not be interpreted as meaning, that the integrity of the special features referred to above has been preserved.

The features of equipment, which qualify it to be classified as explosion-protected, are necessarily in many cases not forming part of its satisfactory operational features. Thus satisfactory operation of apparatus or equipment does not ensure that all the explosion-related features are intact. Thus in most cases there is no warning of failure of these features and the apparatus may continue to operate while having faults, which make it ignition-capable.

The above makes it imperative that a formal and exhaustive documentation procedure should be put in place. This should also cover the requisite procedures for inspection, testing and certification wherever necessary for the following:

- Commissioning and test verification
- Operation
- Storage of flammable material
- Maintenance schedule
- Authorized modifications
- Overhaul and repair.

It should be noted that the required certificate/documentation needs to be issued for the maintenance schedule and authorized modifications, and these should be included in a hazardous area verification dossier (this dossier is mandatory as per the new regulations). Normal overhaul and other activities need also to be suitably documented.

The facility in charge shall nominate a person (or persons) having the appropriate knowledge and experience of hazardous area installations to be responsible for establishing and implementing a system of inspection and maintenance, which shall keep the plant operating safely. Clauses below give guidance on the essential elements of an appropriate system.

14.4.1 Competency and training of personnel

The inspection, maintenance, replacement and repair of apparatus, systems and installations shall be carried out only by personnel whose training has included instruction on the various methods of safeguarding and installation practices and, in appropriate cases, on the general principles of area classification. Appropriate refresher training shall be given to such personnel on a regular basis at intervals, which should not exceed, say, two years, records being kept of the training given.

Where the inspection of apparatus, systems and installations associated with hazardous areas is carried out by a contractor, the contract shall include a statement to the effect that the inspection shall be carried out only by personnel whose training shall have included instruction on the various methods of safeguarding; evidence of which shall be supplied by the contractor.

14.4.2 Basis and frequency of inspection

A planned scheme of regular routine inspection is the basis of effective maintenance, and whilst production requirements need to be considered they shall not result in the postponement of essential inspection and remedial work.

A principal requirement of any scheme is the keeping of records which for this purpose shall as a minimum include for each plant appropriate details of:

- Up-to-date area classification
- All apparatus, systems and installations
- Inspections carried out and faults revealed
- Corrective action taken to remedy such faults.

14.4.3 Type of inspection with specific reference to hazardous area

Three forms/types of inspection which are given hereunder can be adopted as required by plant operation and maintenance personnel.

Initial inspection

All apparatus, systems and installations shall be inspected on initial installation and after modification, in accordance with the 'initial' inspection schedules and the details shall be recorded. These should be carried out as near to commissioning of plant as possible but before flammable or explosive materials are introduced into the plant, so as to avoid any subsequent inadvertent alterations going unnoticed.

We shall now look into some specific guidelines for initial inspection in hazardous area as given hereunder. It should be noted that these are indicative and comprehensive guidelines of manufacturers should be adhered to in addition to the ones given below.

General initial checks and inspection checklist

- **External visual inspection**

    - Match the protection concept as labeled on the apparatus with the area classification of its location.
    - Match the sub-group (in case the protection concept is subject to a sub-group) as labeled on apparatus with the area classification of its location.
    - Match the surface temperature class as labeled on the apparatus with the appropriate explosive gas, liquid or vapor present in the area classification of its location.
    - The apparatus carries the correct circuit (installation) identification. All installed apparatus/enclosures must be tagged so that they are uniquely identified to trace them at a later date ' so as to either ensure or cross-check for correct selection and installation information. This is termed as the inspection or conformance of the individual installation with respect to what has been envisaged at the design stage.
    - The initial external visual inspection to ensure that the apparatus is free of corrosion and/or undue dust accumulation or material (corrosion agents) which could lead to corrosion at a later date.
    - To ensure that all cable glands, stoppers, and their fastening nuts and bolts are secure. The integrity of the earth connection including its proper tightness to the enclosure/apparatus need to be ensured.
    - To ensure that all cable trays, external special guards around the apparatus, if any, are properly secure and undamaged.
    - The cables and conduits are not damaged. In the case where cables lead to the enclosure through sand-filled trenches it is not necessary to remove the sand until physical evidence of any abnormality is noticed.
    - To ensure that the electrical protection is set properly. This is to include checks of fuses, circuit breakers, protection relays, etc. Proper labeling of these devices needs to be ensured such that they are clearly identifiable without the need to touch the devices or the need to remove any devices such as fuses.
    - To ensure that an external application to prevent corrosion is properly applied, such as grease applied to gaskets and joints, tape to glands, etc.

- **Internal inspection**

    - To inspect for undue accumulation of dust, dirt or corrosion agents inside the enclosure. To check for any damages to gaskets, tightness of electrical connections (to guard against the sparking). Being an internal inspection will require proper electrical isolation of equipment and removal of covers as per requirements. Use of proper tools is a prerequisite for checking the tightness.

- Another internal inspection is for checking of the correct installation of lamps/bulbs as per the design. This also necessitates removal of the cover and isolation of electrical apparatus from service and supply.
- In the case where encapsulating material is used in apparatus/enclosures, cable boxes and stoppers, it should be ensured that the filling is proper. The covers need to be removed for a visual inspection inside the enclosure and requisite isolation done. Generally a visual inspection is sufficient as long as no deterioration or abnormality is observed.

- Either by internal or external inspections the motor air-gaps and other radial and axial clearances of motors need to be checked against the manufacturer's design data sheet. This record will help to determine undue wear and tear or distortion, which could be indicative of impending failure in terms of sparks due to rubbing between rotating parts (such as fan blades, bearing grease guards, rotor body, etc.) or any other abnormality.
- The integrity of the apparatus with respect to design parameters needs to be confirmed with respect to a hazardous dossier and it should be ensured that there are no 'unauthorized' modifications. This can be done through internal and external inspections.
- To ensure by inspection and checking that:

  - Potential equalizer system connections are secure and undamaged
  - Earthing connections are undamaged and secure
  - All enclosure fittings are properly bonded to earth and are secure and undamaged.

The initial inspection should also include in addition to the above checks certain other checks, which are specific to the protection concepts in question. Some illustrative checks required to be carried out for important protection concepts are described hereunder.

Flameproof enclosures

- To ensure that enclosures, fixing bolts, enclosure glasses, any glass/metal seals are not damaged by external visual inspection.
- To ensure that the gaps in enclosures are not unduly obstructed by paint or dirt/dust.
- To ensure that the enclosure bolts are of specified tensile strength. Checking the head marking of such bolts can do this.
- To ensure that there are no unacceptable external obstructions to flame paths. In order to achieve this, an examination needs to be made for any external hindrances from the obstructions, which are not part of the enclosure but within the specified distance.
- To ensure that the cable glands and stopper boxes are of the correct type by examining their markings.
- To ensure that the flange gap dimensions are within the maximum specified, by checking with the help of feeler gages. It is to be noted that there is no specified minimum for this criteria. It may not be possible and necessary to

check in the case of cylindrical gaps and spigot joints as they do not get distorted easily.

- An internal inspection of flame paths, internal parts of glasses and glass/metal seals needs to be done for any damage and corrosion effects. Being an internal inspection, the covers need to be removed and requisite isolations also need to be done before inspection.

Intrinsically safe installation

In the case of an internal inspection of an intrinsically safe device or apparatus, isolation is not required, if in the same enclosure:

- There are no non-intrinsically safe circuits at potentials which constitute risks of shock.
  or
- The apparatus is associated apparatus.
- To ensure that there are no unauthorized connections to the potential bonding/equalization system by carrying out internal inspection. In particular multi-circuit junction boxes need to be checked thoroughly.
- To ensure requisite segregation is maintained in multi-circuit junction boxes between intrinsically safe and non-intrinsically safe circuits.
- To ensure that the barrier devices are of correct type and are correctly mounted.
- To ensure that the cables are connected and segregated correctly. Also, the conduits are likewise carrying properly segregated cables and wires.

Pressurized enclosures

- To ensure through external inspection that the inlet and exit ducting is not damaged. Because any gas leakage from the ducting is likely to contaminate either enclosure or exhaust area depending upon the operating system.
- To ensure that the gas or compressed air supply is given to enclosure free from any contamination and of the appropriate quality. This can be ensured by checking for any exhausts, which can contaminate the air in the area of inlet suction of compressor, and/or the specified gas cylinders are only connected to the enclosure.
- To ensure proper/correct operation of pressurizing and flow monitoring devices by checking suitably the operation of interlocking protections.
- To ensure that adequate and requisite flow and pressure is maintained of air or gas, including purging requirement if any.

Oil-immersed and powder-filled apparatus

- To ensure that the oil seals and powder retention covers are in a good condition by examining visually and physically.
- To ensure that there is no leakage of powder or oil from the enclosure.

Inspection after apparatus repair

In the case of repair, adjustment or replacement carried out on any apparatus system or installation, it shall be checked in accordance with the relevant items as per the 'initial' inspection schedule. These checks may be carried out by the person doing the work and need not be recorded.

It has to be ensured that any repair does not change the integrity of the apparatus, vis-à-vis the protection concept as approved and/or certified and documented in the hazardous area dossier of the plant.

Inspection after change in area classification, sub-group or surface temperature classification

This inspection should ensure that the installation, equipment and apparatus are in conformance to the new classification and are appropriate. This needs to be particularly checked for sub-group and temperature classification.

Periodic/routine inspection

All apparatus, systems and installations shall be inspected in accordance with the 'periodic/routine' inspection schedules and the details shall be recorded. These inspections are required to be done to identify the deterioration in installation conditions due to operating conditions or environment or unauthorized modifications.

The nominated person(s) shall determine the frequency of 'periodic/routine' inspections.

Experience has shown that there is a point beyond which increasing the frequency of inspection does not decrease the significant fault rate. The nominated person may extend the interval between 'periodic' inspections if the significant fault rate is not increased consequently.

In extremely adverse conditions, the interval between 'periodic' inspection may be as low as three months but should not normally exceed two years. However in extremely good, stable environmental conditions, the interval between inspections may be extended to four years.

Where the interval between 'periodic' inspections exceeds two years, 'visual' inspections shall be carried out as defined below.

It should be noted that a 'significant fault' is one in which the certified integrity of the apparatus or system (insofar as it affects safety in flammable atmosphere) is impaired and the significant fault rate is the number of items having significant faults expressed as a percentage of the total number of items inspected.

Visual inspection from floor level

The majority of faults on apparatus, systems and installations, which remain undisturbed, are caused by environmental factors and most of these are detectable by a 'visual' inspection from floor level.

Where the 'periodic' inspection interval exceeds two years, all apparatus, systems and installations shall be inspected in accordance with the 'visual' inspection schedules and at an interval not exceeding half that determined for the 'periodic' inspection. The details shall be recorded.

Following a major shutdown in a hazardous area all apparatus, systems and installations in that area should be inspected in accordance with the visual inspection schedules. The details should be recorded.

Guideline for routine or visual inspection

The following can act as guidelines for the routine or periodic or visual inspections of installations:

- To check for deterioration in enclosures, fittings, conduits, cable glands, cable boxes due to corrosion effect.
- To check for undue accumulation of dust and dirt in cable trays, enclosures, conduits, etc. because these can harbor corrosive liquids and solvents.
- To check for any physical damages to enclosures, conduits, cables, cable trays, etc.
- To check for any leakage of oil, powder, sand from apparatus having moving parts.
- To check for looseness in enclosure fittings, mountings, glands, stoppers, etc.
- To check for any deteriorating gaskets thus exposing the components housed in enclosure to harsh environment.
- To check for any excessive vibration at the point of mounting which may lead to loosening of cable or conduit connections in enclosures and rotating equipments.
- To check for the condition of bearings to ensure that no overheating, rubbing or seizing occurs.
- To check for any abnormal leakage or loss in level of oil or powder or sand, indicating that the protection is deteriorating.
- To check for proper functioning of relays and protection, safety devices used to ensure safety of apparatus, equipment and plant installation.
- To check for any loose electrical connection, in particular with equipotential bonding.
- To check for any unauthorized changes, such as in fuse rating or bulbs/lamps.

Inspection procedure

Generally dividing the installations or plant in a geographical manner and then carrying out the appropriate inspections has been found to be more economical than based on a system approach. In checking the installation by systems, it can be time consuming as each system often occupies more than one physical location.

The format of the procedure should be devised such that positive reporting is done. This means compliance of all systems and the full installation with the requirements is ensured. In negative reporting, only non-compliances are reported and a doubt can linger as to whether the full installation has been inspected from every aspect of safety or not.

A suggested method could be:

- Divide the plant into proper logical locations.
- Detail all parts of the electrical installation in sheets of inspection.
- These sheets should detail type of inspection to be carried out on each apparatus or system or installations in the area identified.
- The sheets should clearly specify the method of reporting so that no ambiguity remains in the recording.
- Record compliance and non-compliance separately. This will ensure quick generation of a punch list and its rectification and subsequent re-inspection for compliance.

- A procedure to integrate all such reports and linking them to each part of the installation to allow for verification of the completeness of the inspection process.

Checklist for inspection

We have included as a guidance the typical checklist which can serve as the basis for initiating initial, periodic/routine and visual inspections.

The definition of the words used are given hereunder,

- *ALL*:   To mean all apparatus.
- *SAMPLE*:   To mean the size to be decided by authorized person of the plant.

Inspection schedule for apparatus relying on pressurizing/purging concept of protection

Check that	Types of Inspection		
	Initial	Periodic	Visual
Apparatus is appropriate to area classification	ALL	ALL	ALL
Apparatus surface temperature class is correct	ALL	SAMPLE	NONE
Apparatus carries the correct circuit identification	ALL	SAMPLE	NONE
There are no unauthorized modifications	ALL	SAMPLE	ALL
Earthing connections, including any supplementary earthing connections are clean and tight	ALL	SAMPLE	ALL
Earth loop impedance, or resistance is satisfactory	ALL	SAMPLE	NONE
Lamp rating and type is correct	ALL	SAMPLE	NONE
Source of pressure/purge medium is free from contaminants	ALL	ALL	ALL
Pressure/flow is as specified	ALL	ALL	ALL
Pressure/flow indicators, alarms and interlocks function correctly	ALL	SAMPLE	NONE
Pre-energizing purge period is adequate	ALL	SAMPLE	NONE
Ducting, piping and enclosures are in good condition	ALL	ALL	ALL
No undue external accumulation of dust and dirt	ALL	ALL	ALL

Inspection schedule for apparatus relying on protection concept 'n', 's' or 'e'...

Check that	Types of Inspection		
	Initial	Periodic	Visual
Apparatus is appropriate to area classification	ALL	ALL	ALL
Apparatus group (if any) is correct	ALL	SAMPLE	NONE
Apparatus surface temperature class is correct	ALL	SAMPLE	NONE
Apparatus carries the correct circuit identification	ALL	SAMPLE	NONE
Enclosures, glasses and glass/metal parts are satisfactory	ALL	ALL	ALL
There are no unauthorized modifications	ALL	SAMPLE	ALL
Earthing connections, including any supplementary earthing connections are clean and tight	ALL	SAMPLE	ALL
Earth loop impedance, or resistance is satisfactory	ALL	SAMPLE	NONE
Lamp rating and type is correct	ALL	SAMPLE	NONE
Bolts, glands and stoppers are of the correct type and are complete and tight	ALL	ALL	ALL
Enclosed-break and hermetically sealed devices are undamaged	ALL	SAMPLE	NONE
Condition of enclosure gaskets is satisfactory	ALL	SAMPLE	NONE
Electrical connections are tight	ALL	SAMPLE	NONE
Apparatus is adequately protected against corrosion, the weather, vibrations and other adverse factors	ALL	ALL	ALL
There is no obvious damage to cables	ALL	ALL	ALL
Automatic electrical protection devices are set correctly	ALL	SAMPLE	NONE
Automatic electrical protection devices operate within permitted limits	SAMPLE	SAMPLE	NONE
No undue external accumulation of dust and dirt	ALL	ALL	ALL

Inspection schedule for intrinsically safe systems

Check that	Types of Inspection		
	Initial	Periodic	Visual
System and/or apparatus is appropriate to area classification	ALL	ALL	ALL
System group or class is correct	ALL	SAMPLE	NONE
Apparatus surface temperature class is correct	ALL	SAMPLE	NONE
Installation is correctly labeled	ALL	SAMPLE	NONE
There are no unauthorized modifications (including readily accessible lamp and fuse ratings)	ALL	SAMPLE	ALL
Apparatus is adequately protected against corrosion, the weather, vibration and other adverse factors	ALL	ALL	ALL
Earthing connections are permanent and not made via plugs and sockets	ALL	ALL	ALL
The intrinsically safe circuit is isolated from earth or earthed at one point only	ALL	ALL	NONE
Cable screens are earthed in accordance with the approved drawing	ALL	SAMPLE	NONE
Barrier units are of the approved type, installed in accordance with the certification requirements and securely earthed	ALL	ALL	ALL
Electrical connections are tight	ALL	SAMPLE	NONE
Point-to-point check of all connections	ALL	NONE	NONE
Segregation is maintained between intrinsically safe and non-intrinsically safe circuits in common marshaling boxes or relay cubicles	ALL	ALL	ALL
There is no obvious damage to apparatus and cables	ALL	ALL	ALL
No undue accumulation of dust and dirt	ALL	ALL	ALL

14.5 Maintenance and safe practices

The inspection and maintenance of installations in hazardous areas is necessarily circumscribed by constraints not necessarily applicable in safe areas. Special attention has to be given to the following issues.

The work permit system needs to be installed early on in the organizational procedures such that it becomes part of the culture of the organization and gets embedded in ethos of the people and organization. This requires that no work is done without proper authorization from the manager in charge or the person authorized by him. These rules should be so framed as to keep the safety of persons and equipment always uppermost. Toward this end, some guidelines are described in the following paragraphs.

14.5.1 Procedures for withdrawing equipment from service

The documentation shall clearly define the 'safety tag and lock-out procedures' for the equipment, when:

- It is in a dangerous condition
- It is being worked on
- It has not been completely installed
- It is out of operation for repair or alteration.

There are two main warning systems used in electrical industry for tagging electrical and non-electrical equipment to indicate isolation before work begins. These are,

1. Personal tag
2. Out-of-service tag.

It should be noted that an out-of-service tag is not an indication that the equipment or machinery is safe to work on. Wherever feasible, mechanical locking may be done after isolation.

No person may work on the equipment until a personal danger tag and, where practical, a lock has been placed on equipment isolation. Each person before working on equipment shall be protected by his/her personal tag.

Given hereunder are some general precautions, which must be supplemented with installation-specific procedures:

- To clearly understand and learn local procedure within the installation before starting the work.
- To employ correct testing procedures so that correct isolation, tagging and lock-out of equipment or that part of installation is done.
- Never rely on memory. Test before you touch.

Further, procedures should define the following.

14.5.2 Steps for isolation of equipment in hazardous area

Unless it is part of an intrinsically safe circuit, apparatus containing live parts and located in a hazardous area must not be opened (except as noted below) without isolating all incoming connections including the neutral conductor (Figure 14.1). Isolation in this context means:

- The withdrawal of fuses and links
- The locking-off of an isolator or switch.

It should be noted that if the continuing absence of a flammable atmosphere can be guaranteed by the authority responsible for that area, and a certificate is issued to this effect, essential work for which the exposure of live parts is necessary may be carried out subject to the requirements of the electricity regulations as applicable.

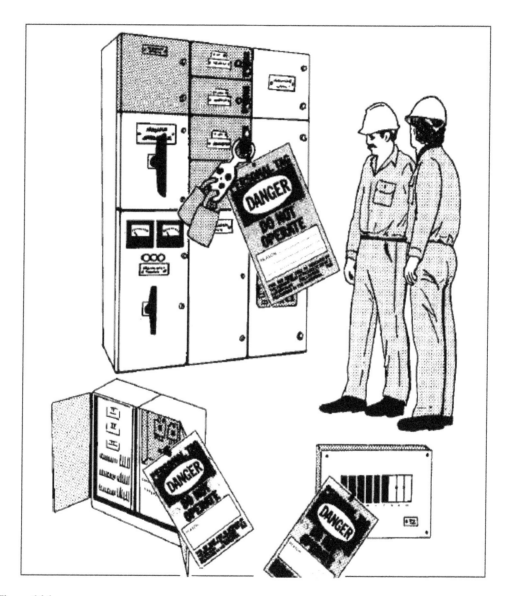

Figure 14.1
Typical tagging shown for safe working

14.5.3　Alterations and repairs to apparatus and systems

Any alteration, which it is necessary to make to apparatus or systems, shall maintain the integrity of the method of safeguarding.

In the case of certified or 'approved' apparatus or systems no alteration which might invalidate the certificate or other documents relating to the safety of the apparatus or system shall be made without appropriate approval which shall be obtained by the nominated person and shall be recorded.

In the case of apparatus or systems which rely on the techniques of segregation, pressurizing, purging or other methods of safe-guarding not covered by the previous paragraphs no alteration shall be made to the safety features without the permission of the nominated person.

Repairs carried out on apparatus or systems shall comply with the requirements of the previous paragraphs. For some types of apparatus, more specific guidance on repairs can be obtained from the manufacturer's documents.

14.5.4　Withdrawal of apparatus from service

Permanent withdrawal

When apparatus in a hazardous area is permanently withdrawn from service, the associated wiring should be removed. Where this is not practicable the exposed conductors shall be correctly terminated in an appropriate enclosure and identified at both ends.

Temporary withdrawal

When apparatus in a hazardous area is temporarily withdrawn from service the exposed conductors shall be either correctly terminated in an appropriate enclosure or solidly bonded together and earthed at one end. In order to avoid a static risk when apparatus is withdrawn from service and the wiring not removed the armoring and conductor shall be earthed at least at one end.

Withdrawal of intrinsically safe devices

When intrinsically safe apparatus or associated apparatus is withdrawn from service the associated conductors shall be bonded together and earthed at the point where intrinsically safe circuits are normally earthed and elsewhere correctly insulated from earth.

14.5.5　Safety assessment of testing

There is a legal 'duty of care' on all industrial personnel that adequate precautions are taken so as not to endanger life or investment during the course of work. The duty is reflected in the standards discussed, in that they require some assessment of work on electrical apparatus in order to ensure that the integrity of protection is not compromised.

Legislation concerned with electrical testing (at any voltage level) applies in some countries where work on any apparatus live requires justification. This include IS. It is easy to demonstrate safe working on the hazardous area side but less easy on the safe area side where no limiting systems are inherent.

It is clear that with IS apparatus, the installation must be covered by a system approach which justifies the safety by validating the interconnection of equipment.

In the same way it may be argued that any work in testing an IS loop is permitted if it is adequately covered by existing systems descriptive documentation. If this is the case then no further precautions need be taken from the point of view of explosion protection because safety has been justified.

If, however, work on an IS loop is not covered by existing documentation, it may be created or ignored. The formal creation of documentation to demonstrate safety is a daunting prospect, which sounds expensive and time consuming. It is hoped that the following discussions will allay those fears and open a way forward to a logical and manageable approach.

The act of ignoring the requirement may result in unacceptable consequences.

14.6 Fault-finding – safety ensured

Having safely isolated the system, equipment or apparatus, it is important that a proper system of fault location be institutionalized. There is no right or wrong way to fault-find on instrument loops and other systems. There are however, safe and potentially unsafe ways, which are worthwhile considering.

The object of this section is to discuss the approach to fault-finding in such a way that guidance can be given on what shall and shall not be done from an explosion-protection safety point of view.

14.7 Insulation testing in hazardous area

The majority of fires and ignitions are triggered by faulty electrical insulation and hence it is essential that integrity of this vital component of electrical installation be maintained (at any cost). Any degradation will not only result in a loss in terms of material and costly repair but also in down-time of production and resultant financial losses. Hence, this section is devoted to the testing and monitoring of the insulation with special reference to hazardous atmospheres.

14.7.1 General

During normal maintenance or fault diagnosis, it may be necessary from time to time to carry out insulation testing. If the test is to be carried out using conventional insulation or pressure test equipment the locations where the test is to be applied and/or *where incentive sparking may occur must be declared free from a flammable atmosphere* by the responsible authority and guaranteed as remaining so for the duration of the test. A certificate to this effect shall be issued.

Insulation testing in hazardous areas should, if practicable, always be carried out in the continued absence of a flammable atmosphere, but where this is not possible any instrument used for testing must be intrinsically safe. It should, however, be appreciated as explained below that with the majority of such intrinsically safe insulation testers danger can arise if the limitations of the certification are not fully observed.

It should be noted that the instruments referred to above be intrinsically safe only in the sense that no danger can result from sparks produced by current generated within the instrument. The effect of external loads, however, cannot be predicted and an intrinsically safe charging current applied to an energy storage circuit such as a length of cable can give rise to incendive discharges. This difficulty has been recognized and the instrument is provided with a discharge circuit, which, if the instrument is used correctly, ensures that the circuit under test is left in a safe condition. This provision, however, gives no protection against a possible discharge within the circuit during the application of a test.

To guard against the dangers that may arise the following precautions shall be observed when using an intrinsically safe insulation tester:

- The 'test' button shall not be operated until the instrument is properly connected to the circuit under test.
- 'Touch' or 'flick' connections are not permitted.
- The instrument **shall not** be disconnected for at least 30 s after releasing the 'test' button.
- Tests on circuits with exposed conductors or connections in a hazardous area and which run between the hazardous and a non-hazardous area should be carried out from the point of exposure and not from the non-hazardous area.

Insulation testers shall not be used on any intrinsically safe system.

14.7.2   Intrinsically safe systems

The insulation testing of intrinsically safe circuits is necessary to confirm that they are earthed at one point only or insulated from earth throughout, whichever of these conditions is required by the original design. The concept of IS permits unprotected sparking to occur in hazardous areas and permits the use of lightly protected cables often carrying several intrinsically safe circuits. Thus the indiscriminate use of a high-voltage test device with a high-energy capability can cause incentive sparking in the hazardous area at normally unprotected contact points or through the invasion of other intrinsically safe circuits resulting from insulation breakdown in the cables.

Furthermore, as most intrinsically safe systems operate at low voltages, a test device can cause damage to the safety features of the apparatus in the system.

In view of the above limitations, insulation testing of intrinsically safe systems or circuits shall only be carried out using a test device specifically approved for connection to such circuits.

14.8   Earthing in hazardous area

The earth loop impedance requirements of electrical circuits to ensure automatic disconnection in the shortest practicable time under fault conditions are detailed in an earlier section. The alternative methods of measurement, which may be used to assess earth loop impedance, are briefly described hereunder. Subsequent clauses give guidance on the application of these methods during 'initial' and 'periodic' inspections and of other related checks at these times.

Special problems associated with the testing of intrinsically safe systems are considered in the clause below.

14.8.1   Assessment of earth loop impedance

According to the circumstances, one or more of three methods is available, as follows:

(i) *Earth loop impedance*: This measurement shall be made using an instrument producing an alternating current of not less than 15 A. The measurement should embrace the total loop including the effective impedance of the supply transformer. Where this is not practicable, allowance shall be made for the impedance of the transformer and any other impedance in the loop not included in the measurement.

(ii) *Earth loop resistance*: This measurement shall be made using an intrinsically safe DC ohmmeter. The measurement should embrace the total loop including the resistance of the supply transformer winding. When this is not practicable it shall be made from a defined point in the circuit as close to the supply transformer as possible and shall include at least the whole of the final sub-circuit. Allowance shall be made for any resistance in the loop not included in the measurement.

(iii) *Earth return resistance*: This measurement shall be made using an intrinsically safe DC ohmmeter. The measurement shall be made from the apparatus enclosure to the transformer-neutral earth point. Where this is not practicable, the measurement should be made from the apparatus to a defined earth reference point if the return resistance between the reference point and the transformer neutral earth point is known.

It should be appreciated that this method assumes the resistance of the live conductors does not change.

14.8.2 Earth checks during 'initial' inspection

All earthing connections, including those of any supplementary earthing conductors, shall be checked to ensure that they are clean and tight.

The earth loop impedance of all circuits operating at or above 240 V, and any other circuits decided upon by the nominated person(s), shall be measured in the manner specified in the clause above and the values shall be recorded.

If it is anticipated that environmental conditions will prevent the use of the above test on subsequent 'periodic' inspections, then either or both of the alternative measurements described in the above clauses shall also be made and the values shall be recorded. The resistance values should be consistent with the impedance values obtained from the tests specified in initial startup. These resistance values shall be used as the basis for comparison with the corresponding resistance measurements made at subsequent 'periodic' inspections.

14.8.3 Earth checks during 'periodic' inspection

All earthing connections, including those of any supplementary earthing conductors, shall be checked to ensure that they are clean and tight.

The earth loop impedance of a representative sample of circuits, decided upon by the nominated person(s) should be measured in the manner specified in the clause above.

Where environmental conditions prevent the use of the above test, the sample measurements shall be made by either of the alternative methods specified in above clauses if the corresponding measurement was made during the 'initial' inspection. If there is a significant increase in the measured value compared with the corresponding 'initial' value, steps shall be taken to ascertain the reason for the change and further measurements shall be made on other circuits of which the sample is representative.

14.8.4 Earth checks of intrinsically safe systems

The requirements for 'initial' and 'periodic' inspection in the above clauses apply for intrinsically safe systems but the following *special requirements* shall also be observed.

Earth loop impedance/resistance of intrinsically safe systems

All mains-powered apparatus associated with intrinsically safe systems shall have their earth loop impedance/resistance measured in the manner specified above. But in situations where the earth is also directly connected to the intrinsically safe circuit (e.g. in some associated non-hazardous area apparatus) the test shall be carried out only if the earth connection to the intrinsically safe circuit is first removed. This precaution is necessary to prevent potentially dangerous voltages or currents from the test device being applied to the intrinsically safe circuit. Where it is not possible to disconnect the common earth from the intrinsically safe system, the test shall be carried out only at a low voltage/current using a device specifically approved for the purpose.

Care shall be taken to ensure that the tests do not affect other intrinsically safe circuits sharing a common earth connection.

Where this may occur, those circuits shall first be disconnected from the common earth, unless the test is carried out using a low-current device specifically approved for the purpose.

It is to be noted that earth loop impedance measurements are not required on the hazardous area part of intrinsically safe circuits because earthing in these circumstances is not associated with the operation of protection devices.

Earth continuity of intrinsically safe circuits

All earth conductors connected to intrinsically safe circuits shall have an earth continuity measurement to the earth reference point for that circuit. The measurement shall be made only with a test instrument specifically approved for the purpose.

The tests required in BS 3807 may be carried out by any responsible organization but it is usual to use the facilities of the SABS test house. In all cases, a certificate of test shall be obtained from the testing authority.

14.9   Handling 'fall out of fire and electrical shock'

No matter how well various safe working practices are implemented and codes and standards followed in installation, operation and maintenance, an occasional hazard of fire and electrical shock cannot be ruled out. Hence we are giving below a few 'safe practices' which if followed will enhance the chances of survival of victims.

A CO_2 extinguisher should be used to extinguish electrical fires.

Generally fallout of any fire could be small or large burns; first- to third-degree burns, and hence let us devote a few lines to understand the precautions that can be taken. This is in no way a substitute for regular statutory training and doctor's advice but to serve maintenance and operation persons in time of emergency.

The seriousness of a burn depends on two factors:

1. The extent of the burned area
2. The depth of the burn.

Shock can be expected from burns involving 15% or more of the body. Burns involving 20% endanger life. Without adequate treatment, burns of over 30% are usually fatal. The depth of the injury determines whether it is a first-, second- or third-degree burn.

First-degree burns are the mildest. Symptoms are slight pain, redness, tenderness and increased temperature of the affected area.

Second-degree burns are more serious. The inner skin may be damaged, resulting in blistering, severe pain, some dehydration and possible shock.

Third-degree burns are the worst of all. The skin is destroyed, and possibly the tissue and muscle beneath it. The skin may be charred, or it may be white and lifeless (from scalds). After the initial injury, pain may be less severe because of the destroyed nerve ends. The person may feel very cold and shivery. Some form of shock will result.

Probably the most important aspect is the extent of the burned area. A first-degree burn covering a large area could be more serious than a small third-degree burn. Sunburn, for example, ranging from mild to serious, is easily obtained, particularly if you are not accustomed to the exposure. If you were to fall asleep while sunbathing, second-, or even third-degree burns of a possibly fatal nature could result.

The most *effective immediate treatment of burns* and of pain is to immerse the affected area in cold water or to apply cold compresses if immersion is impracticable. Cold water not only minimizes pain, but also reduces the burning effect in the deeper layers of the skin. Gently pat, dry the area with lint-free cloth or gauze. Aspirin is also effective for the relief of pain. Continue treatment until no pain is felt when the burn area is exposed to the air.

Burn victims require large amounts of water, which should be slightly salted. Because of the nature of the injury, most burns are sterile. The best treatment for uninfected burns, therefore, is to merely protect the area by covering it with the cleanest (preferably sterile) dressing available. Never apply ointments to a burn. Do not attempt to break blisters or to remove shreds of tissue or adhered particles of charred clothing. Never apply a greasy substance (butter, lard or petroleum jelly), antiseptic preparations or ointments. These may cause further complications and interfere with later treatment by medical personnel.

Electrical shock

The isolation of victim from the live electrical part can be done quickly and safely by carefully applying the following procedures:

Protect yourself with dry insulating material.

Use a dry board, belt, clothing or other available nonconductive material to free the victim from electrical contact.

Do not touch the victim until the source of electricity has been removed.

Once the victim has been removed from the electrical source, it should be determined, if the person is breathing or not breathing, and the following first aid can be administered.

First aid for electrical shock includes the following actions:

- Remove the victim from the source of the shock.
- Check the victim to see if the person is breathing.
- If the victim is not breathing, give artificial ventilation. The preferred method is mouth-to-mouth.
- CPR may be necessary if the heartbeat has stopped, but do not attempt this unless you have been trained in its use.

Obtain medical assistance as soon as possible.

15

Fault-finding and testing

15.1 Fault-finding

There is no right or wrong way to fault-find on instrument loops and systems. There are however, safe and potentially unsafe ways that are of the greatest importance to consider.

The object of this section is to discuss the approach to fault-finding in such a way that guidance can be given on what shall and shall not be done from an explosion-protection safety point of view.

15.2 Fault-finding routine

It is usual for a fault to be investigated by following a logical routine, an example of which is shown in Figure 15.1.

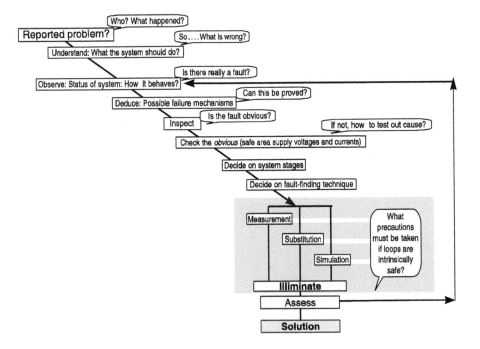

Figure 15.1
Possible fault-finding routine

The approach is described here such that the areas where aspects of safety must be considered can be clearly identified. This is one of a number of possible ways of proceeding; it is not a recommendation of any standard but is distilled from good engineering practice and experience in very general terms.

1. A problem is reported on the plant.

    - Who reported the fault?
    - Will the interpretation of the fault be accurate?
    - What actually happened?
    - Is there really a fault?

2. An understanding of how that part of the plant should perform is necessary.
3. Observation of the current status of the system and how it now behaves.

    - How does this compare with what was originally working?

4. Reason what could cause the observed effect.

    - What has gone wrong? What part of the system is likely to have failed to cause this? Can this be proved?

5. Is the system explosion-protected?

    - What precautions are necessary?

6. Visually inspect to check the obvious.

    - Confirm explosion-protection types.
    - Supply system failure, obvious damage and visible mechanical problems.

7. Decide on 'system' stages.

    - Understand the breakdown of interconnected modules.

8. Decide on a logical fault-finding technique or regime and implement it. Options are probably:

    - Measurement
    - Substitution
    - Simulation or
    - Some combination.

9. Assess the behavior of each stage. Eliminate the stages logically. Narrow down the possibilities to home in on the problem and therefore the likely solution.

    - Does this cure the problem?
    - If not go back to Stage 3.
    - If yes: Effect a permanent repair. Document the failure.

10. Re-commission the system.

    - Inspection is required in order to verify; safety is not compromised.
    - Return to an operational mode.

The concern with Stages 5–8 is that the devised test must be safe in that it cannot involve the compromise of any safety system, unless under adequately controlled conditions. Such systems are concerned with plant operation safety as well as explosion protection.

The modularity of instrument loop system components enable substitution techniques to be used efficiently in order to cure the fault quickly, thereby minimizing 'down-time'. As systems become more complex, however, it is not possible to substitute complete systems as this is too costly and takes too long. This aspect of maintenance should be considered during the design of installations.

Most large organizations operate a formal fault reporting system into which the recommendations or requirements of the codes of practice can be built into. The inspection requirement should become an automatic reaction to any work done on hazardous area apparatus. The recording of the actions taken in the fault-reporting system required by the code of practice can be incorporated into the plant documentation. Many think that there must be separate and therefore duplicated systems but there is no evidence in any of the known codes to suggest that these cannot be effectively integrated. Compliance with ISO 9000 documentation encourages integration and therefore distribution of useful and safety-related information.

15.3   Safety assessment of testing

There is a legal 'duty of care' on all industrial personnel that adequate precautions are taken so as not to endanger life or investment during the course of work. The duty is reflected in the standards discussed, in that they require some assessment of work on electrical apparatus in order to ensure that the integrity of protection is not compromised.

Legislation concerned with electrical testing (at any voltage level) applies in some countries where work on any apparatus live requires justification. This does include IS. It is easy to demonstrate safe working on the hazardous area side but less easy on the safe area side where no limiting systems are inherent.

It is clear that with IS apparatus, the installation must be covered by a system approach which justifies the safety by validating the interconnection of the equipment.

In the same way it may be argued that any work in testing an IS loop is permitted, provided that it is adequately covered by existing systems descriptive documentation. If this is the case then no further precaution need be taken from the point of view of explosion protection because safety has been justified.

If, however, work on an IS loop is not covered by existing documentation, it may be created or ignored. The formal creation of documentation to demonstrate safety is a daunting prospect, which sounds expensive and time consuming. It is hoped that the following discussions will allay those fears and open a way forward to a logical and manageable approach.

The act of ignoring the requirement may result in unacceptable consequences.

15.4   Test equipment

There is a great variety of test equipment available on the market. It may seem obvious to point out that test equipment for hazardous area use will be either certified or uncertified.

If equipment is uncertified then, interpreting the standards and codes of practice, it cannot be carried into the hazardous area unless:

- It remains electrically isolated from any source of supply including that from internal energy-storing devices, such that sparking or heating through power dissipation cannot occur under any circumstance and
- The temperature of any part of the equipment is and remains below the ignition temperature of the expected hazard or

- There are acceptable levels of safety in the assurance that there will be no hazardous atmosphere present for the duration of the attendance of the equipment. If this is the case then operational electrical equipment may be taken into a designated hazardous area. It should also be ascertained that if the presence of the hazard were to be detected, then adequate allowances would be made for the equipment to be suitably isolated and/or removed.
- IS-certified test equipment may take the form of specifically designed test instruments such as that made by Fluke or AVO.

Older IS test equipment available, such as that manufactured by Metrohm, Megger, and Edgecome Instruments, was certified to BS 1259: 1958 (which is now obsolete). This standard did not consider the consequences of interconnection of such equipment into an existing 'certified system'. They are not compatible under arrangements where systems considerations are fundamental to safety, without some further assessment.

There are very few test equipment devices certified to the equivalent of IEC 60079-11 because the apparatus cannot be provided with a 'blanket' systems certificate. This is a good example of where the systems descriptive document is most helpful.

The document can be written in such a way that assessment of loops on a plant can list which loops may be connected to the test equipment. More importantly the documentation must state how this is to be done and with what limitations. This is not difficult to do and can be incorporated into design and plant safety documentation with little extra effort.

The test equipment may have a safety description that is non-energy storing and a global document can state that this may be added into any loop in the current measuring mode without prejudice to the existing certification.

Voltage measurement may be performed through a high series resistance and with a low-fault voltage contribution; it can be argued that the energy output is sufficiently low to be added temporarily into any loop without cause for concern.

Typical safety description values for IS multi-meters on the voltage ranges are 2 V 2 mA. It is not difficult to justify this power/energy contribution level.

15.4.1 Insulation testing

Insulation testing is a requirement of all standards. The test used is for the circuit to withstand 500 V AC rms for 1 min, as discussed in the installations section. This is an internationally adopted IEC recommendation. Testing at this voltage in safe area conditions, i.e. where there is no risk of the presence of ignitable vapors, is acceptable with some obvious precautions taken by the person performing the testing owing to the risk of shock at the high-voltage level.

The cable capacitance of relatively short runs at this high voltage can store a significant charge. Sudden discharge of the stored energy can cause ignition in a surrounding explosive atmosphere. The use of high-voltage tests in hazardous atmospheres must therefore be done under well-controlled conditions.

Insulation testing performed on cables running through the hazardous area should be undertaken with a procedure that minimizes the risks. Such a procedure may be to inspect for damage throughout the length of the cable prior to the test being performed. If no damage is visible then it is likely that the cable insulation is acceptable. If it fails the failure will be within the sheath and it will be unlikely to cause an explosion.

Where the cable cannot be inspected because of the routing, it is preferable to do a low-voltage test first. The application of a high voltage after this is acceptable.

15.4.2 Low-voltage insulation test

Low-voltage testing may be done by the technique shown in Figure 15.2. The principle is one of applying a low-voltage source and monitoring the current taken with an adequately sensitive measuring instrument. Most multi-meters have a resolution to 1 µA, which is acceptable for this sort of testing.

A variable supply, which can be raised from 0 V to some higher value, is useful for this duty.

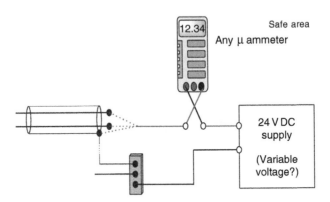

Figure 15.2
Safe area testing at low voltage

Where a cable of an instrument loop runs through a hazardous area, as in Figure 15.3, the procedure for carrying out this test, for example, on a conventional two-wire transmitter system, is as follows:

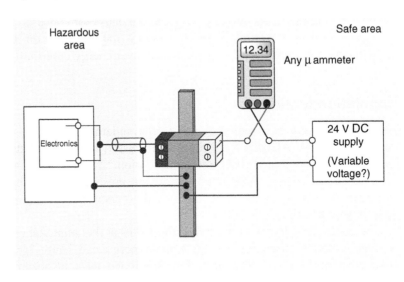

Figure 15.3
Low-voltage insulation testing with a barrier

1. Disconnect the field cables to be tested from its normal associated apparatus interface.
2. Disconnect the cabling at the field end in the hazardous area. Ensure the conductors are left open circuit from each other and from any earth; temporarily insulate.

3. Apply the highest voltage from the power supply to the safe area terminal of the barrier channel without permitting measurable leakage currents through the barrier. The supply will need to be set to a voltage just less than the $V_{WORKING}$ of the barrier channel used. Note the reading of the ammeter (it should be zero).

4. Connect the field cable and note any change in the reading. Core-to-core and each core-to-screen measurements are normally required.

5. If satisfactory, bridge the input terminals (+ve and −ve) of the transmitter.

6. Connect one of the cores of the cable to the bridged terminals of the transmitter and note the meter reading.

7. Calculate the system leakage resistances by Ohm's law.

The insertion of a 28 V safety description barrier (as shown) will enable low-voltage measurements of cable insulation resistance as well as apparatus 'structure to electronics' insulation of hazardous area certified apparatus to be checked while being installed. The justification for safety is explained in Section 15.5.

15.4.3 The 500 V AC test

The high-voltage test to instrumentation should be performed as shown in Figure 15.4. This is preferably carried out in the safe area. If necessary to do so in the hazardous area, please refer to Section 15.4.2.

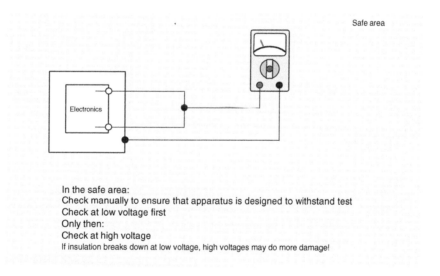

In the safe area:
Check manually to ensure that apparatus is designed to withstand test
Check at low voltage first
Only then:
Check at high voltage
If insulation breaks down at low voltage, high voltages may do more damage!

Figure 15.4
High-voltage testing of instruments

The instruction manual and certification should first be consulted to ascertain that the instrument would withstand the test. Some older devices do not have the same quality of isolation that modern equipment can withstand. Some simple apparatus devices such as load cells require a careful evaluation before proceeding with the test. The terminals of the apparatus should be short-circuited. If breakdown occurs in part of the circuit then further damage could be done to the electronics.

Check at low voltages first before applying the full test. Record the results for future reference.

15.5   Use of uncertified test apparatus

Where there is a need to perform testing on instrument loops, which enter the hazardous area through interfaces, several precautions must be taken. Equipment introduced on to the safe area side of an interface must not compromise the conditions of being supplied by or generating greater than the U_M rating of the interface (normally 250 V AC). Additionally, the device should be fed from a double-wound mains transformer. Oscilloscopes with CRTs operate at higher than 250 V, say 30 kV EHT, and so there are schools of thought which suggest that surge protection should be fitted upstream of the barrier. The use of a LCD display or a PC-based system will reduce the risks.

Barrier characteristics are simply determined by voltage, current and resistance relationships. Provided that the working voltage is not exceeded and the series resistance characteristics are allowed for in the measurement, it is acceptable to interpose a barrier between an uncertified item of test equipment and the hazardous area loop.

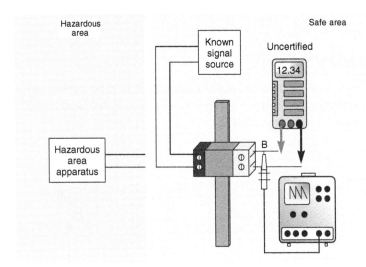

Figure 15.5
The use of barriers with test equipment

In Figure 15.5, the use of uncertified safe area equipment may be used through a barrier. The effect of the introduction of the barrier may be observed and measured prior to connection to the hazardous area circuit. This comparison will enable the measurement to be corrected by any effect observed.

If the barrier (or isolator) is added into a loop for the specific purpose of making some measurement then its inclusion must be assessed and deemed safe (by being adequately covered by a systems certificate). With some thought this is not difficult to do. In the example in Figure 15.5, a current and waveform is being measured on an oscilloscope. The interface is correctly system-certified together with the hazardous area apparatus. The barrier introduced has a safety description of 1 V 100 mA. It can be argued that if a fault occurred on the barrier in the safe area, the total energy contribution would be the equivalent of the maximum permitted by 'simple apparatus'. Simple apparatus may be added into any loop without the need to re-certify the system. The input is non-energy storing and therefore cannot jeopardize safety in that way.

If, for example, the waveform being monitored was the propagation response of the isolating interface or the switch-on surge characteristics of the loop, the resistor would be chosen such that the sensitivity of the oscilloscope was suitable. Refer to Figure 15.6.

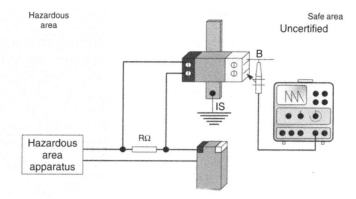

Figure 15.6
Current measurement waveforms using an oscilloscope

Suppose that the expected current was 30 mA then a 1 Ω resistance would develop 30 mV. The $R_{\text{E-E}}$ of the barrier at 10 Ω would not prejudice the signal into the high impedance of the scope. Leakage currents would not attenuate this voltage developed if the MTL 751 barrier were used.

15.5.1   Multi-meters

Hand-held multi-meters, which are IS-certified, are increasingly available but are expensive compared to non-certified devices. Whether they are certified or not there are certain modes in which they must not be used. Figure 15.7 shows a meter bridging the interface. This should not be done under any circumstance with the hazardous area wiring connected.

If the meter is certified, it may be operated in the hazardous area. Connection to circuits must be assessed to ensure compliance with the systems certificate of the loop. If a certified meter is not used, the person performing the tests is fully liable for the consequences of his actions.

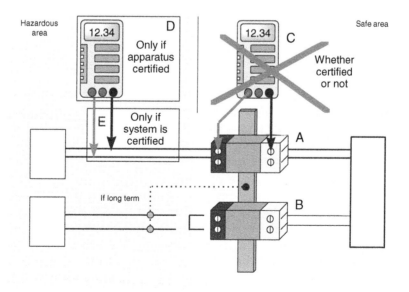

Figure 15.7
Use of multi-meters

15.5.2   Testing using standard IS apparatus

The lack of range and raised cost of purpose-built certified IS test equipment supports the sensible option that instrument engineers can look to the existing range of certified apparatus available and realize its usefulness in testing and fault-finding functions.

IS apparatus can be used for safe area applications and so there is no reason why IS apparatus cannot be used for general test purposes. Figure 15.8 shows how loop-powered indicators are especially versatile with very low insertion loss of loop voltage and are certified as non-energy storing. Some indicators may be externally powered by a second loop (see the applications section). These devices are capable of being added into loops with insertion losses in the order of a few millivolts.

The use of IS apparatus as permanent test equipment must be documented and identified. It would need to be inspected more rigorously and at shorter intervals than normal IS apparatus because of its specialized use.

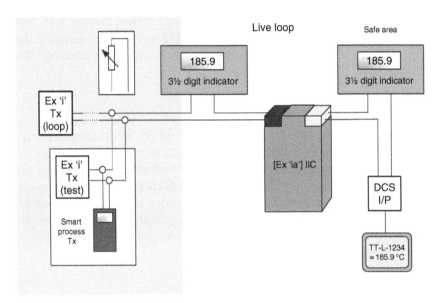

Figure 15.8
Using standard certified apparatus to fault-find

A smart transmitter with a hand-held controller becomes a very practical test set up for calibrating analog loops. It offers high accuracy and resolution with good stability. It is easy to set up and operate. Safety is unaffected, provided that initial selection of the chosen test apparatus is compatible with all installed loops. This is of course a luxury to have. The same effect can be achieved with a suitable variable resistance, claiming 'simple apparatus' for safety purposes.

There are many and devious ways of using apparatus to test and fault-find on hazardous area apparatus and loops. The rules are that the test arrangement should be demonstrably safe, being covered by system certification principles and should be worked on by competent persons, with adequate training and supervision. The precise way forward must be left to the individual with the proviso that the standards use the recommendation to 'seek expert advice' where there is any doubt on the acceptability of a proposed test technique.

In many cases initial fault-finding can be done with a simple equipment. This is to be recommended. Open circuits and short circuits may be applied on the hazardous area side of any powered analog loop in order to create large current swings for the purpose of loop identification and simple testing. Such a test is most unlikely to cause any damage because current limiting is inherent in IS circuits. Needless to say that the safe area equipment must be placed in a state where the under- or overrange signals will not cause havoc (for example, in a control or shutdown system).

Monitoring of plant bonding may also be performed with IS apparatus. In Figure 15.9, an interface is used as an IS power supply. This would need to be an isolating interface such as a solenoid/alarm driver, which is continually energized to provide a current through an earth loop.

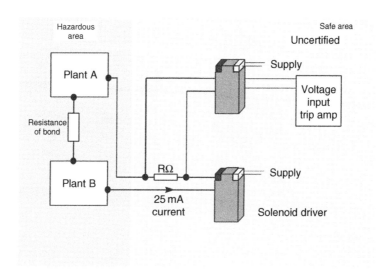

Figure 15.9
Monitoring an earth bond between two structures

The sensing system monitors the level of current, which falls if the bond breaks. This arrangement can be implemented in many ways, depending on the requirements of what is to occur if the bond resistance goes above certain levels. Normally a voltage-input trip amplifier will operate an alarm. The system can be used for preventing a tanker loading system from operating without adequate earthing connection for static hazard risks.

The current injection and monitoring principle is a useful tool in fault-finding. The system certificate provided with the solenoid driver for simple apparatus use normally covers this application because the trip amplifier and the connections through earth are deemed simple apparatus. One connection to earth is permitted with isolating interfaces.

15.6 Interface testing

In this section the testing of the different types of IS interface will be examined. The testing discussed concerns functional operation, as the integrity of the safety protection cannot be verified without access to internal components.

15.6.1 Barrier testing

It is relatively easy to test barriers. The practical approach is to measure the R_{E-E} (the channel resistance) (see Figure 15.10).

This is quoted by the manufacturer as a maximum resistance; it verifies that the safety fuse is intact. It can only exceed this value by being at 'infinity' or open-circuit, owing to the fuse being blown. In practice it is likely to be a few ohms less than the maximum R_{E-E} because component tolerances are be taken into account. A $300\,\Omega$ barrier will have a maximum end-to-end resistance of $340\,\Omega$ but the expected measure value will be approximately $320\,\Omega$. This may be done with any low-current measuring device such as an AVO or almost any digital multimeter.

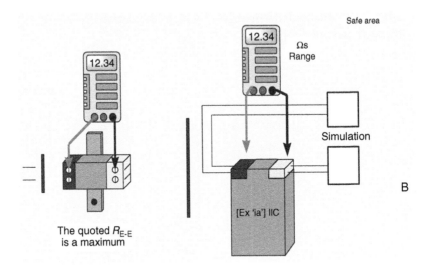

Figure 15.10
Interface testing

Note: Testing must not be carried out on barriers using instruments designed to test for low resistance continuity by applying a high current.

On some barriers there are external (non-safety) fuses which are designed to blow in advance of the safety fuse. It may be this that has been damaged and replacing it with the correct type from the manufacturer will reclaim the barrier. Occasionally, the internal safety fuse may blow instead of the external fuse, owing to the unfortunate overlap mismatch of characteristics.

Diode failure cannot be easily verified. The likelihood of failure is acceptably low. A current may be injected to the safe area (provided that it is below the level of the safety fuse rating) to check the approximate voltage of the Zener diode chains in parallel. Individual diodes or chains cannot be tested because the barrier is encapsulated. Barriers are therefore not repairable.

Damage to barriers is easy to sustain and is most likely to occur during commissioning. They are highly reliable once correct operation has been established.

The cause of barrier fuses blowing will not be owing to a faulty barrier but most likely because of faults, albeit long term or short term on other parts of the associated instrument loop, possibly on either side of the barrier.

Note that periodic IS interface removal and testing is not recommended by manufacturers and by some standards/codes of practice.

15.6.2   Isolator testing

It is not possible to test isolating interfaces by any external measurement in the same way that barriers can be.

Certain obvious things can be checked, such as external (non-safety) fuses, if fitted. The application and wiring should be confirmed as correct. Thereafter, the supply current drawn can give some indication of possible problems.

The easiest way of testing is to simulate the function as depicted in Figure 15.10. It is recognized that this is not always practical as it may take some considerable time to set up and perform. Substitution may well be the quickest way of proving a loop. There should be no problem in substituting a device of the same manufacturer and model number, although it is recommended that procedures encourage the routine of checking the salient details.

However, care should be taken to establish that if a unit from a different manufacturer is used on an installed loop, that it is 'no less safe' than that of the original device. In practice, this too should not cause a problem because most isolator manufacturers use industry standard safety descriptions for the common applications.

It is useful to know that the replacement interface is new and is known to work properly. Many hours of time have been wasted by a damaged stray interface being returned to a spares stock and drawn on the false assumption that it is new.

Isolating interfaces may be returned to the manufacturer or his agent for testing. If the device is electrically damaged, the manufacturer will be able to discover why. If there is a problem with the unit then the manufacturer will be grateful for the opportunity to put it right and to save face. If the problem is concerned with the application then depending on the damage done, as dictated by the state of internal protection devices, the manufacturer can guide the user as to what to look for in sorting out the cause.

If the change from one device type to another is temporary, then documentation of the change will be short term. Where the change is permanent, loop drawings, other schedules and records will require updating with new details, termination numbers, etc.

Substitution of barriers to isolators (and vice versa)

From a safety point of view, provided that the chosen interface devices have the same safety description and the hazardous area apparatus permits the other device to be used, unlimited by certification or special conditions, the only constraint is an operational one. The equipment in the safe area must be able to operate with the alternative device. There are not many instances where exchange is of benefit but this is a 'frequently asked question'.

15.6.3   Associated apparatus

Where 'associated apparatus' is manufactured and certified specifically to operate with hazardous area mounted apparatus, the ability to test is dependent on possessing adequate knowledge of safety and operational parameters. Some simulation can be attempted if the manufacturer's manual gives enough detail. The interfacing technique may be via shunt diodes or by isolation. If the documentation calls specifically for an IS earth, then the technique will be using the equivalent of internal barriers. Beware that there is some associated apparatus that requires an 'earth' connection but the purpose is not clearly defined. Some clarification of this should be sought.

Intelligent and informed guesses can be used to infer a basic mode of operation in order to discern if there is a problem. Otherwise the manufacturer's 'informed approach' must be sought.

15.6.4   Physical damage to apparatus

If IS apparatus suffers physical damage then repair can only be undertaken by the original manufacturer. In repairing the device it must be restored to the original condition on which the application for certification was made. Extensive damage will pose a risk that subtle changes may have occurred in components, which cannot therefore guarantee them to perform to their original design specification. The whole sub-assembly may be treated in this way and the manufacturer/user commercial agreement may not run to the cost of repair. Replacement is therefore the only option.

Barriers are encapsulated and are therefore not repairable. Isolator cases are sacrificial such that any attempt to remove the box results in damage seen during initial return inspection. Other apparatus has PCBs that are coated with a conformel coating varnish where it will be apparent if any attempted repair has been undertaken since manufacture. These techniques are useful in preventing the uninitiated from 'unauthorized interference'.

15.7 Certified apparatus

The testing of IS-certified apparatus which can be performed on site whilst installed will not reveal the condition or status of the explosion protection. Testing is not a specific requirement of the codes of practice where the test is not able to check the protection integrity. Unnecessary testing is not substantially useful and can cause more harm than good.

There are certain safety-related tests to which apparatus may be subjected, such as insulation tests. It is debatable as to how often this should usefully be done. The 'self-revealing fault' nature of instrumentation is best relied upon to indicate problems, particularly where the alternative approach of removing and testing actually poses a greater risk than damage or incorrect reinstallation may result in. If the apparatus is correctly functioning then the likelihood of the IS protection having failed is accepted as highly unlikely, owing to their close relationship in the design.

The 500 V AC insulation test is a normal requirement and may be tested for on a type/sample basis. The code of practice gives guidance on this. The method of execution is discussed in Section 15.4.1.

15.7.1 Simulation testing

Where required, simulation testing should be realistic. It is highly recommended that the IS apparatus be functionally tested when connected into actual operating conditions and including appropriate interfaces. The operational and safety characteristics must not be confused, when ascertaining correct operation. Diagnostic equipment provided by the manufacturer should be checked for use in hazardous areas. Older battery-operated apparatus has limits of certification to IIB apparatus groups or possible Ex 'ib', such that connection in Zone 0 circuits is not permitted.

15.7.2 System testing

System testing invariably involves the simultaneous operation of many safety systems of which explosion protection is just one. Care should be taken to understand the various scopes of these preventative measures within the context of pieces of equipment working together. In the modification of one aspect of an installation, due regard must be given to its effect on other parts of the installation. Apparatus and systems may be evaluated to comply with other safety assessments. For example, those carried out to the IEC 61508 standard.

IEC 61508 is entitled, 'Functional Safety of Electrical/Electronic/Programmable Electronic Systems'. Its intention is to provide an internationally accepted way of determining overall reliability and plant safety. It operates by the quantification of safety integrity levels desired on an installation and matched to the 'SIL' values awarded by the design of equipment proposed for use on the installation.

Substitution of equipment must not reduce the design safety levels. This is unlikely at the sub-function levels of IS apparatus, unless the instruments carry a TUV safety approval or the equivalent.

15.8 IS apparatus repair procedure

The IEC has recently adopted the BEAMA/AEMT (British Electrical and Allied Manufacturers' Association/Association of Electrical Machinery Traders) document entitled: Code of practice for the repair and overhaul of electrical apparatus for use in potentially explosive atmospheres (other than mining applications or explosive processing and manufacture).

This has been published as IEC 60079-19: 1997: Repair and overhaul for apparatus used in hazardous atmospheres (other than mines or explosives).

The intent of the original document was to guide the industry when repairs to certified apparatus were to be undertaken. Both repairers and users needed to agree on terms of reference, and the trade associations were well placed to promote and mediate in the preparation of this code.

Whilst it was originally and primarily designed to deal with the mechanical approach to repair on motors and high-power applications, it does give useful guidance on the documentation and identification of repaired apparatus.

There are general conditions applying to all types of protection and then further specific recommendations for each of the types.

15.8.1 Definitions

Some important definitions are:

- *Serviceable condition*: A condition which permits a replacement or reclaimable component part to be used without prejudice to the performance or explosion-protection aspects of the apparatus, with due regard to the certification requirements as applicable, in which such a component part is used.
- *Repair*: An action to restore a faulty apparatus to its fully serviceable condition and its compliance with the relevant standard.
- *Overhaul*: An action to restore to fully serviceable condition, an apparatus which has been in use or in storage for a period of time but which is not necessarily faulty.
- *Maintenance*: Routine actions taken in accordance with the recommendations of ... (a relevant Code of Practice) to preserve the fully serviceable condition of the installed apparatus.
- *Component part*: An indivisible item from an assembly of such parts that form an apparatus.
- *Modification*: A change to the design of the apparatus that may affect material, fit, form or function.

15.8.2 General conditions

The general conditions are divided into the interests of the three organizations, which may be involved or affected by the repair.

1. *The manufacturer*: The code recommends that documentation and data suitable for the repair and/or overhaul should be available. Additionally, spare parts, which affect the means by which the apparatus complies with the appropriate safety certification, must be identified.

   Where sufficient and accurate information on apparatus is unavailable, for whatever reason, then the 'repair' cannot be undertaken because the repairer cannot know what aspects may affect the safety of the apparatus.

2. *The user*: The code recommends that the user should acquire certification and related documentation as part of the original contract of supply. The user should notify the repairer of the fault, the history of the apparatus and any special requirements of the original specification. (Re-installation after repair must follow the appropriate Code of Practice when the apparatus is to be re-commissioned.)

   The repairer needs to know what he is dealing with.

3. *The repairer*: In essence, the code advises that 'the repairer's attention be directed to the need to be informed of, and to comply with the relevant explosion-protection standards. . . . Training and refresher training should be provided to personnel involved in repair work.

   Recommendations and reminders of other issues affecting testing of apparatus, spare part authenticity and safety, identification of repairs and QA procedures are given. This includes reminders of legal responsibilities.

15.8.3 Identification of repairs

If IS apparatus has been repaired, normally by the manufacturer, records will be kept by the repairer, clearly identifying the device. To depict that a repair has been carried out on apparatus, a 'Triangle-R' symbol shown in Figure 15.11 is used. It appears the opposite way up purely for ease of identification between the IEC and BEAMA/AEMT codes of practice.

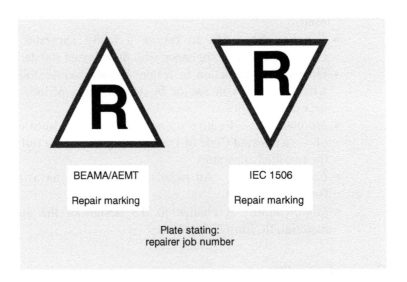

Figure 15.11
Repair marking

15.8.4 Repair of IS apparatus

The code provides clear guidelines in respect of mechanical parts such as enclosures, cable entries and termination systems. Any repair performed must not result in a reduction in the degree of protection provided by the original design.

It states that electrical components must be replaced by types, which are obtained only from the manufacturer (of the apparatus using these components). This is because the devices may be specially selected for the purpose. Such components as Zener diodes will inevitably fall into this category.

Opto-couplers, for example, may be specially designed, constructed and additionally certified (as 'component-certified ... /U') for a particular purpose, so only these types may be used.

Fuses are normally closely specified to be of a particular characteristic and rupturing capacity. Only that type can be used to protect other devices and circuits.

Faulty relays should be replaced and no attempt to repair them should be made. This also applies to mains or supply transformers. Any overcurrent protection device is embedded and may have been damaged if a fault has occurred, causing the transformer to fail.

Only batteries of a type specified by the apparatus manufacturer may be used as replacements. Encapsulated battery assemblies, which contain built-in internal current-limiting resistance, must be replaced entirely.

Encapsulated components, including actual shunt diode safety barriers, which are found to be defective, are 'non-repairable' and must be replaced.

The repair of printed circuit boards is considered possible by this code. The warning is given that varnished boards will need two coats of the prescribed type of varnish to be applied in the correct manner.

The code states that no attempt should be made to 'reclaim' components on which IS depends. The code defines reclamation as a means of repair by restoration. This is not practicable, given the nature of electronic components but the code does not amplify this more. The industry views that restoration of any safety component after possible damage may have weakened the component and therefore its reliability cannot be accepted.

The IS apparatus manufacturing industry generally takes the view that attempts at repair other than by their own facility render the apparatus unreliable in failure mode and will refuse to attempt to repair. The established installation and maintenance codes state that 'the same informed approach as that of the manufacturer' is required for any such work on certified apparatus. They argue that control over the handling of repairs is lost if others make attempts.

15.8.5 Modifications

Modifications are permitted by the code.

16

ATEX Directive

16.1 General

The concept of a globalized single market is one of the great achievements of our time. To realize this dream, conducive conditions need to be created such that the goods, services, capital and labor can circulate freely, and thus providing a foundation for prosperity to all continents as we move toward the twenty-first century.

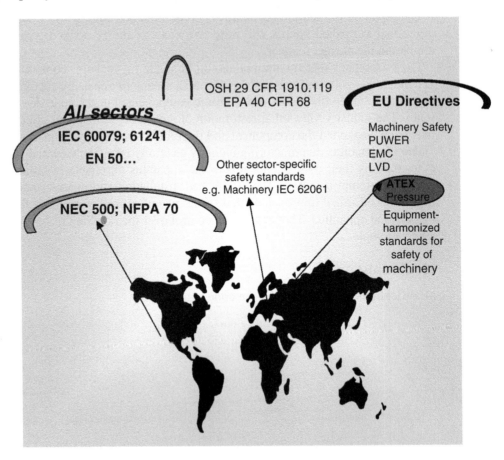

Figure 16.1
The legislation and standards for hazardous area standards

The EU, being one of the earliest industrialized zones, has contributed a lot toward standardization in the field of hazardous area (Figure 16.1). It has developed original and innovative instruments to remove the barriers to free circulation of goods. Among these, the new approach to product regulation and the Global Approach (IECEx scheme) to conformity assessment are notable. Since 1987 some 20 directives, adopted on the basis of the New Approach and the Global Approach, have progressively come into force.

The common thread amongst these complementary approaches is that they limit public intervention to what is essential and leave business and industry the greatest possible choice on how to meet their public obligations.

Progressive globalization obviously also includes industries such as large-scale chemicals, pharmaceuticals, petroleum, and gas extraction and processing, and many other industries directly or indirectly involved with hazardous areas. Hence, in the long term, it will be unacceptable for globally active companies to have to conform to different safety standards in different areas of the world. The development of internationally valid Explosion-Protection Directives and Standards will therefore continue to accelerate.

What are the EC guidelines for hazardous area?

Two EC guidelines have been issued which form the basis of the new system of European explosion protection, also known as 'New Approach':

1. Guideline 94/9/EG derived from Article 100a of the EWG agreement, which describes the basic principles for improving technical equipment for workers' protection and for the user, and
2. The guideline based on Article 137a of the EWG agreement, which governs the measures required to protect workers in potentially explosive atmospheres.

To whom they are applicable?

Whereas the first is aimed principally at the manufacturers of explosion-protected electrical equipment and the notified bodies, the second guideline is mandatory for operators of electrical equipment in hazardous areas.

Users will, however, find the ATEX guidelines extremely useful. Despite the discrepancies, the EU Directive represents 'enormous progress' as numerous difficult points are clarified for the official testing authorities, manufacturers and operators. The new approval concept ensures that innovative manufacturers and users have, as long as the safety level is at least equivalent, great scope for new and further technological developments.

European Directive 94/9/EC provides a new legal framework for placing explosion-protected equipment, components and protective systems on the market and has been in effect now for some time.

For the first time, not only electrical but also non-electrical equipment and autonomous protective systems are included. The supplemental European Directive 1999/92/EC (ATEX 137), which is directed to employers and includes provisions for protecting personnel from potential explosions, has been recently adopted.

By July 2003, all equipment and protective systems intended for use in potentially explosive atmospheres within the EU had to bear the '*CE marking*' in accordance with the ATEX Directive 94/9/EC.

What is this ATEX Directive?

The ATEX Directive 94/9/EC is a directive adopted by the EU to facilitate free trade in the EU by aligning the technical and legal requirements in the member states for products intended for use in potentially explosive atmospheres. The full text of the Directive was published in the Official Journal of the European Communities No. L 100, dated 19 April 1994 (Figure 16.2).

The period known as 'transitional period' lasted until 30 June 2003, when the ATEX Directive 94/9/EC canceled all the old approaches.

One question, which crops up in everybody's mind, be it manufacturer or operator of a plant, is

'Why do I have to do all this to comply with the directive?'

Probable answer could be –

The countries that make up the European Economic Area adopt Directives as their national laws. The ATEX Directive is the law in every country in the European Community. And part of the law states that all products for use in explosive atmospheres must meet the Directive.

Another equally compelling reason is that of economics. The purpose of harmonizing provisions in the area of explosion protection is to remove obstacles related to the free trade of goods, expand the market, afford Europe more economic clout and reduce competitive disadvantages.

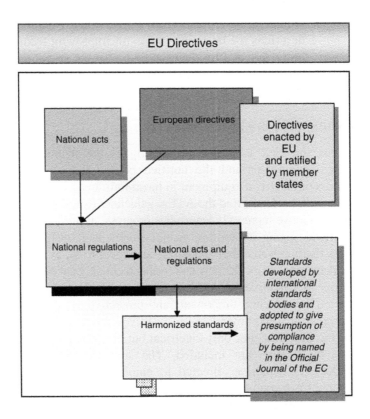

Figure 16.2
EU directive and regulations

Pertinent, reasonably oriented criteria in the application of the provisions are necessary to prevent additional costs accruing to manufacturers and users without a suitable gain in safety. The opportunities and risks of the new directive are discussed in this chapter.

At the end of the transition period, manufacturers in the EU had to comply with the relevant national regulations, in line with the ATEX Directive. All equipment and protective systems intended for use in potentially explosive atmospheres (as well as the specified devices), which are made or sold in the EU, including imports, will have to:

- Satisfy wide-ranging health and safety requirements
- In some cases, be subject to type-examination by a notified body (NB)
- In many cases be subject to conformity assessment procedures by a NB
- Carry CE marking and information (generally about the manufacturer).

The ATEX Directive replaces the old 'Explosive Atmospheres and Gassy Mines Directive' (76/117/EEC) and is intended to bring products covered by the old directive into line with the other, so called, 'New Approach' CE-marking Directives.

Let us examine first what the Old Approach Directives are.

'Old Approach' Directives

The Potentially Explosive Atmospheres Directive, 76/117/EEC (18 December 1975), along with 79/196/EEC (6 February 1979) and all of its amendments (mentioned below), form a working document. This directive applies to electrical equipment capable of use in a potentially explosive atmosphere employing certain types of protection.

Types of protection included are:

- Flameproof enclosure 'd'
- Pressurized apparatus 'p'
- Powder filling 'q'
- Oil immersion 'o'
- Increased safety 'e'
- Intrinsic safety 'i' and
- Encapsulation 'm'.

The 'Old Approach' Directive allows the 'Ex' Mark to be placed on products that comply with its requirements.

This directive was in effect only until 30 June 2003.

To have total protection, the industry has to look at other directives based on whether it is an underground industry or hazards are from gas or dust. The other relevant directives for other than mining, in additions to the above two are,

- Commission Directive 84/47/EEC of 16 January 1984 – replaced the Annexure I and II of Directive 79/196/EEC
- Commission Directive 88/571/EEC of 10 November 1988 – another replacement of the Annexure I of Directive 79/196/EEC
- Council Directive 90/487/EEC of 17 September 1990 – added encapsulation 'm' and intrinsically safe electrical systems 'i' protection concept in Article 1 and amended Annexure I of Directive 79/196/EEC
- Commission Directive 94/26/EEC of 15 June 1994 – another replacement of the Annexure I of Directive 79/196/EEC
- Commission Directive 97/53/EC of 11 September 1997 – another replacement of the Annexure I of Directive 79/196/EEC.

Similarly there are Old Approach Directives for mining equipment, as given hereunder,

- Council Directive 82/130/EEC of 15 February 1982 – apply to electrical equipment for use in underground parts of mines susceptible to firedamp which may be endangered by firedamp
- Commission Directive 88/35/EEC of 2 December 1988 – amendment to Annexure A, B and C of the Directive 82/130/EEC
- Commission Directive 91/269/EEC of 30 April 1991 – another amendment to Annexure A and C of the Directive 82/130/EEC
- Commission Directive 94/44/EEC of 19 September 1994 – amendment to Annexure A of the Directive 82/130/EEC
- Commission Directive 98/65/EEC of 3 September 1998 – amendment to Annexure A and addition of two annexes to it in the Directive 82/130/EEC.

It is obvious that any 'user' not very current with the contents of the directives is likely to get thoroughly confused. All the to criss-crossing of amendment and at the end-users were at a loss to find most economical means of safety protection and at some point safety was the casualty.

Hence, in order to have a single directive covering whole gamut of plausible scenarios this new approach directive has been enshrined in the EU.

'New Approach' or ATEX Directive

On 1 March 1996, a transitional period began for the implementation of the ATEX Directive (94/9/EC). This directive applies to electrical and non-electrical equipment/components and protective systems intended for use in potentially explosive atmospheres. The ATEX Directive became mandatory on 1 July 2003.

The fundamental underlying principles enshrined in ATEX Directive are,

- Prevent the formation of explosive atmospheres
- Prevent the ignition of unavoidable explosive atmospheres
- Control the effects of unavoidable explosions.

The ATEX Directive sets out the essential requirements that products must meet and defines procedures,

- For the evaluation of a product's design
- For manufacture (production) based on equipment groups and categories
- For the conformity assessment that manufacturers must undertake before affixing the CE marking to them.

Equipments located outside potentially explosive atmospheres are also covered by the ATEX Directive if the equipment includes,

- A safety device
- Controller
- Regulatory device
- Required for the safe function of equipment
- Required for protective systems with respect to the risk of explosion.

Differences between Old and New Approach

The change from old to new directive is not merely an update, but a major change in philosophy, extending to cover under normal operating conditions,

- All equipment and protective systems, which may be used in areas endangered by potentially explosive atmospheres created by the presence of flammable gases, vapors, mists or **dusts**
- Both electrical and mechanical equipments
- Constructional features
- Methods of testing
- Certification requirements and procedures
- Requirements for safety-related devices (flame arrestors, suppression systems, etc.) and safe area equipment
- Additional quality system requirements
- The need to produce a 'technical file'.

'USE' Directive

Along with the ATEX Directive EU has also issued a new directive for 'The Protection of Workers at Risk from Potentially Explosive Atmospheres' (1999/92/EC), on minimum requirements for improving the safety and health protection of workers potentially at risk from explosive atmospheres (15th individual Directive within the meaning of Article 16(1) of Directive 89/391/EEC) – commonly known as the **'Use' Directive**. This directive also became mandatory under EU law from 1 July 2003.

The essential requirements under this directive are that sites shall document evidence where potentially explosive atmospheres (gas or dust) may develop by carrying out:

- Risk analysis
- Area classification
- Site inspections.

This 'Use' Directive ensures that only ATEX-certified electrical, mechanical and safety-related systems are installed in potentially explosive atmospheres.

Some other salient features of this directive are

- The requirement to create documentation on explosion protection with a comprehensive risk assessment
- The duty to protect workers
- The safety strategy in force at the operator's premises
- The determination of the risk posed by the explosion danger
- The classification of areas (Zone 0, 1, 2, 20, 21, 22)
- The requirements relating to qualifications of employees working in the hazardous areas
- The procedure for giving approvals to work
- Regular testing
- The criteria governing the selection of working substances and installation materials for the different zones
- The identification of the areas
- The criteria for the authorization of work in the various zones.

Explosion protection is of particular importance to safety of workers. The consequences of explosions are,

- Endangering the lives and health of workers as a result of the uncontrolled effects of flame and pressure, and
- Consumption by workers of noxious reaction products along with oxygen in the ambient air.

The establishment of a coherent strategy for the prevention of explosions requires that organizational measures complement the technical measures taken at the workplace.

Documentation

As mentioned above, Directive 89/391/EEC requires the employer to be in possession of an assessment of the risks to workers' health and safety at work. This directive specifies that the employer is to draw up an explosion-protection document, or set of documents, which satisfies the minimum requirements laid down in this directive. In addition to this, the employer is to keep it up to date.

The explosion-protection document may be part of the assessment of the risks to health and safety at work and should include,

- The identification of the hazards
- The evaluation of risks
- The definition of the specific measures to be taken to safeguard the health and safety of workers at risk from explosive atmospheres.

An assessment of explosion risks may be required under other community acts; whereas, in order to avoid unnecessary duplication of work, the employer is allowed, in accordance with national practice, to combine documents, parts of documents or other equivalent reports produced under other community acts to form a single 'safety report'.

Places where explosions may occur, the sign in Figure 16.3 is to be displayed.

Warning sign for places where explosive atmosphere
may be present

Figure 16.3
The sign to be displayed where explosive atmosphere is likely to occur

Directive implementation

The ATEX Directive came into effect on a voluntary basis on 1 March 1996. However, from 1 July 2003, it became necessary for all products intended for use in potentially explosive atmospheres, to comply with this directive.

16.2   Definitions

ATEX   'Atmospheres explosibles' is French for 'potentially explosive atmospheres'. The Directive 94/9/EC is frequently called the 'ATEX-Directive'. In other parts of the world, areas with potentially explosive atmospheres are often called hazardous locations.

CE The CE marking is a distinctive community mark of the EU. It is a marking that signifies declaration by the responsible party (usually the manufacturer) that a product is compliant with all appropriate European Union New Approach Directives, such as the ATEX, low voltage, EMC, Machinery Directives and quality assurance, third-party assessment, depending on the directives.

Product It covers equipment, protective systems, devices, components and their combinations.

Notified body A notified body is a testing laboratory and certification organization. Every national government notifies (or, designates) one or more testing/certification lab as an agency able to issue certificates pertaining to the directives.

Categories of equipment 'The Old Approach' Directives only dealt with electrical equipment covered by EN 50014-5039. The scope was clearly Zone 1, the comparable situation in a mine. Because the new Directive 94/9/EC covers also Zone 2 and Zone 0 equipment, and as the mystery rules of the commission did not allow the explicit use of the definitions of zones (this is a matter for another general directorate), the categories are created.

16.3   Scope of the ATEX Directive

The objective of Directive 94/9/EC is to ensure free movement for the products to which it applies in the EU territory. Therefore the directive, based on Article 95 of the EC Treaty, provides for harmonized requirements and procedures to establish compliance.

The directive notes that to remove barriers to trade via the New Approach, provided for in the Council Resolution of 7 May 1985(17), essential requirements regarding safety and other relevant attributes need to be defined by which a high level of protection will be ensured. These essential health and safety requirements (EHSRs) are listed in Directive 94/9/EC at Annex II. The essential requirements must be met by equipment and protective systems intended to be used in potentially explosive atmospheres.

The essential requirements fall into three groups:

1. Common requirements
2. Requirements for equipment
3. Requirements for protective systems.

After 30 June 2003 products could be placed on the market in the EU territory, freely moved and operated as designed and intended in the expected environment only if they comply with Directive 94/9/EC (and other relevant legislations).

Atmospheres covered

An explosive atmosphere, for the purposes of Directive 94/9/EC, is defined as a mixture, of flammable substances in the form of gases, vapors, mists or dusts with air; under atmospheric conditions;

... in which, after ignition, the combustion spreads to the entire unburned mixture.

It has to be noted that in the presence of dust not always the whole quantity of dust is consumed by the combustion.

An atmosphere which could become explosive due to local and/or operational conditions is called a potentially explosive atmosphere. It is only this kind of potentially explosive atmosphere which products falling under the Directive 94/9/EC are designed for.

It is important to note, that products are not covered by Directive 94/9/EC (27) where they are intended for use in or in relation to atmospheres which might potentially be explosive, but one or more of the defining elements as above are not present.

Product and equipment covered

The word 'equipment' is used instead of 'apparatus'. It should be noted that Directive 94/9/EC provides for the first time essential health and safety requirements for non-electrical equipment intended for use in potentially explosive atmospheres and is to include both electrical and non-electrical equipments.

Equipment, as defined in Directive 94/9/EC, means

- Machines
- Apparatus
- Fixed or mobile devices
- Control components and instrumentation thereof
- Detection or prevention systems, which, separately or 'jointly', are intended for

    - The generation
    - Transfer
    - Storage
    - Measurement
    - Control
    - Conversion of energy
    - The processing of material.

and which are capable of causing an explosion through their own potential sources of ignition.

It should be noted that intrinsically safe equipment is included in the scope of the directive.

The following is a partial list of equipment, which as an example is subject to Directive 94/9/EC, Article 1.2, due to being a potential source of ignition:

- Electrical equipment and apparatus (motors, lamps, switches, measuring devices, gas detection devices, etc.)
- Ventilators, fans, compressors
- Pumps
- Forklift trucks
- Fast-running mechanical machinery
- Centrifuges
- Drive belts
- Mixers
- Hoisting gear.

Electrical equipment

Directive 94/9/EC does not define 'Electrical Equipment'. However, because such equipment is subject to its own conformity assessment procedure it may be useful to provide a definition, which has been generally accepted by the majority of Member States, as follows:

Electrical equipment Equipment containing electrical elements is used for the generation, storage, measurement, distribution and conversion of electrical energy, for controlling the function of other equipments by electrical means or for processing materials by the direct application of electrical energy. It should be noted that a final product assembled using both electrical and mechanical elements may not require assessment as electrical equipment provided the combination poses no additional risk.

Examples: A *pump* (non-electrical) is assessed under the appropriate conformity assessment procedures and is then connected to an *electric motor* (electrical equipment), which has already been assessed. As long as the combined equipment poses no additional hazards, no further assessment for the electrical part is necessary.

If the same pump and electric motor have not been through the appropriate conformity assessment procedures and are connected, then the resulting product is to be regarded as electrical equipment and the conformity assessment should treat it as such.

Protective systems

'Protective systems' mean design units, other than components, which are intended to halt incipient explosions immediately and/or limit the effective range of explosion flames, and include items that prevent an explosion that has been initiated from spreading or causing damage.

They include,

- flame arrestors
- quenching systems – water trough barriers
- pressure-relief panels or explosion-relief systems (using e.g. bursting discs, vent panels, explosion doors, etc.)
- fast-acting shut-off valves
- extinguishing barriers.

to name but a few. Products are to be categorized by the level of protection that they offer against the risk of them becoming a potential source of ignition of an explosive atmosphere.

Some examples of devices falling under protection systems are:

- A power supply feeding an intrinsically safe (Ex 'i') measurement system used for monitoring process parameters.
- A pump, pressure-regulating device, backup storage device, etc. ensuring sufficient pressure and flow for feeding a hydraulically actuated safety system (with respect to the explosion risk).
- Overload protective devices for electric motors of type of protection EEx 'e' 'Increased safety'.
- Controller units in a safe area, for an environmental monitoring system consisting of gas detectors distributed in a potentially explosive area, to provide executive actions if dangerous levels of gas are detected.
- Controller units for sensors temperature, pressure, flow, etc. located in a safe area, for providing information used in the control of electrical apparatus, used in production or servicing operations in a potentially explosive area.

Components

A component means any item essential to the safe functioning of equipment and protective systems but with no autonomous function.

Components intended for incorporation into equipment or protective systems, which are accompanied by an attestation of conformity including a statement of their characteristics and how they must be incorporated into products, are considered to conform to the applicable provisions of Directive 94/9/EC. Ex-components, as defined in the European Standard EN 50014, are components in the sense of the ATEX Directive 94/9/EC as well. Components must not have the CE marking affixed unless otherwise required by other directives (e.g. the EMC Directive 89/336/EEC).

Examples:

- Terminals
- Push button assemblies
- Relays
- Empty flameproof enclosures
- Ballasts for fluorescent lamps
- Meters (e.g. moving coil)
- Encapsulated relays and contactors, with terminals and/or flying leads
- Test reports.

Assemblies

From the term 'jointly' in the definition above it follows that an assembly, formed by combining two or more pieces of equipment, together with components if necessary, has to be considered as a product falling under the scope of Directive 94/9/EC, provided that this assembly is placed on the market and/or put into service by a responsible person (who will then be the manufacturer of that assembly) as a single functional unit.

Such assemblies may not be ready for use but require proper installation. The instructions (Annex II, 1.0.6) will have to take this into account in such a way that, compliance with Directive 94/9/EC is ensured without any further conformity assessment, provided the installer has correctly followed the instructions.

In the case of an assembly consisting of different pieces of equipment as defined by Directive 94/9/EC which were previously placed on the market by different manufacturers these items of equipment have to conform with the directive, including being subject to proper conformity assessment, CE-marking, etc. The manufacturer of the assembly may presume conformity of these pieces of equipment and may restrict his own risk assessment of the assembly to those additional ignition and other relevant hazards (as defined in Annex II), which become relevant because of the final combination. If additional hazards are identified a further conformity assessment of the assembly regarding these additional risks is necessary. Likewise, the assembler may presume the conformity of components, which are accompanied by a certificate, issued by their manufacturer, declaring their conformity.

However, if the manufacturer of the assembly integrates parts without a CE marking into the assembly (because they are parts manufactured by himself or parts he has received from his supplier in view of further processing by himself) or components not accompanied by the above-mentioned certificate, he shall not presume conformity of those parts and his conformity assessment of the assembly has to cover those parts as required.

Appendix H may be referred for further details.

Installations

A common situation is that pieces of already-certified equipment are placed on the market independently by one or more manufacturer(s), and are not placed on the market by a single legal person as a single functional unit. Combining such equipment and installing at the user's premises is not considered as manufacturing and thus does not result in equipment; the result of such an operation is an installation and is outside the scope of Directive 94/9/EC. The installer has to ensure that the initially conforming pieces of equipment are still conforming when they are taken into service. For that reason he has to carefully follow all installation instructions of the manufacturers. The directive does not regulate the process of installation. Installing such equipment will generally be subject to legal requirements of the member states. An example could be instrumentation consisting of a sensor, a transmitter, a Zener barrier and a power supply if provided by several different manufacturers installed under the responsibility of the user.

Product type

A product type is a unique design that is sufficiently well defined that items manufactured to that definition would be in compliance with the requirements applicable to the product type. The concept of product type is of particular importance where products are to be manufactured in volume and where the cost of a NB verifying the conformity of each item produced would be prohibitive.

In compiling the technical file, the manufacturer must prepare a set of documents, normally in the form of drawings that specify every feature of the product that could affect conformity with the essential requirements. Different features will be addressed by different requirements. For example the materials of which the product is made must have,

- Adequate strength
- Stability
- Resistance to the operating conditions (including corrosion, heat and ultraviolet light, for example) defined by the manufacturer
- The dimensions of component parts must have suitable tolerances that the required degree of fit will be achieved.

The sample(s) presented for inspection and testing must be in conformity with the specified design features. Where a particular test, for example mechanical impact, is to be applied, the sample(s) must represent the worst-case condition such as the minimum wall thickness of an enclosure. The sample(s) must also be sufficiently representative of the production methods to be used, not to adversely affect the outcome of the tests. If a plastic case has been fabricated from sheet materials for testing purposes and the final product is to be moulded, it may be necessary to repeat some tests on the moulded article if there is a possibility that the moulding process may give rise to weaknesses such as flow lines.

Exclusions of products and equipment

The following product types, which have no own potential sources of ignition, are not subject to 94/9/EC:

- Medical devices intended for use in a medical environment
- Equipment and protective systems where the explosion hazard results exclusively from the presence of explosive substances or unstable chemical substances
- Products for use in the presence of explosives
- Equipment intended for use in domestic and non-commercial environments where potentially explosive atmospheres may only rarely be created, solely as a result of the accidental leakage of fuel gas

- Sea-going vessels and mobile offshore units together with equipment on board such vessels or units
- Military equipment
- Personal protective equipment covered by Directive 89/686/EEC
- Means of transport, i.e. vehicles and their trailers intended solely for transporting passengers by air or by road, rail or water networks, as well as means of transport in so far as such means are designed for transporting goods by air, by public road or rail networks or by water. Vehicles intended for use in a potentially explosive atmosphere shall not be excluded, the equipment covered by Article 223 (1) (b) of the Treaty.

For example, hereunder is a illustrative list of equipment and apparatus not covered under the directive,

- Containers, reactors, evaporators, columns and their parts
- Tanks
- Tubes
- Conduit installation parts (nozzles, apertures, reducers)
- Steam traps
- Cabling
- Filters
- Jet mills
- Mechanical components
- Fittings (hand valves, stands and racks, taps, flaps, latches)
- Mechanical measuring equipment (mechanical flow-meters, level measuring devices with mechanical indicators).

Groups and categories

All the equipments which are covered under this directive are divided into traditional groups of,

- Group I – mining
- Group II – non-mining.

Group I – mining equipment

The categories of equipment for gassy mines are defined as hereunder,

Equipment Category	Protection	Comparison to Current IEC Classification
M1	2 levels of protection; or 2 independent faults	Group I
M2	1 level of protection based on normal operation	Group I

The M1 Category of equipment is designed to operate continuously in explosive atmosphere and must be,

- Equipped with additional special means of protection. It must remain functional with an explosive atmosphere present.
- So constructed that no dust can penetrate it.
- So constructed that the surface temperatures of equipment parts are kept clearly below the ignition temperature of the foreseeable air/dust mixtures in order to prevent the ignition of suspended dust.
- So designed that the opening of equipment parts, which may be sources of ignition, is possible only under non-active or intrinsically safe conditions. Where it is not possible to render equipment non-active, the manufacturer must affix a warning label to the opening part of the equipment. If necessary, equipment must be fitted with appropriate additional interlocking systems.

Some features of equipment coming under Category M2 which are required to *operate in a potentially explosive atmosphere* are,

- The equipment is intended to be de-energized in the event of an explosive atmosphere.
- Equipment must be so designed that the opening of equipment parts, which may be sources of ignition, is possible only under non-active conditions or via appropriate interlocking systems. Where it is not possible to render equipment non-active, the manufacturer must affix a warning label to the opening part of the equipment.

'*Mines cannot be zoned*' like a petrochemical plant. There is no distinct source of release and gas is emitted by,

- Most of the walls
- Floors
- Ceilings
- Main product (coal) during transport.

Consequently mines have two different situations, neither depending on the place, nor depending on the time.

Sometimes the gas (methane) content in the air (at those places where measuring instruments are installed or where somebody uses a portable instrument) is below the permissible value, normally fixed by the inspecting bodies at levels between 1 and 2%, well below the LEL. Of course at other places at the same time there may be other, higher concentrations, even higher than LEL.

This increase of the gas from source (or decrease of the ventilation) normally is recognized very soon in a properly operated mine. Then the power is switched off and the machinery is stopped. During the time before switching off, the electrical equipment (Category M2) – which is designed using identical principles as Zone 1 equipment – of course may operate in explosive atmosphere; this is the intended use. After switching off the M2 equipment, there is a need for instrumentation for safety purposes and communication. This equipment needs to meet criteria similar to Zone 0 equipment. Consequently it is named 'Category M1'.

Category M1/M2 equipment and coal dust

The M1 and M2 equipment are not only safe in explosive methane atmosphere, they are also safe with respect to coal dust layers and coal dust clouds.

Coal dust explosions in underground mines normally are triggered by an ignition of a methane explosion. But also coal dust alone is able to form an explosive atmosphere, which can be ignited by ignition sources with sufficient energy (e.g. high-power arcing, wrong use of explosives, hot spots, smoldering nests in coal dust layers). Consequently since 1977 the CENELEC standards (which since that time contain basic requirements for 2D and M2 equipment) required what we call now M2 equipment to meet IP 54 requirements and to limit the maximum external surface temperature to 150 °C. For flameproof (American english: explosion proof) electrical equipment the gaps, which do not transmit a methane explosion, are also safe for a coal dust explosion.

A 200 m gallery for coal dust explosions was installed in Dortmund 1911. With this set-up for the first time it could be demonstrated, that with coal dust alone (without any methane) a self-propagating explosion is possible. The gallery is still in use.

The CENELEC standard EN 50014 is slightly amended and re-introduces special requirements concerning electrostatic ignition risks, and EN 50020-2 is supplemented by a separate standard for intrinsically safe systems for Group I. This is in line with the original approach in Directive 82/130/EEC, 'electrical equipment for use in underground parts of mines susceptible to firedamp which may be endangered by firedamp'.

Group II – non-mining equipment

Equipment intended for use in other than Equipment Group I, i.e. non-mining places that are liable to be endangered by explosive atmospheres is defined as hereunder,

Equipment Category	Protection	Comparison to Current IEC Classification
1	2 levels of protection; or 2 independent faults	Group II, Zone 0 (gas) Zone 20 (dust)
2	1 level of protection based on frequent disturbances; or equipment faults	Group II, Zone 1 (gas) Zone 21 (dust)
3	1 level of protection based on normal operation	GrouSp II, Zone 2 (gas) Zone 22 (dust)

As explained above within the ATEX Directive, equipment is classified into following categories according to the level of risk in its intended area of use.

Category 1 (1G, 1D)

The equipment under this category is intended for high-risk areas where an explosive atmosphere is present for long periods. The equipment is ignition safe in normal operation, also with foreseeable and rare malfunctions present.

Equipment intended for use in Zone 0 would normally be classified under ATEX as Category 1. This would include equipment designed to meet the IS 'ia' standard.

Category 2 (2G, 2D)

The equipment under this category is intended for medium-risk areas where an explosive atmosphere may occur under normal operating conditions. The equipment is ignition safe in normal operation, also with foreseeable malfunctions present.

Equipment intended for use in Zone 1 would normally be classified as ATEX Category 2. Equipment complying with the following protection standards is included in Category 2.

- Intrinsic safety 'ib'
- Flameproof enclosure 'd'
- Increased safety 'e'
- Purged and pressurized 'p'
- Encapsulated 'm'
- Oil-filled 'o'
- Powder-filled 'q'.

Category 3 (3G, 3D)

The equipment under this category is intended for areas where an explosive atmosphere is only likely under abnormal circumstances i.e. Zone 2 areas.

Equipment intended for use in Zone 2 would normally be classified as ATEX Category 3. This would include equipment designed to the non-incentive 'n' standard. The category equipment will have the following distinctive features:

- Distinctive community mark
- CE marking
- New type 'n' standard with many additional tests and requirements
- Quality audits not required
- NB involvement is not required.

Verifying conformity with the ESHRs

The New Approach Directives contain so-called 'essential requirements' regarding safety and health. The times have gone in which one tried to regulate down to the last detail, which made new innovations difficult or even obstructed them. It is a paradox that now the missing details of the EC Directives in European Standards are a fact. Which leads to the arguable conclusion that products manufactured on that basis are lawfully in conformance.

The intention is to regulate the use of products in explosive atmospheres by a 'Council directive on minimum requirements for improving the safety and health protection of workers potentially at risk from explosive atmospheres' which will be an individual directive within the meaning of Article 16 of Directive 89/391/EEC and based on Article 138 of the EC Treaty (19).

In general, the use of such products in potentially explosive areas has to be monitored as part of the surveillance activity undertaken by the competent authorities in the member states.

Associated with each equipment category, whether for electrical or mechanical equipment, is a complex conformity assessment procedure defined by Directive 94/9/EC and entailing various degrees of effort. The following outlines the main features of the conformity assessment procedures for electrical and mechanical equipment, apparatus, machines and internal-combustion engines, and the documents that are generated from the procedures.

From the point of view of the operator, it must be ensured that the conformity assessment procedure is appropriately performed and not overdone. For example, the manufacturer of non-electrical equipment of Category 2G (suitable for Zone 1) is responsible for internally controlling the manufacturing process and providing the relevant documents to a NB. The manufacturer can thus presume equipment conformity to the directive without involving a testing house.

Unfortunately, the new legal basis and allegedly unclear liability issues cause uncertainty amongst manufacturers. The result is that they are either not at all inclined to submit EC conformity declarations or they contact notified bodies (who are not unhappy about acquiring additional clients).

The users and their organizations should immediately enter into a 'trialog' with their equipment suppliers and with notified bodies in order to avoid negative developments. The attempt should be made to convince manufacturers to avoid unnecessary EC type-examination procedure and to initiate self-certification (manufacturer's declaration) as far as this is possible.

The table below summarizes the intended place of use and the verification procedure associated with each ATEX category of equipment.

ATEX Equipment Category	Intended Place of Use	Conformity Assessment Modules Described in Directive
Category 1G/1D	Explosive atmosphere (Zone 0/Zone 20)	NB performs – Type examination (B), plus either Quality Assurance (D) or Product verification (F)
Category M1	Gassy mines, operating in an explosive atmosphere	
Protective system	Anywhere	
Category 2G/2D electrical equipment or internal combustion engine	Potentially explosive atmosphere (Zone 1/Zone 21)	NB performs – Type examination (B), plus either Product Quality Assurance (E) or Conformity to type (C)
Category M2 electrical equipment or internal combustion engine	Gassy mines, de-energized if the atmosphere changes from potentially explosive to explosive.	
Category 2G/2D non-electrical equipment	Potentially explosive atmosphere (Zone 1/Zone 21)	Manufacturer performs – Internal control of production (A)
Category 2M non-electrical equipment	Gassy mines, de-energized if the atmosphere changes from potentially explosive to explosive	(Manufacturer's Technical File)
Category 3G/3D electrical, or non-electrical equipment	Potentially explosive atmosphere (Zone 2/Zone 22)	Manufacturer performs – Internal control of production (A) (Manufacturer's Technical File)
All of the above categories	As appropriate to the category	NB performs – Unit verification (G) of each item manufactured

Note: The letters (A), (B), (C), (D), (E), (F) and (G) are modules described in the Directive.

The ATEX Directive now means that far more equipment will require certification. This relates mainly to the need for mechanical equipment and protective systems to comply with the directive. Presently there are over 70 new standards being prepared by CEN committees specifically for these types of equipment.

All the equipment to be used in hazardous atmospheres under this directive will have to bear distinctive marking of category. The marking of the equipment with the category will help the end-user with their selection of the equipment in that it identifies which zone it can safely be installed in. This is a major improvement over the old cryptic marking system that only listed the protection concepts used in the design of the equipment. This meant that the user of the equipment had to be familiar with all eight recognized protection concepts and furthermore had to know which of them was suitable for a particular type of zone.

Harmonized standards

CENELEC and CEN technical committees are recognized as the bodies competent to adopt harmonized standards, which follow the general guidelines for cooperation between the CEC and those two bodies, signed on 13 November 1984 (Figure 16.4).

CENELEC is the European Committee for Electro-technical Standardization.

CEN is the European Committee for Standardization.

For the purposes of this directive, a harmonized standard is a technical specification (European Standard or harmonization document) adopted by one or other of those bodies, or by both, at the prompting of the Commission pursuant to Council Directive 83/189/EEC of the 28 March 1983 providing for a procedure governing the provision of information on technical standards and regulations and pursuant to the general guidelines referred in the Directive.

CENELEC and CEN are European standardization bodies recognized as competent in the area of voluntary technical standardization and listed in Annex I of the Union Directive 98/34/EC (replacing 83/189/EEC) concerning 'the information procedure' for standards and technical regulations. Together they prepare European Standards in specific sectors of activity. When these standards are prepared in the framework of the 'new approach' directives, they are known as 'harmonized standards' and will be cited in the Official Journal. Products manufactured in accordance with these standards benefit from a 'presumption of conformity' to the essential requirements of a given directive.

Existing standards may be suitable in their existing form or may require amendment in order to address the essential requirements. Harmonized standards may originate as existing standards or be created at the specific request of the CEC.

CEN or CENELEC carries out the process of creation or amendment as appropriate. The work is assigned to the relevant committee or, if none exists, a new committee is set up. The member bodies of CEN or CENELEC will appoint the members of the committee. The member bodies are the national standards bodies of each member country. In practice the appointment of members is delegated to the corresponding committee at national level. The chairman and secretary of the committee are appointed following consultation with the members.

The process of creating a standard starts with the adoption of the item on the work program of the committee. A draft is prepared, usually by a small working group or one of the members and is circulated to national committees for their comments. The comments are discussed at a meeting of the committee and a decision taken on either amending the draft for further discussion, or putting the draft forward.

The national committee considers comments before sending the official national comments to the European Committee. The European Committee considers the comments and decides either to put the document out for voting or to refer the draft back for further consideration.

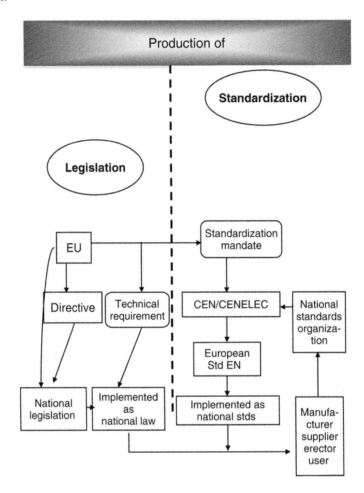

Figure 16.4
Generation of harmonized standards

Once the document is sent out for voting the national committees must decide whether to accept or reject the document. The accepted document is translated into English, French and German by the European Committee and sent to the CEC for them to accept as a harmonized standard according to the mandate. Provided that the document is acceptable, the CEC will publish the number and title in the OJEC, together with a reference to the directive to which it relates.

The published standards are produced by each national standards body in their own language and as a direct transposition of the text into a national standard.

There is an agreement between CENELEC and the IEC to coordinate their standards work. Thus, if an IEC standard already exists or is in preparation, it will be used as the starting point for a CENELEC standard. In this way products produced in Europe should be acceptable in other countries where IEC standards are used. Equally, products from these countries should comply with.

The preparations of the standards (EN 50 series) covering electrical products have been completed by CENELEC and are now included in the Official Journal. The standards for mechanical products and protective systems are still either under development or under approval.

Category 1 and M1 equipment

Category 1 and M1 require the highest levels of explosion protection, because they are the types of equipment intended to remain operational in a continuously present explosive atmosphere. Even though no constructional details are given in the directive the basic requirements as explained earlier are,

- With even one fault the equipment is safe against ignition
- To have two independent protection types applied in such a way that if one fails the other will protect against ignition.

CEN and CENELEC have produced the following harmonized standards to enable notified bodies to issue certificate of conformity:

- For Category 1G equipment it is EN 50284: 1999
- For Category M1 equipment it is EN 50303: 2000.

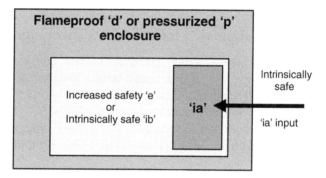

Figure 16.5
Principle of Category 1 and M1 equipment

The majority of equipment in this category is likely to be of the type as indicated above. Here, as shown in Figure 16.5 either increased safety 'e' or intrinsically safe 'ib' apparatus is located inside a flameproof 'd' enclosure or pressurized 'p' enclosure. This provides two independent means of protection and to ensure the integrity of the protection concept the power supply to such equipment is through cables carrying intrinsically safe 'ia' circuits.

Category 2 and M2 equipment

The protection concepts for Category 2 and M2 electrical equipment are already well known and are being used. They have been listed in European Standard EN 50014 – general requirement. The others as discussed during the course are,

- *EN 50018*: flameproof 'd': This type of protection being used for high/heavy current (power) switchgear.
- *EN 50019*: increased safety 'e': This type of protection being used for lighting fittings, traction batteries, etc.
- *EN 50020*: intrinsic safety 'ib': This type of protection being used for gas monitoring, communications, telephones, control circuits, etc.

As mentioned before internal combustion engines are the new type of equipment, which will be covered under this new directive. These diesel engines, when used in powering locomotives and free steered vehicles, have been traditionally of flameproof type. But now new standards have been produced to allow ATEX conformity. These are:

- *EN 1834-1: 2000*: Reciprocating internal combustion engines – safety requirements for design and construction of engines for use in potentially explosive atmospheres – Part 1 – Group II engines for use in gas and vapor atmospheres.
- *EN 1834-2: 2000*: Reciprocating internal combustion engines – safety requirements for design and construction of engines for use in potentially explosive atmospheres – Part 2 – Group I engines for use in underground workings susceptible to firedamp and/or combustible dust.
- *EN 1834-3: 2000*: Reciprocating internal combustion engines – safety requirements for design and construction of engines for use in potentially explosive atmospheres – Part 3 – Group II engines for use in flammable dust atmosphere.

New series of standards for non-electrical equipment

Traditionally so many other equipments are being used in potentially explosive atmosphere for centuries. These could include the mechanical de-watering pumps used in mines to the modern pneumatically or hydraulically driven equipment, like pumps.

CEN is now producing a new series of standards, which are similar to the ones in existence and has allotted a 'symbol letter' to each type of protection so as to allow them to be identified with a ignition-protected electrical equipment, as given hereunder.

EN 134463-1 – basic methodology and requirements

This is similar to EN 500014 and includes many of the requirements covered therein. Important thing to note is that it formally applies it for the first time to non-electrical equipment.

This places responsibility on manufacturers to,

- Assign a 'maximum surface temperature' to equipment
- Take into account the restrictions on use of exposed light metals, like alloys containing aluminum, magnesium, titanium and zirconium
- Avoid the risk of electrostatic discharge
- Perform impact tests on the parts that could have their ignition protection hampered in case of impact damage
- Mark the equipment with ATEX Group, category and ignition-protection symbol
- Submit technical file (as described below).

Further it provides for self-certification for some of the non-electrical equipment and detailed procedure is given therein.

EN 134463-2 – flow restricting enclosure 'fr'

Part 2 is an ignition-protection concept equivalent to the electrical equipment 'restricted breathing enclosure' described in EN 50021 and is only considered suitable for Category 3 equipment.

EN 134463-3 – flameproof enclosure 'd'

Part 3 is an already familiar concept of protection for containing the explosion and flames from surrounding atmosphere for electrical equipment. This is identical to electrical standard EN 50018 and as ignition concept is suitable for,

- Category 2 equipment
- Category M2 equipment
- Category 3 equipment.

EN 134463-4 – inherent safety '?'

Not much work has been done and no letter allotted. The prime reason being, the inability to fix threshold limit of mechanical power level where ignition can occur.

EN 134463-5 – constructional safety 'c'

Part 5 is an already familiar concept of protection for electrical 'increased safety'. It requires better engineering techniques to be employed. This is identical to electrical standard EN 50019 and as ignition concept is suitable for non-electrical,

- Category 2 equipment
- Category M2 equipment
- Category 3 equipment.

EN 134463-6 – control of ignition sources 'b'

Part 6 is a new ignition concept based on the age-old method of removing the source of energy or power to equipment whenever explosive atmospheric conditions occur. This as ignition concept is suitable for non-electrical,

- Category 2 equipment
- Category M2 equipment
- Category 3 equipment.

EN 134463-7 – pressurization 'p'

Part 7 is an already familiar concept of protection for keeping in an enclosure the source of ignition and then pressurize it with air or an inert gas. This is identical to electrical standard EN 50016 and as ignition concept is suitable for non-electrical,

- Category 2 equipment
- Category M2 equipment
- Category 3 equipment.

EN 134463-8 – liquid immersion 'k'

Part 8 is a familiar concept of protection based preventing ignition sources from becoming effective by immersing them in a liquid. Here water can be used as a protective liquid. This is based on electrical standard EN 50015 and as ignition concept is suitable for non-electrical,

- Category 2 equipment
- Category M2 equipment
- Category 3 equipment.

Technical file

The technical documentation to be submitted by manufacturer to NB is termed as 'technical file'. A technical file is a dossier of information specifying a product in sufficient detail for it to be manufactured in accordance with the requirements of the ATEX Directive and containing the evidence that the product conforms to those requirements. The ATEX Directive does not use the term 'technical file' but refers to 'technical documentation'. It has become common in the field of EU directives to refer to such an item as a technical file.

Annexes III, VIII and IX of the ATEX Directive specify that the technical file shall contain:

- A general description of the product
- Design and manufacturing drawings and layouts of components, sub-assemblies, circuits, etc.
- Descriptions and explanations necessary for the understanding of the drawings and layouts and the operation of the product
- A list of harmonized or other standards that have been applied in full or in part
- For aspects where standards have not been applied, descriptions of the solutions that have been adopted to meet the essential requirements of the directive
- Results of design calculation examinations carried out.

Certification and CE marking

The CE marking is intended to facilitate the free movement of products within the EU by signifying that essential health and safety requirements have been met.

The CE marking comprises the symbols CE together with such other information as may be required by the EU directives, which apply to a particular product. For the ATEX Directive, the symbols CE must be accompanied by the following:

- Name and address of manufacturer
- Designation of series or type
- Serial number, if any
- Year of construction.

The specific marking of explosion protection ⟨Ex⟩ followed by the symbol of the equipment group and category,

- For equipment-group II, the letter 'G' (concerning explosive atmospheres caused by gases, vapors or mists) and/or
- The letter 'D' (concerning explosive atmospheres caused by dust).

Furthermore, where necessary, they must also be marked with all information essential to their safe use.

The manufacturers normally affix the CE marking to the products. The manufacturer's legally appointed representative in the EU might affix the CE marking where products are manufactured outside the EU. However, the representative would then be taking legal responsibility for verifying the conformity of the products with the requirements of the relevant directives. In that case the representative would have to comply with the conformity assessment procedures, including, where required, type examination and quality modules.

For the ATEX Directive, the CE marking must be affixed to each item of equipment or to each protective system. The CE marking must not be affixed to components which do not of themselves comply with all relevant requirements but which must be combined with other parts in order to comply.

Conversions of existing certification

The vast majority of equipment certified before 1994 will not comply with the latest harmonized standards for the ATEX Directive and, manufacturers who do not consider the design implications now are liable to be caught out in the very near future. It is imperative that before putting your equipment up for certification you ensure that you are now compliant with all of the 'latest' relevant Harmonized European Standards and the Essential Health and Safety Requirements (EHSR) of the directive.

Conformity assessment procedure

Quality requirements

The annexe to directive specifies quality assurance requirements, known as ATEX Quality Modules. These modules are issued to the product manufacturer by the notified bodies. The issuing of the modules is dependent upon the manufacturer achieving a satisfactory level of quality control, which is determined by an external audit performed by the NB. Maintaining the quality module will be dependent upon a periodic audit program, i.e. Surveillance under the responsibility of the NB, which again is carried out by the NB.

There are two quality modules specified in the directive as hereunder,

1. *Production Quality Assurance*: Which applies to equipment in Categories 1 and M1, and to protective systems
2. *Product Quality Assurance*: Which applies to electrical equipment and internal combustion engines only, in Categories 2 and M2.

The directive requires the quality assurance system to address the following points:

- Quality objectives, organizational structure, responsibilities and powers of management with regard to equipment quality
- Manufacturing, quality control and quality assurance techniques, processes and systematic actions that will be used
- Examinations and tests which will be carried out before, during and after manufacture and frequency with which they will be carried out
- Verification and testing of each piece of equipment shall be carried out as set out in the relevant standard(s) in order to ensure their conformity with the type as described in the EC-type-examination certificate and the relevant requirements of the directive
- Quality records (inspection reports, test data, calibration data, qualifications of personnel, etc.).

These points will be covered by a quality system complying with ISO 9002: 1994.

A manufacturer's perspective

The very fundamental reason for formulation of ATEX Directive is that there should be common certification in EU so that manufacturers can have much wider market and users can have wider choices for safety products without compromising on the safety of plant and personnel. We will now look into this aspect of ATEX from a manufacturer's perspective.

Placing ATEX products on the market

This means the first making available, against payment or free of charge, of products, in the EU market, for the purpose of distribution and/or use in the EU. It is the manufacturer's responsibility to ensure that each and all of his products comply with the directive, where these fall under the scope of the directive.

'Making available' means the transfer of the product, that is, either the transfer of ownership, or the physical hand-over of the product by the manufacturer, his authorized representative in the EU or the importer to the person responsible for distributing these onto the EU market or the passing of the product to the final consumer, intermediate supplier or user in a commercial transaction, for payment or free of charge, regardless of the legal instrument upon which the transfer is based (sale, loan, hire, leasing, gift or any other type of commercial legal instrument). The ATEX product must comply with the directive at the moment of transfer.

Putting ATEX products into service

This means the first use of products referred to in Directive 94/9/EC in the EU territory, by its end-user.

Manufacturer

This is the person responsible for the design and construction of products covered by Directive 94/9/EC with a view to placing them on the EU market on his own behalf.

Whoever substantially modifies a product resulting in an 'as-new' product, with a view to placing it on the EU market, also becomes the manufacturer.

The manufacturer has sole and ultimate responsibility for the conformity of his product to the applicable directives. He must understand both the design and construction of the product to be able to declare such conformity in respect of all applicable provisions and requirements of the relevant directives.

Manufacturing of ATEX products for own use: Whoever puts into service products covered by the directive, which he has manufactured for his own use, is considered to be a manufacturer. He is obliged to conform to the directive in relation to putting into service.

Authorized representative

This is the person or persons expressly appointed by the manufacturer by a written mandate to act on his behalf in respect of certain manufacturer's obligations within the EU. The extent to which the authorized representative may enter into commitments binding on the manufacturer is restricted by the relevant articles of the directive and determined by the mandate conferred on him by the latter.

As an example, he could be appointed to undertake the testing in the EU territory, sign the EC declaration of conformity, affix the CE marking and hold the EC declaration of conformity and the technical documentation within the EU at the disposal of the competent authorities.

The quality assessment system of the authorized representative/responsible person will not be subject to assessment by a NB, but the quality assessment system of the real manufacturer will be. It would not be reasonable to assess a quality assessment system of a person who is not producing the product and might only be a trading agent.

Other persons responsible for placing on the market

Where neither the manufacturer, nor the authorized representative is established within the EU, any other person resident in the EU who places the product on the EU market has obligations under the scope of the directive. The only obligation is to keep available the necessary documentation at the disposal of the competent authorities for ten years after the last product has been manufactured. In their capacity as 'person responsible for placing on the market' they are not entitled to assume other responsibilities, which are solely reserved to the manufacturer or his authorized representative (e.g. signing the EC declaration of conformity).

Why should manufacturer subject his product to ATEX new approach directive?

The manufacturers of explosion-protection equipment/apparatus may plan to meet the ATEX requirements based on the following criteria:

- Would their products be sold and used within the EU (including the country of origin) after 2002?
- Are their products intended for use in conjunction with potentially explosive atmospheres?
- Do their products contain potential ignition sources, e.g. electrical, mechanical, thermal?
- Do their products have an explosion control function, e.g. flame arresters, shut-off valves, quenching systems, relief panels?
- Do their products provide a control function for equipment or systems in a potentially explosive atmosphere?
- What range of flammable gases, vapors, mists or dusts will your products be used with?
- What is the likelihood that the flammable substances will be present in the air surrounding these products?
- What explosion safety standards are applicable to their products?
- To what extent have those standards been deemed to satisfy the ATEX requirements?
- To what extent will the existing certification of their products satisfy the ATEX requirements?

The answers to these questions will help manufacturers to establish whether the ATEX Directive applies to their products and if so which of its requirements they will have to meet.

If you are a manufacturer of products intended for use in potentially explosive atmospheres or operate, a plant that produces such areas, then the time to start working toward satisfying the requirements of the ATEX and 'Use' Directives was well before July 2003. Manufacturers and users must accept that 2003 has come and gone, otherwise they will have put themselves into the position of not being able to sell their products or operate their plants from the cut-off date. The notified bodies already had relatively long lead times for product approval in place which only increased as the implementation date of the directive approached. Additionally, a significant number of products that were covered by approvals issued prior to 1994 would require some redesign work in order to satisfy the latest approvals standards. Therefore, the time to start the process of gaining compliance with the directive if you are not to be left behind has long passed.

Approval lead times

As the directive will require the use of a NB for equipment intended for use in Zone 0 (an area where an explosive atmosphere is continually present, for example inside liquid fuel storage tanks) and Zone 1 (an area where an explosive atmosphere will be present in normal operation but not continually – for example re-fueling areas) classified areas, there will probably be a very long lead time for certification, as all existing certified equipment will have to be re-certified to the ATEX Directive if the manufacturer wishes to continue selling it in to Europe.

Self-certification

Products that are intended for use in Zone 2 will not require the involvement of a NB for approval. (This is an area where an explosive atmosphere will be present only under abnormal conditions for short periods, for example solvent storage areas. The solvent is normal contained in sealed drums however, if a drum is damaged during the process of moving it then a potentially explosive atmosphere could exist for a short period.) These products can be self-certified or if the in-house resources are not available, then third-party approvals from a suitable test house can be used.

Internal control of production is a process whereby the manufacturer, on his own responsibility, carries out the necessary work to ascertain that products, which he places on the market, comply with the requirements of the directive. Internal control of production applies to equipment in Category 2 and M2, which is neither electrical equipment nor internal combustion engines, and to Equipment Category 3. It is specified in Annex VIII of the Directive.

16.4   Bird's eye view of ATEX Directive

This directive consists of four chapters having sixteen articles and eleven annexes, as described hereunder. The Annexes of the ATEX Directive detail the specifics pertaining to the required conformity assessment procedures based on the equipment groups/categories.

Chapter I

This describes the scope of the directive. It also gives the requirements to be met for placing on the market the equipment for hazardous area and freedom of movement of such products, components, etc. This comprises of seven articles (1–7).

Chapter II

This describes in brief the conformity assessment procedures required to be met. Article 8 covers this.

Chapter III

This describes the requirement for CE conformity marking system and consists of Article 10.

Chapter IV

All other provisions are covered under this chapter and termed, Final provisions. It consists of 6 Articles (11–16).

Annex I

Criteria determining the classification of equipment-groups into categories

- Equipment-group I
- Equipment-group II.

Annex II

This annex describes the essential health and safety requirements relating to the design and construction of equipment and protective systems intended for use in potentially explosive atmospheres.

It is the responsibility of the NB to verify compliance to the essential safety requirements (ESRs) as part of an EC-type examination (see below).

It is to be noted that technological knowledge, which can change rapidly, must be taken into account as far as possible and be utilized immediately. The table below gives the requirements/obligations as given in various clauses of this annexure for guidance of manufacturers and users.

Common Requirements for Equipment and Protective Systems	
Clause	**Subject Covered**
1.0	General requirements
1.0.1	Principles of integrated explosion safety
1.0.3	Special checking and maintenance conditions
1.0.4	Surrounding area conditions
1.0.5	Marking
1.0.6	Instructions
1.1	Selection of materials
1.2	Design and construction
1.2.1	Adequate technological knowledge of explosion protection
1.2.2	Components to be incorporated
1.2.3	Enclosed structures and prevention of leaks
1.2.4	Dust deposits
1.2.5	Additional means of protection
1.2.6	Safe opening
1.2.7	Protection against other hazards
1.2.8	Overloading of equipment

(Continued

Clause	Subject Covered
1.2.9	Flameproof enclosure systems
1.3	Potential ignition sources
1.3.1	Hazards arising from different ignition sources
1.3.2	Hazards arising from static electricity
1.3.3	Hazards arising from stray electric and leakage currents
1.3.4	Hazards arising from overheating
1.3.5	Hazards arising from pressure compensation operations
1.4	Hazards arising from external effects
1.4.1	Safe function in changing external conditions
1.4.2	Withstanding attack by aggressive substances
1.5	Requirements in respect to safety-related devices
1.5.1	General requirements for safety devices
1.5.2	Safety device failure
1.5.3	Restart lockouts
1.5.4	Control and display units
1.5.5	Devices with a measuring function
1.5.6	Checking accuracy and serviceability
1.5.7	Safety factor
1.5.8	Risks arising from software
1.6	Integration of safety requirements relating to the system
1.6.1	Manual override
1.6.2	Emergency shutdown
1.6.3	Hazards arising from power failure
1.6.4	Hazards arising from connections
1.6.5	Placing of warning devices as parts of equipment
2.0	Requirements applicable to equipment in Category M of equipment – group I
2.0.1	Category M1
2.0.2	Category M2

Clause	Subject Covered
2.1	Requirements applicable to equipment in Category 1 of equipment – group II
2.1.1	Category 1G
2.1.2	Category 1D
2.2	Requirements for Category 2 of equipment – group II
2.2.1	Category 2G
2.2.2	Category 2D
2.3	Requirements applicable to equipment in Category 3 of equipment – group II
2.3.1	Category 3G
2.3.2	Category 3D
3.0	Supplementary requirements in respect of protective systems
3.0	General requirements
3.0.1	Dimensioning of protective systems
3.0.2	Design and positioning of protective systems
3.0.3	Functioning in event of power failure
3.0.4	Outside interface
3.1	Planning and design
3.1.1	Characteristics of materials
3.1.2	Withstanding shock waves
3.1.3	Accessories
3.1.4	Reactions caused by pressure in peripheral systems
3.1.5	Pressure-relief systems
3.1.6	Explosion suppression systems
3.1.7	Explosion decoupling systems
3.1.8	Protective systems

Annex III

Module EC-type examination

This module describes that part of the procedure by which a NB ascertains and attests that a specimen representative of the production envisaged meets the relevant applicable provisions of the directive.

EC-type examination is a process carried out by a NB to verify that a product type conforms to the relevant essential requirements. The process is specified in Annex II of the directive. The NB will examine the dossier of technical information (known as the technical file) supplied by the manufacturer and conduct such inspections and tests as may be required to show that the product type complies with the requirements stated by the manufacturer. The NB is required to verify compliance with the directive. This includes verifying that the manufacturer meets the essential safety requirements of the directive, evaluation to the EN 500xx series of standards, and the continuing verification such as quality assurance, product verification, or conformity to type (see below). The NB may advise on the applicable requirements but it is ultimately for the manufacturer to decide the intended use of the product. The EC-type examination process culminates in a certificate issued by the NB, attesting that the type as defined in the technical file complies with the directive.

Annex IV

Module: production quality assurance

The NB verifies compliance with a design and manufacturing quality assurance program, i.e. EN 29002 (ISO 9002). The quality registration should be with the NB. If it is not, the NB must review/audit the manufacturer's quality registration.

Annex V

Module: product verification

The NB evaluates each sample of a product as the manufacturer produces it. The NB is involved from start to finish in the manufacturing process. This specifically applies to Equipment Categories 1 and Ml and to protective systems.
    The directive requires:

- The manufacturer to ensure that the manufacturing process guarantees conformity of the equipment with the type described in the EC-type examination certificate. The manufacturer or his authorized representative in the EU to affix the CE marking to each piece of equipment.
- The NB to examine and test each item of equipment to verify conformity with the type as described in the EC-type examination certificate.
- The NB to affix its identification number to each approved item and provide a certificate of conformity covering each approved item.

Annex VI

Module: conformity to type

This module describes that part of the procedure whereby the manufacturer or his authorized representative established within the community ensures and declares that the equipment in question is in conformity with the type as described in the EC-type examination certificate and satisfy the requirements of the directive applicable to them. The manufacturer or his authorized representative established within the community shall affix the CE marking to each piece of equipment and draw up a written declaration of conformity.
    This specifically applies to electrical equipment and internal combustion engines in Equipment Categories 2 and M2.

The directive requires the manufacturer to:

- Ensure that the manufacturing process assures compliance of the manufactured products with the type described in the EC-type examination certificate.
- Carry out tests under the responsibility of a NB to confirm the conformity of each item manufactured with the certified type.
- Affix the CE marking to each item that has been found to be in conformity.
- Affix the NB's identification number to each item that has been found to be in conformity, under the responsibility of the NB.
- Periodic auditing of the capability.
- Examination of inspection and testing records for each item produced.

Annex VII

Module: product quality assurance

The manufacturer shall operate an approved quality system for the final inspection and testing of equipment and shall be subject to surveillance as specified. The identification number of the NB responsible for surveillance shall accompany the CE marking.

The directive requires the quality assurance system to address the following points:

- Quality objectives, organizational structure, responsibilities and powers of management with regard to product quality
- Examinations and tests that will be carried out after manufacture
- Means to monitor effective operation of the system
- Quality records (inspection reports, test data, calibration data, qualifications of personnel, etc.).

Note that this module requires each piece of equipment to be examined and tested as appropriate to confirm conformity with the requirements of the directive.

The NB verifies compliance with a manufacturing quality assurance program, i.e. EN 29003 (ISO 9003). The quality registration should be with the NB. If it is not, the NB must review/audit the manufacturer's quality registration.

Annex VIII

Module: internal control of production

This module describes the procedure whereby the manufacturer or his authorized representative established within the community shall affix the CE marking to each piece of equipment and draw up a written declaration of conformity. The manufacturer self-declares compliance with all provisions of the directive.

The NB involvement is for manufacturers of equipment in Category 2 and M2, which is neither electrical equipment nor internal combustion engines. In that case the manufacturer must deposit a copy of the technical file with a NB. The NB will hold the file for access by the responsible authorities at their request. Any changes to the technical file as a result of changes to the product or the introduction of variants must be copied to the NB so that they can add the amendment to the file. The NB will not examine the contents of the file nor be responsible in any way for the conformity of the product with the requirement of the directive.

Annex IX

Module: unit verification

The manufacturer declares compliance with the directive and the NB verifies compliance, conducting tests as necessary. This annex is primarily intended for one-time evaluations or unique products. This may be applied at the option of the manufacturer as an alternative to any other conformity assessment modules.

Annex X

CE marking

The CE conformity marking shall consist of the initials 'CE' taking the form in Figure 16.6.

'CE' conformity marking

Figure 16.6
CE conformity marking

If the marking is reduced or enlarged, the proportions given in the above graduated drawing must be respected.

The various components of the CE marking most have substantially the same vertical dimension, which may not be less than 5 mm.

This minimum dimension may be waived for small-scale equipment, protective systems or devices.

EC declaration of conformity

- Assess the conformity of the equipment with the essential requirements specified in the directive
- Draw up the technical documentation described in the info sheet on the technical file
- Check that each piece of equipment conforms to the design specified in the technical file
- Affix the CE marking to each conforming product
- Draw up a declaration of conformity
- Retain the declaration of conformity and the technical file for at least 10 years after the last piece of equipment was manufactured
- Update the technical file to cover changes to the equipment
- In some cases send a copy of the technical file to a NB.

Annex XI

Minimum criteria to be taken into account by member states for the notification of bodies

This gives guidelines for nomination of notified bodies. In nut shell, NB, its director and the staff responsible for carrying out the verification tests shall not be the designer, manufacturer, supplier or installer of equipment, protective systems, or devices referred to in Article 1 which they inspect, nor the authorized representative of any of these parties. This does not preclude the possibility of exchanges of technical information between the manufacturer and the body.

16.5 Summary

Major aspects on introduction of the ATEX Directive are summarized hereunder:

- *Degree of harmonization*: Total harmonization is mandatory.
- *Scope*: All equipment and protective system intended for use in potentially explosive atmospheres and auxiliary items
- *Areas of use*: All potentially explosive atmospheres – flammable gases, vapors, mists or dusts – with specific exceptions like medical and home environments, ships and public transport.
- *Combustible dust hazards*: The essential health and safety requirements cover technical requirements for equipment and protective systems where the risk arises from combustible dust.
- *Equipment categories*: Equipment is divided into two groups:

    1. Coal-mining (two sub-categories)
    2. Others (three subcategories).

- *Technical requirements*: This is a 'New Approach' Directive. It specifies the essential health and safety requirements (EHSRs), which must be met, but accepts compliance with other harmonized standards.
- *Conformity assessment*: Adopts various procedures ranging from EC-type examination to control of production by the manufacturer. Following briefly sums up the assessment guidelines as per directives:

    - *Categories 1 and M1 protective systems*: EC-type examination and either production quality assurance or product verification.
    - *Categories 2 and M2 electrical equipment and internal combustion engines*: EC-type examination and either product quality assurance or conformity to type.
    - *Categories 2 and M2 non-electrical equipment*: Internal control of production and deposit technical file with NB.
    - *Category 3*: Internal control of production.

As an alternative to the above procedures, unit verification can be used as a route to conformity.

- *Marking*: The 'CE conformity marking' is mandatory. In special cases, the ID number of the NB is also required. Other markings may be required.
- *Documentation*: Manufacturers are required to maintain technical records of all matters relating to explosion safety of equipment. All quality-system documentation must also be available.

Appendix A

IEC series standard titles for explosive atmospheres

IEC	Date	Title
60079-		*Electrical apparatus for explosive gas atmospheres*
0	1983	General Requirements Amendment No. 1 (1987) Amendment No. 2 (1991)
1	1990	Part 1: Construction and verification test of flameproof enclosures of electrical apparatus Amendment No. 1 (1993)
1A	1975	First Supplement Appendix D: Method of test for ascertainment of maximum experimental safe gap
2	1983	Part 2: Electrical apparatus, type of protection 'p'
3	1990	Part 3: Spark test apparatus for intrinsically safe circuits
4	1975	Part 4: Method of test for ignition temperature
4A	1970	First Supplement *Note*: This supplement applies also to the second edition of 1975
5	1967	Part 5: Sand-filled apparatus. First edition (1967) incorporating the first supplement (1969)
6	1996	Part 6: Oil-immersion, 'o'
7	1990	Part 7: Increased Safety, 'e' Amendment No. 1 (1991) Amendment No. 2 (1993)
10	1995	Part 10: Classification of hazardous areas
11	1991	Part 11: Intrinsic Safety, 'i'

IEC	Date	Title
12	1978	Part 12: Classification of mixtures of gases and vapors according to their maximum experimental safe gaps and minimum igniting currents
13	1982	Part 13: Construction and use of rooms or buildings protected by pressurization
14	1996	Part 14: Electrical installations in hazardous areas (other than mines)
15	1987	Part 15: Electrical apparatus with type of protection 'n'
16	1990	Part 16: Artificial ventilation for the protection of analysers(s) houses
17	1996	Part 17: Inspection and maintenance of electrical installations in hazardous area other than mines
18	1992	Part 18: Encapsulation, 'm'
19	1993	Part 19: Repair and overhaul for apparatus used in hazardous atmospheres (other than mines or explosives)
20	1996	Part 20: Data for flammable gases and vapors relating to the use of electrical apparatus
61241-		*Electrical apparatus for use in the presence of combustible dust*
1-1		Part 1: Electrical apparatus protected by enclosures Section 1: Specification for apparatus
1-2		Part 1: Electrical apparatus protected by enclosures Section 2: Selection, installation and maintenance
2-1		Part 2: Test methods Section 1: Methods for determining the minimum ignition temperatures of dust
2-2		Part 2: Test methods Section 2: Method for determining the electrical resistively of dust in layers
2-3		Part 2: Test methods Section 3: Method for determining minimum ignition energy of dust/air mixtures
3		Part 3: Classification of areas where combustible dust are or may be present
60529		*Degrees of protection provided by enclosures (IP Code)*
60050 (426)		*International electro technical vocabulary* *Chapter 426: Electrical apparatus for explosive atmospheres*
61779-		*Electrical apparatus for the detection and measurement of flammable gases*
1		Part 1: General requirements and test methods

(Continued)

IEC	Date	Title
2		Part 2: Performance requirements for Group I apparatus indicating a volume fraction up to 5% methane in air
3		Part 3: Performance requirements for Group I apparatus indicating a volume fraction up to 100% methane in air
4		Part 4: Performance requirements for Group II apparatus indicating up to 100% lower explosive limit
5		Part 5: Performance requirements for Group II apparatus indicating a volume fraction up to 100% gas
6		Part 6: Guide for the selection, installation, use and maintenance of apparatus for the detection and measurement of flammable gases

Appendix B

Listing of IS standards and codes of practice by country

Country	Certifying Authority	Standard	Code of Practice Specific to IS	Notes
Argentina	INTICITEI	IAP CA4.00		IEC adopted
Australia	SA	AS 2380.7-1987		IEC accepted
Brazil	CEPEL			CENELEC adopted
Canada	CSA	C 22.2 No.157	CEC Part 1 Sect 18/App F	IEC 79- becoming adopted from 1/1/98
China	NEPSI	GB 3836-1/7		Own standards
CIS (Russia)	VNIIVE, ISZWE	GOST 22782.5-78		IEC accepted CENELEC accepted
Czech Republic	FTZU	CSN 33 0380		
Denmark	DEMKO	EN 50 020 + 039		
France	LCIE	EN 50 020 + 039		
Germany	PTB, TUV	EN 50 020 + 039	VDE 0165	VDE 0171
Holland	KEMA	EN 50 020 + 039		
Hungary	BKI	MSZ 4814/7-77		IEC accepted
India	IS			IEC accepted
Italy	CESI	EN 50 020 + 039		
Japan	TIIS			Own standards. Was RIIS

(Continued)

Country	Certifying Authority	Standard	Code of Practice Specific to IS	Notes
Korea	KRS			
Norway	NEMKO	EN 50 020 + 039		
Poland	KDB	PN-84/E 08107		IEC accepted
Romania	ISM	STAS 6877/1-86 & 4-87		
South Africa	SSA	SABS 549: 1996	SABS 0108	Shell documents for IEC 79 Adoption
Switzerland	SEV	EN 50 020 + 039		
UK	SSS, BASEEFA	EN 50 020 + 039	BS 5345	Was BS 5501
USA	FM, UL	FM 3610-Entity	RP 12.1 and 12.6	NFPA NEC A.500-504

Notes: Countries may have more than one testing authority.
If unspecified, no local codes of practice were operated at the time of compilation.

Appendix C

IEC 79-17 Ex 'i' inspection schedule

Check that:	Grade of Inspection		
	Detailed	Close	Visual
A Apparatus			
1 Circuit and/or apparatus documentation is appropriate to area classification	●	●	●
2 Apparatus installed is that specified in the documentation – fixed apparatus only	●	●	
3 Circuit and/or apparatus category and group correct	●	●	
4 Apparatus temperature class is correct	●	●	
5 Installation is clearly labeled	●	●	
6 There are no unauthorized modifications	●		
7 There are no visible unauthorized modifications		●	●
8 Safety barrier units, relays and other energy-limiting devices are of the approved type, installed in accordance with the certification requirements and securely earthed where required	●	●	
9 Electrical connections are tight	●		
10 Printed circuit boards are clean and undamaged	●		
B Installation			
1 Cables are installed in accordance with the documentation	●		●
2 Cable screens are earthed in accordance with the documentation	●		●
3 There is no obvious damage to cables	●	●	
4 Sealing or trucking, ducts, pipes and/or conduits is satisfactory	●		
5 Point-to-point connections are all correct			●
6 Earth continuity is satisfactory (e.g. connections are tight and conductors are of sufficient cross-section)	●	●	
7 Earth connections maintain the integrity of the type of protection	●		

(Continued)

Check that:	Grade of Inspection		
	Detailed	**Close**	**Visual**
8 The intrinsically safe circuit is isolated from earth or earthed at one point only (refer to documentation)	●		
9 Separation is maintained between intrinsically safe and non-intrinsically safe circuits in common distribution boxes or relay cubicles			
10 As applicable, short-circuit protection of the power supply is in accordance with the documentation	●		
11 Special conditions of use (if applicable) are complied with	●		
12 Cables not in use are correctly terminated	●		
C Environment			●
1 Apparatus is adequately protected against corrosion, weather, vibration and other adverse factors	●		●
2 No undue external accumulation of dust and dirt	●	●	● ●

Appendix D

CENELEC members

Countries in the EC CENELEC membership

United Kingdom	Norway
Ireland	Sweden
France	Finland
Germany	Austria
Spain	Switzerland
Italy	Greece
Luxembourg	Denmark
Belgium	Netherlands
Portugal	Iceland
Associates	
Slovakia	Czech Republic
Poland	Hungary

Appendix E

IP code

The Ingress Protection Code IEC 529 specifies the degree of protection provided by enclosures to levels of solids and water as depicted in the code list.

Note that 'water' is used as the test for liquid ingress. Enclosures may suffer a greater degree of penetration from other liquids such as solvents where corrosion or reactions may also result. Assessment of ingress by other liquids shall therefore be determined by testing.

Code	Solids	Code	Water
0	No protection against ingress of solid foreign bodies	0	No protection
1	Protection against inadvertent contact with live or moving parts by the human body	1	Protection against drops of condensed water
2	Protection against contact by fingers	2	Protection against drops of liquid
3	Protection against contact by objects of thickness greater than 2.5 mm	3	Protection against rain
4	Protection against contact by objects of thickness greater than 1 mm	4	Protection against splashing liquid
5	Complete protection against contact with live or moving parts plus harmful deposits of dust	5	Protection against water jets
6	Complete protection against contact and ingress of dust	6	Protection against conditions on ships decks
		7	Protection against immersion in water
		8	Protection against indefinite immersion in water

Appendix F

Standards reference

Protection Technique	Australian Standard	CENELEC Standard	IEC Standard
General	AS 2380.1	EN 50014	IEC 60079-0
Ex 'd'	AS 2380.2	EN 50018	IEC 60079-1
Ex 'p'	AS 2380.4	EN 50016	IEC 60079-2
Ex 'e'	AS 2380.6	EN 50019	IEC 60079-7
Ex 'i'	AS 2380.7	EN 50020	IEC 60079-11
Ex 'n'	AS 2380.9	EN 50021	IEC 60079-15
Ex 'm'	AS 2431	EN 50028	IEC 61271-1
DIP	AS 2236		IEC 61241-1
IP	AS 1939	EN 50529	IEC 60529

Appendix G

Familiarization with electricity

This appendix should be read in conjunction with Chapter 2 and serves to complement the definitions and explanations given therein. It also gives an overview of our daily interface with electricity, to a person other than an electrical engineer.

The technical term electricity is the property of certain particles to possess a force field, which is neither gravitational nor nuclear. To understand what this means, we need to start simply looking at things around us.

Everything, from water and air to rocks, plants and animals, is made up of minute particles called atoms. They are too small to see, even with the most powerful microscope. *Atoms consist of even smaller particles called protons, neutrons and electrons.* The nucleus of the atom contains protons, which have a positive charge, and neutrons, which have no charge. Electrons have a negative charge and orbit around the nucleus. An atom can be compared to a solar system, with the nucleus being the sun and the electrons being planets in orbit (Figure G.1).

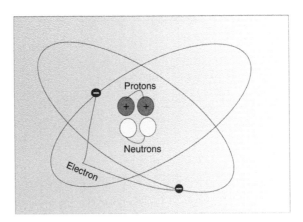

Figure G.1
Parts of an atom

Electrons can be freed from their orbit by applying an external force, such as movement through a magnetic field, heat, friction or a chemical reaction.

A free electron leaves a void, which can be filled by an electron forced out of its orbit from another atom. As free electrons move from one atom to another, an electron flow is produced. *This electron flow is the basis of electricity.*

The cliché, 'opposites attract', is certainly true when dealing with electrical charges. Charged bodies have an invisible electrical field around them. When two like-charged bodies are brought close together, they repel each other. When two unlike charged bodies are brought closer together, their electrical fields work to attract.

The flow of free electrons in the same general direction from atom to atom is referred to as current and it is measured in *amperes* ('amps' or 'A'). The number of electrons that flow through a *conductor*'s cross-section in 1 s determines amps. Current can be expressed in a number of different ways, such as:

Quantity	Symbol	Decimal
1 mA	1 mA	1/1000 A
1 A	1 A or 1 amp	1 A
1 kA	1 kA	1000 A

We have already seen what is a current. Further to it we come across in our daily life two types of electric current, AC and DC current:

1. Direct current (DC) flows in one direction.
2. Alternating current (AC) flows in two directions because the source voltage changes from positive to negative.

Similarly, electrical resistance is the opposition of a conductor to the flow of electrons. It is the ratio between the potential difference (V) across a conductor and the resulting current (I) that flows. The unit for electrical resistance is the ohm, which is represented with a 'Ω'.

Resistance is based on four factors:

1. The material of the conductor
2. The length of the conductor; the longer the conductor the greater the resistance
3. The conductor's cross-sectional area; the thicker the conductor the less the resistance
4. Temperature.

Ohm's law
This law is the very basis of all electrical engineering calculations and defines the relationship between voltage, current and resistance.

Current is directly proportional to voltage and inversely proportional to resistance

Formula: Pressure (volts) = Current (amps) × Resistance (ohms)

Now let us look at some other aspects connected with electricity supply, such as –

- Design of electrical installation – including architectural
- Estimation of load
- Generation of electricity
- Transmission of electrical supply
- Distribution of electrical supply
- Utilization of electrical energy, etc.

The architectural dimension

Electrical installations are vital in all modern buildings and industries. Provision has to be made not only for adequate space for the installed equipment, but also in allocating the cabling raceways. The space required for raceways needs careful consideration firstly for access for repairs and maintenance, and secondly because the bending radius of heavy cables is surprisingly large. Metal-cored cabling has high density, and cable-trays are often required.

The designer has to specify the switches of lighting and the positioning of power outlets. Where communication equipment is involved, electrical interference has to be avoided. There is much legislation to be complied with, particularly in dealing with the separation of electrical power from water and gas supplies.

Generally, adding extra electrical capacity during the construction or refurbishment program is cheaper than trying to add it later.

Assessing electrical needs

The process of establishing the electrical load has two stages, the first simpler than the second.

The initial task is to tabulate all of the equipment proposed to be installed in the building, including plant and office equipment. Once the total load is established, the second task is that of determining what the maximum demand at any one time will be. This will govern the size of the main distribution plant, as well as affecting the basis of calculation of the electricity bills in future. Load scheduling and automatic load shedding should be investigated as a means of minimizing peak demand.

The design of electrical supply and reticulation systems demands the employment of determining the specification and translating the organizational needs. Before setting out to specify a user's electrical installation, the user's business and the specific requirements, which it might have, must be understood. There are three strategic questions, which any business must address:

1. *Continuity of supplies*:   As discussed above, the importance of continuous operation of the business to the user must be assessed. Continuity of business is dependent upon the continuity of electrical power, and so it is an essential early consideration, which will influence the budget considerably.
2. *Growth*:   Allowance for the probable areas of growth must be made.
3. *Flexibility*:   Finally 'flexibility' and the 'churn' factor has to be taken into account. 'Flexibility' is determined in this context by the ability of a building to absorb plant and machinery and its ability to adapt to change. The value of in-built flexibility is measured in the reduced cost of each change multiplied by the number of changes against the annual value of the additional costs in providing the upgraded system.

Supply

In our day-to-day life we come across electrical systems' constituents like utility-owned pole-mounted over-head conductors, motor control centers, power distribution centers, pump-motor sets, local control panels, cables, bus-ducts, etc.

The local utility company, via four metal cables will supply electricity for most non-domestic buildings. In dense urban centers the cables are likely to be underground, otherwise overhead, carried by poles (refer Figure G.2). Environmental concerns will probably increase the number of remote area power supplies (RAPS) conferring immediate self-sufficiency for sub and urban domestic, and other small buildings.

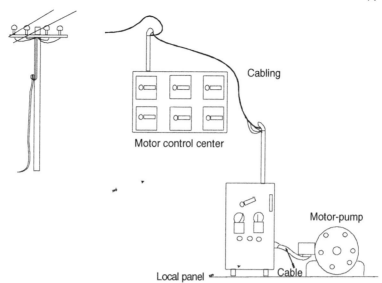

Figure G.2
A typical electrical installation

Generation

Most electrical supplies are generated by power stations. Without subsidies it would also be the most expensive form of energy, since its generation requires the use of other fuels and involves a 30% efficiency loss in doing so, and a further 30% loss in transmitting it to the user. Fuels like coal, uranium, etc. are burnt to generate steam, which drive turbines and generator sets. These stations are a major cause of carbon emission.

The power is transmitted via over-head transmission towers or cables (above ground or under-ground) at very high voltage (>15 000 V), which has to be vastly reduced by transformation to be usable for nearly all applications (Figure G.3).

Transformation

The high voltage is reduced by a large heavy static device known as transformer, which consists of a ring magnet with conductor coils wound around opposite segments. The ratio of turns on the two coils will determine the transformation ratio. The electromagnetic process involved produces fluctuating magnetic fields, heat (loss) and noise.

Transformers are installed either indoors or outdoors. The transformer housing (usually referred to as a substation) becomes the property of the utility company and must be designed according to their specifications. The generation of electro-magnetic interference, noise and heat has to be accounted for by the designer as does access for maintenance and when the transformer needs replacing.

Electrical intake

A heavy cable or cables will then be run from the transformer to a distribution frame. This might be in the switch-room of a large building. In any event, a heavy cable or a number of cables connect the incoming supply with the main electrical switchboard in the building. These cables have a diameter of over 50 mm and are therefore very difficult to bend and require a minimum fair radius of about 600 mm.

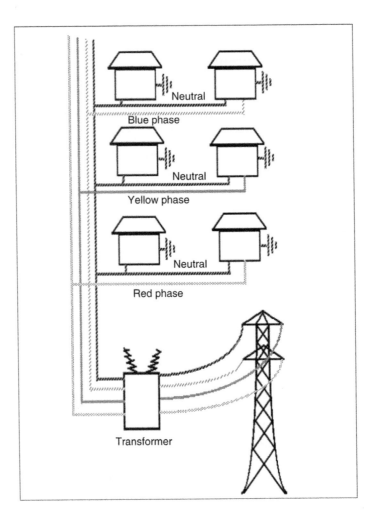

Figure G.3
Distribution of power

A rule of thumb for determining the bending radius of a Cable is,

$$\text{Bending radius} = 12 \times \text{Cable diameter}$$

Fuses

Between the incoming power supply and the switchboard, the utility company will install fuses. There are a number of types, but they generally consist of a metal conductor of lower melting point than the cables on either side. When the electrical power flows through conductors, they heat up. The fuse is designed to melt long before the cables. There will be a fuse or instantaneous trip device on all three live cables in a three-phase power supply.

Metering

Beyond the fuses will be the utility company's consumption meters, such that all power drawn will register. Meters are becoming increasingly sophisticated like digital, electronic, etc. Still the classical meter, of the type where a wheel driven at a speed proportional to the current drawn, increments a series of dials graduated in different orders of magnitude, is being used extensively.

Switchgear

The switchgear should be as near to the incoming electrical cables/overhead wires as possible. The physical size of the switchgear is important and depends upon the maximum power that will be drawn through it. If the switches are too small, there is a danger that the electrical power might arc between terminals, or they might melt. The larger the switchgear the more expensive it is, as it is usually made out of non-ferrous metal.

The switches will connect the supply with a distribution board. On this board will be the terminals from the incoming supply bus bars, from which a number of separately fused take-off points, will be tapped. These take-offs will lead to sub-distribution boards placed strategically around the plant and building. These sub-boards also have further fused tap-offs, each of which will supply a circuit. The capacity of the fuse will determine the power that each final circuit can take, and the rating of the cabling to be provided to each one.

Earthing and lightning protection

Electricity only flows when there is a difference in electrical potential between two points. To ensure that there is no PD between accessible surfaces of appliances and earth, electrical appliances are always 'earthed'; that is, they are electrically connected to a large object of low electrical potential – the earth.

If the potential of the earth fluctuates, then the current flowing through circuits will also fluctuate. This happens when lightning strikes nearby causing lights to dim. Lightning has a very high voltage and can in fact cause the electrical flow to reverse if the earth is raised to a higher potential than the supply. Depending where the lightning strikes, transformers and other electrical gear can explode, if fuses and diodes do not protect them. The concentration of electrical power in lightning can cause damage and fire. Buildings ought to be protected by a metal strap running from its highest point to an earthing point in the ground to protect against it.

Three-phase supply

The advantage of three-phase power is safety in that there is no resultant current if all the three transmission lines are shorted together simultaneously (Figure G.4). Each phase is 240 V; connecting two phases together 120° apart produces 415 V. A range of equipments, which uses 415 V supplies, is discussed further.

Motors, fans and pumps

Most industrial equipment involving heavy electric motors require three-phase power. When they start up, the power drawn might be of a magnitude greater than when they run at operating speed, hence lights dim. To prevent this power loss to other items of equipment, it is usual to provide 'capacitance' in the three-phase electrical system.

It is a good practice for each heavy equipment item to have its own fuse, isolation switch and emergency cutout.

Lifts, cranes, BMUs and hoists

Lifts are driven by electric motors, and installations usually have their own electrical switchboard. The intermittent usage because of frequent stopping and starting has led to the development of systems, which rather than varying the voltage to the motors, vary the frequency.

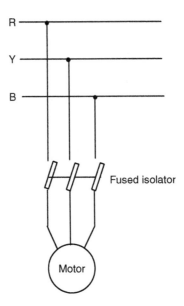

Figure G.4
Three-phase power

When a rheostat is used to vary the voltage, the resistance of the devices is increased, thus converting more of the power to heat. To avoid this waste it is more efficient to just 'chop' the full power into small bursts whose frequency is altered.

The power drawn by heavy motors has already been mentioned. Motors that are designed for constant stopping and starting are referred to as 'shunt-wound'. Normal motors are known as 'series-wound'. These two types of motors have different speed to torque characteristics.

Chillers and refrigeration

Apart from lifts, the most frequent requirement for three-phase supplies in buildings is the air conditioning plant. Motors power the refrigeration equipment, and fans and pumps circulate the chilled media. Larger installations will generally have a separate electrical switchboard.

Single-phase supply

The power supply to most desktop equipment is generally known as 'small power'. The power requirements of industries include small power, and the power is needed for lifts, lights, process equipment and utilities. Until perhaps ten years ago small power was a minor element in the total requirement. As IT equipment spread into offices small-power demand has grown substantially and has become a significant part of the total power requirement in non-air-conditioned office buildings.

In estimating the initial small-power demand for a new building, offices, industries for a known process, it is reasonable to extrapolate from the existing usage. Planners should base their estimates on the measured average power demands of the equipment to be used, and should never rely on the nominal loads.

Lighting

Commercial lighting schemes will draw much heavier currents and a consultant should design circuitry. The current drawn by 50 industrial light fittings would be considerable in comparison to normal household fittings. Hence, it is needed to enlist the help of a consultant.

Small power

Small means the 15 amp supply which powers industrial and office equipment. The supply should take the form of 30 amp fused ring circuits leading from and returning to the local switchboard. Each circuit should take no more than 10 switch socket outlets. Multiple adapters should be discouraged.

Emergency systems

Industrial and commercial premises have to be provided with emergency systems, such as lighting and escapes signage. These items generally have short-term battery packs, which are constantly charged from a 12 V transformer from the mains. When mains power is lost (or disconnected) the lights remain lit. Alternatively, these fittings can be connected by pyro cables, which have a high fire rating.

Other essential equipment, like fire pumps and fireman's lifts, have their own specially protected power supply and reticulation.

Circuitry is dependent on country. In some countries, such as South Africa, Australia, India and UK earth leakage protection is mandatory for small power requirements.

Clean power

It is a practice to provide special switch sockets giving access to a special clean power distribution network. The power might come from an UPS, or a system, which simply 'cleans' the main line power.

UPS

As organizations become increasingly dependent upon their computers, the necessity arises for fail-safe operation, as even the best electricity supply is not good enough for the computer environment. Momentary losses of supply are sufficient to cause computers to 'crash'. The risk is particularly high, for example – in the case of EFTPOS – electronic fund transfer from point of sale – systems, where the cash registers in the shops are linked directly to the shoppers' bank. EFTPOS systems in shops are but one of the organizations whose profitability is seriously affected by the inability to trade. The loss is not just the short term, but also the greater damage done to goodwill. The buyer is presented with the need, not just the opportunity, to find an alternative vendor with whom they might remain. Here, a failure of the computer/communication systems curtails retailing activity, and cost shops thousands of dollars per minute in lost sales. One solution is the installation of an uninterruptible power supply (UPS) (Figure G.5).

An UPS is precisely what the name suggests: a machine, which continues to provide electrical power without any noticeable glitch in the quality or continuity, despite disruption or noise in the mains. Cheaper UPSs may provide no more than a supply in a square sine wave form that will be suitable only for the simplest of computers.

Uninterruptible power supply consist of a bank of batteries (depending upon capacity requirement), which is charged by the electricity mains. The batteries feed an invertor, which converts the battery DC power into clean and reliable AC power. In case of a breakdown, the batteries maintain power for the short space of time, as necessary for the stand-by generators, to come on-stream and assume the electrical supply. At the very worse, the UPS allows the computers to be shutdown in an orderly fashion. Typically, some old types of lead acid batteries emit hydrogen, which has to be vented to a safe place, as it is explosive. Now a days we are getting maintenance-free and fully sealed batteries, which are being generally used.

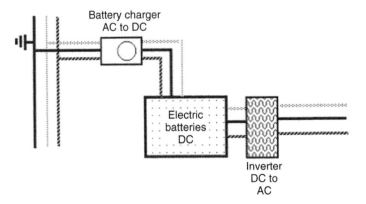

Figure G.5
Uninterruptible/emergency power supply

Wiring

Unlike the flow of water through a pipe, the amount of power conducted through an electric cable is not limited by physical size. As power flows, the conductor offers resistance and this produces heat. The greater the power flow the more heat is produced, and this is often the cause of fires. It is therefore important to match the cable capacity to the highest loads in the plant to which it might be subjected.

Single cables tend to be quite small and flexible. However, multiple cables bulk up rapidly, and require cable trays and dedicated runs to thread them through buildings.

The proliferation of cabling becomes a major problem in large and moderately sized buildings. This is due to three reasons:

1. The increase in the number of electrical devices
2. The increase in the type of devices and
3. The frequent addition and changes made to the equipment inventory.

This surreptitiously increases the power demand through existent cabling harnesses, hence the fire risk. In this respect, the use of multiple adapters is a potential hazard.

Effect of 'switching' on incoming electrical supply

The 'backwash' created by untoward events, like the random switching in and out of heavy loads or supplies, as well as local machinery, can produce interference in the power supply, as shown in Figure G.6. These transient 'glitches' in the supply are sufficient to disrupt sensitive computers and communications equipment.

It may be noticed that there are a few transient surges (Figure G.6), which are even higher than the V_{max} and these will repeatedly stress the insulation. Proper allowance for these need to given as per relevant codes and practices.

This aspect needs to be addressed by designer, while selecting the electrical equipment for hazardous area because any failure of insulation and resultant flash-over would be disastrous to the installation and may lead to explosion, given the conducive environment conditions.

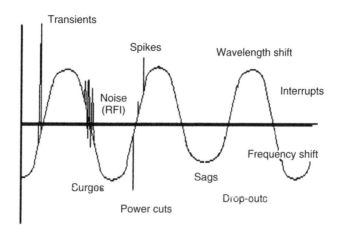

Figure G.6
Interference to power supply

Physiological effects of electricity

The physiological effects of electricity are 'primarily' related to the amount of current. The effects of current on the body are measured in milliamps,

- 1 mA = 0.001 A
- 1 A = 1000 mA.

Examples of effects based on current levels

- The 'Let Go Threshold' is that current above which one cannot let go of the conductive surface being grasped, and ranges from 6–10 mA. At this current level there is normally involuntary contraction of both extensor and flexor muscle groups. The flexors in the hand are stronger, thereby preventing the person from letting go.
- Painful shock with muscular contraction at 10–30 mA.
- Lung paralysis at 30–50 mA. Asphyxia occurs when the passage of continuous current through the chest cavity causes the chest muscles to constantly contract, thereby interfering with breathing. Respiratory arrest occurs when the current flows through the medulla of the brain at the base of the skull.
- Possible ventricular fibrillation at 50–100 mA. Ventricular fibrillation is an uncoordinated, abnormal contraction of the ventricular muscle fibers of the heart. This abnormal contraction causes the heart to quiver and the blood circulation ceases. Ventricular fibrillation can occur when the current flows through the chest or between an arm and a leg. At a current greater than 100 mA ventricular fibrillation is often fatal.

Appendix H

ATEX Directive

Situation:
1. Parts:
   Assembly is
   composed of
	Equipment, protective systems, devices (Article 1.2) all CE-marked, and components accompanied by a written attestation (Article 8.3) (parts with proven conformity)		Equipment, protective systems, devices (Article 1.2), including non-CE-marked, and components not accompanied by a written attestation (Article 8.3) (parts without proven conformity)	

2. Configuration:
   Assembly is
   placed on the
   market as
	Exactly defined configuration(s)	A 'modular system' of parts, to be specifically selected and configured to serve a specific purpose, maybe by the user/installer	Exactly defined configuration(s)	A 'modular system' of parts, to be specifically selected and configured to serve a specific purpose, maybe by the user/installer

3. Result:
   Manufacturer may
   presume
   conformity for
	All parts	All parts	Only parts with proven conformity	Only parts with proven conformity

4. Conformity Assessment (CA)	CA has to cover the whole configuration regarding all risks, which might arise by the interaction of the combined parts, with respect to the intended use	CA has to cover at least those of the possible and useful configurations, which are assessed to be the most unfavorable regarding all risks, which might arise, by the interaction of the combined parts, with respect to the intended use	CA has to cover: • All parts without proven conformity regarding all risks, and • All configuration(s) regarding all risks which might arise by the interaction of the combined parts, both with respect to the intended use	CA has to cover: • All parts without proven conformity which are part of the 'modular system', regarding all risks, and • At least those of the possible and useful configurations, which are assessed to be the most unfavorable regarding all

				risks which might arise by the interaction of the combined parts, both with respect to the intended use
5. Information to be provided (a) by EC-Declaration of conformity (b) by instructions for installation and use	(a) Identification of all parts forming the assembly (b) Instructions for installation and use, sufficient to ensure that resulting assembly complies with all relevant EHSRs of Directive 94/9/EC	(a) Identification of all parts forming the 'modular system' (b) Instructions for the selection of parts, to be combined to fulfil the required purpose, and instructions for installation and use, sufficient to ensure that resulting assembly complies with all relevant EHSRs of Directive 94/9/EC	(a) Identification of all parts forming the assembly (b) Instructions for installation and use, sufficient to ensure that resulting assembly complies with all relevant EHSRs of Directive 94/9/EC	(a) Identification of all parts forming the 'modular system' (b) Instructions for the selection of parts, to be combined to fulfil the required purpose, and instructions for installation and use, sufficient to ensure that resulting assembly complies with all relevant EHSRs of Directive 94/9/EC

Appendix I

Properties of combustible compounds

Compound	Ignition Temp	Temp Class	Gas Group	Compound	Ignition Temp	Temp Class	Gas Group
Acetaldehyde	140	T_4	IIA	Butadiene	430	T_2	IIB
Acetic acid	485	T_1	IIA	Butane	365	T_2	IIA
Acetone	535	T_1	IIA	Butanol	340	T_2	IIA
Acetylacetone	340	T_2	IIA	Butene	440	T_2	IIB
Acetyl chloride	390	T_2	IIA	Butyl acetate	370	T_2	IIB
Acetylene	305	T_2	IIC	Butylamine	(312)	T_2	IIA
Acrylonitrile	480	T_1	IIB	Butyldigol	225	T_3	IIA
Allyl chloride	390	T_1	IIA	Butyl methyl ketone	(530)	(T_1)	IIA
Allylene	–	–	IIA	Butyraldehyde	230	T_3	IIA
Ammonia	630	T_1	IIA	Carbon disulfide	100	T_5	–
Amphetamine	–	–	IIA	Carbon monoxide	605	T_1	IIB
Amyl acetate	375	T_2	IIA	Chlordimethyl ether	–	–	IIA
Amyl methyl ketone	–	–	IIA	Chlorobenzene	637	T_1	IIA
Aniline	617	T_1	IIA	Chlorobutane	(460)	(T_1)	IIA
Benzene	560	T_1	IIA	Chloroethane	510	T_1	IIA
Benzaldehyde	190	T_4	IIA	Chloroethanol	425	T_2	IIA
Benzyl chloride	585	T_1	IIA	Chloroethylene	470	T_1	IIA
Blue water gas	–	T_1	IIC	Chloromethane	625	T_1	IIA
Bromobutane	265	T_3	IIA	Chloropropane	520	T_1	IIA
Bromoethane	510	T_1	IIA	Coal tar naphtha	272	T_3	IIA

Compound	Ignition Temp	Temp Class	Gas Group	Compound	Ignition Temp	Temp Class	Gas Group
Coke oven gas	–	–	IIB	Dioxolane	–	–	IIB
Cresol	555	T_1	IIA	Epoxypropane	430	T_2	IIB
Cyclobutane	–	–	IIA	Ethane	515	T_1	IIA
Cyclohexane	259	T_3	IIA	Ethanol	425	T_2	IIA
Cyclohexanol	300	T_2	IIA	Ethanolamine	–	–	IIA
Cyclohexanone	419	T_2	IIA	Ethoxyethanol	235	T_3	IIB
Cyclohexene	(310)	(T_2)	IIA	Ethyl acetate	460	T_1	IIA
Cyclohexylamine	290	T_3	IIA	Ethyl acrylate	–	–	IIB
Cyclopropane	495	T_1	IIB	Ethylbenzene	431	T_2	IIA
Decahydronaphthalene	260	T_3	IIA	Ethyldigol	–	–	IIA
Diacetone alcohol	640	T_1	IIA	Ethylene	425	T_2	IIB
Diaminoethane	385	T_2	IIA	Ethylene oxide	440	T_2	IIB
Diamyl ether	170	T_4	IIA	Ethyl formate	440	T_2	IIA
Dibutyl ether	185	T_4	IIB	Ethyl mercaptan	295	T_3	IIA
Dichlorobenzene	(640)	(T_1)	IIA	Ethyl methyl ether	190	T_4	IIB
Dichloroethane	440	T_2	IIA	Ethyl methyl ketone	505	T_1	IIA
Dichloroethylene	(440)	(T_2)	IIA	Formaldehyde	424	T_2	IIB
Dichloropropane	555	T_1	IIA	Formdimethylamide	440	T_2	IIA
Diesel fuel	220	T_3	IIA	Hexane	233	T_3	IIA
Diethylamine	(310)	(T_1)	IIA	Hexanol	–	–	IIA
Diethylaminoethanol	–	–	IIA	Heptane	215	T_3	IIA
Diethyl ether	170	T_4	IIB	Hydrogen	560	T_1	IIC
Diethyl oxalate	–	–	IIA	Hydrogen sulfide	270	T_3	IIB
Diethyl sulphate	–	–	IIA	Isopropylnitrate	175	T_4	IIB
Dihexyl ether	185	T_4	IIA	Jet fuels	220	T_3	IIA
Di-isobutylene	(305)	(T_2)	IIA	Kerosene	210	T_3	IIA
Dimethylamine	(400)	(T_2)	IIA	Methaldehyde	–	–	IIA
Dimethylamine	370	T_2	IIA	Methane (firedamp)	595	T_1	IIA
Dimethyl ether	–	–	IIB	Methane (industrial)	–	T_1	IIA
Dipropyl ether	–	–	IIB	Methanol	455	T_1	IIA
Dioxane	379	T_2	IIB	Methoxyethanol	285	T_3	IIB

(Continued)

Compound	Ignition Temp	Temp Class	Gas Group	Compound	Ignition Temp	Temp Class	Gas Group
Methyl acetate	475	T_1	IIA	Petroleum	220	T_3	IIA
Methyl acetoacetate	280	T_3	IIA	Phenol	605	T_1	IIA
Methyl acrylate	–	–	IIB	Propane	470	T_1	IIA
Methylamine	430	T_2	IIA	Propanol	405	T_2	IIA
Methylcycohexane	280	T_3	IIA	Propylamine	(320)	(T_2)	IIA
Methylcycohexanol	295	T_3	IIA	Propylene	(455)	(T_1)	IIA
Methyl formate	450	T_1	IIA	Propyl methyl ketone	505	T_1	IIA
Naphtha	290	T_3	IIA	Pyridine	550	T_1	IIA
Naphtalene	528	T_1	IIA	Styrene	490	T_1	IIA
Nitrobenzene	480	T_1	IIA	Tetrahydrofuran	(260)	(T_3)	IIB
Nitroethane	410	T_2	IIB	Tetrahydrofurfuryl alcohol	280	T_3	IIB
Nitromethane	415	T_2	IIA	Toluene	535	T_1	IIA
Nitropropane	420	T_2	IIB	Toluidine	480	T_1	IIA
Nonane	205	T_3	IIA	Toluol	535	T_1	IIA
Nonanol	–	–	IIA	Triethylamine	–	–	IIA
Octaldehyde	–	–	IIA	Trimethylamine	(190)	(T_4)	IIA
Octanol	–	–	IIA	Trimethylbenzene	470	T_1	IIA
Paraformaldehyde	300	T_2	IIB	Trioxane	410	T_2	IIB
Paraldhyde	235	T_3	IIA	Turpentine	254	T_3	IIA
Pentane	285	T_3	IIA	Xylene	464	T_1	IIA
Pentanol	300	T_2	IIA				

Appendix J

Important changes to AS/NZS 2381.1

The active part being played by Standards Australia in IEC global harmonization and subsequent adoption of the same by Australian and New Zealand standard bodies has brought about certain changes in earlier version of AS 2381.1. This listing only highlights the major changes and there could be numerous others. In our opinion these will have major impact on manufacturers and users of the electrical equipment in explosive atmosphere.

The main changes from the 1991 and 1999 edition are listed below. Clause numbers are the ones used in the 1999 edition:

Clause No.	Change
1.4.47	In a fundamental shift, Australia has now aligned with the IEC and there are three dust zones (20, 21 and 22) rather than the old Class II. As a result Class I has also been removed so there are no longer Class I Zone 1, Class I Zone 2 areas, etc. The gas hazardous areas are now simply called Zone 0, Zone 1, Zone 2 (which is really what the lazy ones amongst us in O&G and petrochem have done for years). People with dust hazards will now have to quantify the Zone and this will affect what technique can be used. With the advent of AS 61241.1.1 replacing AS 2236, DIP will now come in 6 forms (DIP A20, DIP A21, DIP A22, DIP B20, DIP B21 and DIP B22)
1.6	The contents required in the verification dossier have now been added to this standard rather than only residing in the other parts. If the dossier is not kept on site, the site must have a document which states who the owners are and where the dossier is located
1.7	Perhaps the biggest change that will affect users is that personnel designing, constructing, maintaining, testing and inspecting installations in accordance with the standard must be competent. Competency may be demonstrated by meeting the EEHA Standards or the equivalent

(Continued)

Clause No.	Change
1.9.9	The dangers of radiation from optical equipment passing through hazardous areas are now included. Recommendations regarding power and flux are given in Appendix G
1.11.3	Rewording has clarified the issue of working on equipment that is still alive. Normal HOT WORK procedures that include initial and periodic checks for flammable gas should meet the requirements
Table 2.1	Ex 'o' and Ex 'q' have been added to the table showing which equipment is allowed in a certain zone. They are recognized as Zone-1 techniques
Table 2.2	The T rating of gases is now to be taken from IEC 60079-20 rather than NFPA 325M
2.4.5	A new clause is added covering the use of enclosures with an internal source of release of flammable gas. The type of protection required is determined by the hazardous area surrounding the enclosure, and the extent and type of release that can occur within the enclosure
2.4.6	All motors supplied by VSDs must now have additional protection to ensure temperature limits are not exceeded. The requirements vary depending upon technique. For Ex 'e', the motor, converter and protection device must be type tested together. Ex 'd' and DIP motors can be tested per Ex 'e' or they can have embedded temperature sensors which will trip the motor on high temperature. Ex 'n' motors can be tested per Ex 'e' or they can have their temperature rise assessed by calculation for the specified duty
2.5.3.2	For dusts, the temperature differential between cloud/layer ignition temperature and equipment temperature has been increased to 75 °C (from 50 °C). However, this may be reviewed soon, as it is not in accordance with the IEC Standards
2.5.4	With the introduction of three zones for dusts, and the replacement of the Australian DIP standard with an IEC clone, the equipment allowed in dust areas has been revised. Very basically, any equipment in Zone 20 must be approved for Zone 20 use (standards still being developed). The old DIP (to AS 2236), DIP A20 and 21, DIP B20 and 21, Ex 'i', Ex 'p' and Ex 'm' can all be used in Zone 21. In Zone 22, all the above are OK plus DIP A22 and B22 if the dust is non-conductive. Problems will occur here because IEC allows manufacturers to self-certify DIP A22 and B22 but Australia will still require testing station certification
2.6	With Australia now subscribing to the IEC Ex scheme the certification requirements for equipment have changed. Electrical equipment used in a hazardous area must now comply with either the relevant IEC or the relevant Australian Standard. Proof of compliance is via a certificate of conformity issued under the AUS Ex or IEC Ex scheme. Equipment certified to an alternative standard, but shown to provide equivalent safety, can be accepted by the regulatory authority

Clause No.	Change
2.7	A new clause has been added with a flowchart to show the selection rules for existing or repaired equipment
3.2.4	A new clause now requires the user to consider the impact of safe area equipment upon safety in a hazardous area (VSDs, protection devices)
3.2.5	The installation rules in the standard now apply to I.S. too, except where noted otherwise. Generally, if the rule is in place to stop an incendive circuit creating a spark, the rule will not apply to I.S.
3.3.1	The need to earth metal parts of ELV has now been emphasized in the standard. This used to be lost in AS 3000
3.3.2	Requirements for earthing conductor type/locations and their termination have been expanded to cover LV and ELV. In Zone 2, earth conductors can now be run externally, with certain restrictions
3.3.3	Spare conductors must now be earthed at least at one end
3.4.2	Recommendations are now given for the temporary equipotential bonding of equipment (e.g. pail or vehicle loading)
3.5	The generally unknown and ignored detailed requirements regarding earth fault protection have been removed. Now Ex equipment must be protected so that they are quickly disconnected in the event of overcurrent or fault. AS/NZS 3000-2000 gives requirements for disconnect times of electrical equipment under fault. Precautions must now also be taken to prevent motors single phasing
3.6	The electrical isolation requirements have been updated to allow the use of procedures in lieu of locking-off. This now enables the isolation of instruments via disconnected terminals when actioned under adequate permitting
3.7	In a new requirement, an emergency off switch/es must be available which enables the isolation of electrical supplies to the hazardous area
Table 3.1	The table showing allowable cable types has been updated to include braiding designed for mechanical protection. The user needs to be aware that in some cable standards the braiding is NOT deemed to be mechanical protection. The writer's opinion is that braiding is inferior to SWA as it offers less protection and corrodes easily at the cable gland. Cable trunks, trenches, etc. have been added but they must be installed so that they do not create a vapor passage medium
3.8.13	For multistrand cables, the ends must be protected against separation (e.g. lugs, terminal type, etc.)

(Continued)

Clause No.	Change
3.8.15.2	The thread engagement requirements for conduits have been changed from a prescribed axial length in AS 2381.1 to the quantity required by the protection device to maintain Ex protection
3.8.15.4	The distance a conduit seal has to be installed on a conduit entry to a flameproof enclosure has been reduced from 500 to 450 mm. The writer assumes that this is to align with the USA, however, the logic is flawed as IEC and SAA Ex 'd' equipment are not tested with conduit. Ultimately this clause will be changed, probably to place the seal at the entry of the flameproof enclosure. The writer's preference is that there is to be no conduit between the enclosure and the seal because this section will see explosion pressure and the conduit thread is usually site cut. Keep all conduits 'after' the seal. Only use Ex 'd' fittings to join the seal to the enclosure
3.14	A section on specific occupancies has been added giving new rules for electric ovens, fuel dispensing, finishing processes, medical agents, lab fume cupboards and secondary batteries in building
4.3	Inspection requirements are now given in Part 1, rather than referring the reader to the specific parts for each technique. Some points of interest are: The use of 'continuous supervision' in lieu of periodic inspections is raised. This matter is receiving IEC consideration. So far, I am far from convinced the periodic inspections can be replaced. The frequency between inspection has been extended to a maximum of four years without expert advice (was three years). One year for portable equipment
5.2	The old standard said that no alterations that might invalidate the certificate may be made without approval of the authority. This clause has been removed, but a new sentence has been added stating that modifications must be carried out in accordance with AS 3800
5.10	A new clause links AS 2381.1 to AS 3800. Overhaul and repair shall comply with AS 3800
App A and B	Two new appendices regarding O/S-certified equipment. Applicable to NZ only
App E	A new appendix giving guidance on using Ex equipment when there could be internal releases of flammables
App G	A report detailing recent work on optical energy creating ignition

(Source: Alan Wallace of Inlec Engineering)

Appendix K

Practical exercises for hazardous area course

K.1   Introduction

The following is a list of exercises to be worked through in the class setting.   The exercises are designed to be carried out in conjunction with the information presented in the book. Due to pressure of time, not all these exercises will be necessarily carried out. The exercises are not in a specific order.

Please read the information fully before proceeding.

K.1.1   Exercise titles

1. Area classification of your workplace – discussion in class of a few case studies of participants
2. Area classification of a fuel station
3. Brain teasers on hazardous areas
4. A case study for additions or modification in hazardous area of a plant
5. A case study in hazardous area classification of equipment
6. Area classification of a process plant – a schematic
7. Hazardous area classification for 'a fixed roof tank'
8. Exercise on protection concepts and inspection
9. Hazardous area classification for compressor housing
10. Exercise on protection concepts Ex 'd'
11. Switch status input system calculations
12. Solenoid valve operational characteristics.

K.2 Exercises

K.2.1 Exercise 1 Area classification of your workplace

1.1. Think of your work area. Make a list of the 'fuels' that are present that could be a hazard.

1.2. When, where and how are these substances present?

1.3. What could cause these substances to ignite? Be specific.

1.4. Describe the ventilation of the area.

1.5. Given all the above information, what do you think the area classification of the work area would be?

K.2.2 Exercise 2 Area classification of a fuel station

From Figure Ex. 2.1 answer the following:

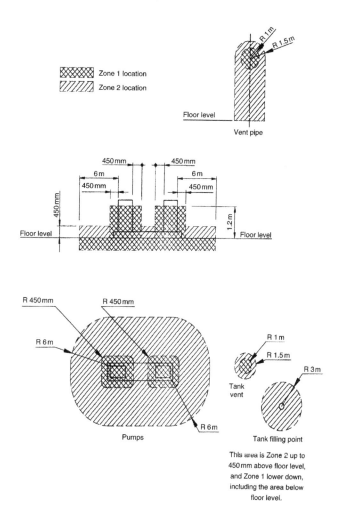

Figure Ex. 2.1
From SABS 0108: 1995

2.1. Make a list of the allowable protection methods for Zone 1.
2.2. The area classification is for a fuel station.

    (a) What types of equipment will typically be used?
    (b) What would the typical Ex rating be for each of these pieces of equipment?

2.3. Fuel and diesel fuel are mixtures of several chemical compounds. How do you think complex chemicals such as these are classified?
2.4. If you consider the situation inside the storage tanks, what do you think the area classification would be? Why?
2.5. What type of electronic sensing device/equipment would you recommend to determine the amount of fuel left in the tank?

K.2.3   Exercise 3   Brain teasers on hazardous areas

3.1. Name several factors that could determine the frequency of maintenance and inspection of equipment in hazardous areas.

3.2. Why are light metal alloys, such as magnesium, not allowed in hazardous areas?

3.3. How would you determine whether a flameproof enclosure is damaged?

3.4. How would you repair intrinsically safe devices?

K.2.4 **Exercise 4 A case study for additions or modifications in hazardous areas of a plant**

Eric Patterson just left the office of the plant manager of the petrochemical company that he works for. He had been given a task that he felt uneasy about. The company did a study and found that it could be economically viable to purify and then sell the cyclohexane that is produced as a by-product from one of the other processes in the plant.

Being a chemical engineer, he knew that cyclohexane is highly flammable and that he had to be careful in considering the safety of the personnel and the plant in the design of the process. He did feel, however, that his knowledge of dealing with this type of hazard was limited and decided to give a friend from the university a call.

Linda Watson was one of the recognized experts in the field of hazardous areas and arranged to visit Eric at the plant to discuss the problem.

The meeting was convened in Eric's office a week later with the following people present: Linda, Eric, two other process engineers, an electrical engineer and the chief control system engineer.

4.1. If you were Linda what would the first things be that you would advise Eric to do?

4.2. Make a schematic/block diagram of such a plant.
   The following are the basics of the process:

   - Unpurified cyclohexane is pumped into the plant from the other process and stored into large tanks.
   - The solution is heated to 100 °C and another chemical is added that binds to the metals that are present in the solution.
   - The impurities are heavy and will drop to the bottom of the tank.
   - The impurities are now extracted and drained away from the tank.
   - The purified cyclohexane is now pumped into storage tanks from where it will be pumped to tanker vehicles or other containers to be transported to their customers.

4.3. Do an area classification for your plant. What shortages do you notice? What else may be needed?

4.4. Select equipment for use in your plant.

4.5. How would you do the electrical installations in your plant when considering the plant is fully automated and controlled from a central location?

4.6. How would you address earthing in your plant when taking into account that a fair number of lightning strikes occur each year?

4.7. Draw up a maintenance and inspection schedule.

Additional information

Cyclohexane is an IIA gas with an ignition temperature of 285 °C.

K.2.5   Exercise 5   A case study of hazardous area classification of equipment

Since many delegates will be familiar with the basic design and construction of motor vehicles, this exercise has been specifically chosen to prompt discussion on the concepts of area classification. For the purposes of instruction, the delegates are asked to examine how the 'industrial practice' would be applied.

It should be noted that extensive vehicle testing is performed to meet exacting international standards for fire safety. Motor vehicle manufactures have agreed standards, which are suitable for these specialized conditions of risk.

Objectives

Petrol-driven motor vehicles such as the typical domestic car comprise mechanical and electrical systems that generate heat and energy in close proximity to the fuel, petrol. Since the layout of most cars is somewhat similar and known to many delegates, in this hypothetical exercise, you are required to area classify the vehicle shown in Figures Ex. 5.1 and Ex. 5.2.

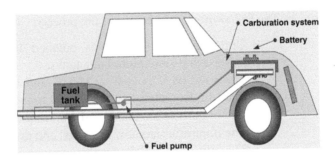

Figure Ex. 5.1
Side elevation of car

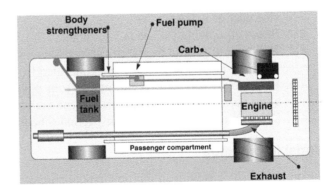

Figure Ex. 5.2
Plan view of car

It is suggested that you work in small teams and brainstorm some ideas. The task is complete when the sketch of the car has been clearly marked with adequate information on the 'hazard' and the area classification assigned to the vehicle in its various conditions of use. The team should agree on the conclusions. Indicate any areas you wish to designate Zones 0, 1 or 2.

Supplementary instructions

It is appreciated that you may have insufficient information to complete the task. Where you require more information, note this need adjacent to a relevant part or area.

The vehicle detail shown is diagrammatic and is specifically for the purposes of this exercise. The 'area classification team' may augment the drawing with additional information.

Any assumptions made should be recorded by appropriate annotations.

Method of approach

Use the principles and terminology outlined in Chapter 3 of this book. Assess the risk by taking account of the nature of the hazard. Consider likely scenarios and postulate possible situations to be considered.

How would the modern car be area classified according to industrial practice? Assuming the layout to be as shown in Figures Ex. 5.1 and Ex. 5.2 (plan and elevation), what should be considered? What other information do you need to know?

Consider the four likely operational situations:

1. Parked in a garage
2. Parked on the road
3. Low speed
4. High speed.

Your ideas should list the methodology used including:

- The identification of all sources of ignition (electrical and mechanical).
- The identification of all locations and routes (storage and piping) of fuel.

Petrol flammability characteristics

Flashpoint	$-40\,°C$
Ignition temperature	$310\,°C$
Vapor density	1.24
Ignition energy	IIA
Vapor dispersion in still air	4 m in diameter from source of release. 0.5 m high from the level ground

You are urged to consider the following situations:

- The vehicle is first started (from cold).
- The engine running but the vehicle is stationary.
- The car moves slowly, say, below 10 mph/16 kph.
- Higher speeds above this value.
- Accidents of varying degrees of severity.
- Long-term storage.
- Fuel leaks and spillage.

What other influences are present on the behavior of the fuel and/or vapor?
How do you think a car can be made 'safe' from risk of explosion protection?

K.2.6   Exercise 6   Area classification of a process plant – a schematic

This exercise has been specifically chosen to prompt discussion on the protection concepts once area classification has been done. For the purposes of instruction, the delegates are asked to examine how the 'industrial practice' would be applied.

Objectives

The schematic of a process area is given below. Indicate type of protection concepts you will choose for each of equipment in various zones to make it compliant to standards. The economics of selection should not be forgotten.

Supplementary instructions

It is appreciated that you may have insufficient information to complete the task. Where you require more information, note this need adjacent to a relevant part or area.

You may augment the drawing with additional information.

Any assumptions made should be recorded by appropriate annotations.

The method of protection needs to be marked for the following equipment and apparatus (Figure Ex. 6.1):

- Lighting
- Enclosure – suggest the best method for protection of various apparatus housed in it
- Valve position sensor
- Motor
- Motor starter

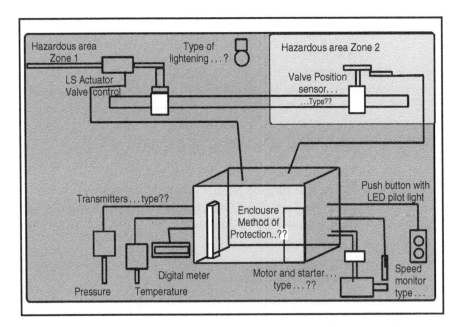

Figure Ex. 6.1
The hazardous area ... application of protection technique

- Speed monitor
- Push button with pilot light.

Method of approach

Use the principles and terminology outlined in this course. Consider likely scenarios and postulate possible situations to be considered.

K.2.7   Exercise 7   Hazardous area classification for 'a fixed roof tank'

Since many delegates will be familiar with the basic design and construction of a process plant, this exercise has been specifically chosen to prompt discussion on the concepts of area classification. For the purposes of instruction, the delegates are asked to examine how the 'industrial practice' would be applied.

Objectives

The tank is fixed roof type construction. It is supposed to be handling gas, which is lighter than air (Figure Ex. 7.1).

Indicate any areas you wish to designate Zones 0, 1 or 2 in the areas already defined in the drawing.

Supplementary instructions

It is appreciated that you may have insufficient information to complete the task. Where you require more information, note this need adjacent to a relevant part or area.

You may augment the drawing with additional information.

Any assumptions made should be recorded by appropriate annotations.

You should also try to estimate the extent of area falling under the influence of the zone classified.

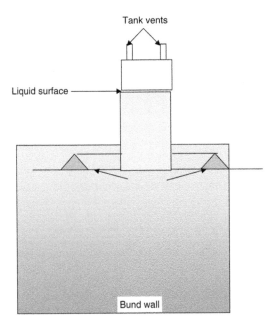

Figure Ex. 7.1
The elevation of the tank with bund wall all around

Method of approach

Use the principles and terminology outlined in Chapter 3 of this book. Assess the risk by taking account of the nature of the hazard. Consider likely scenarios and postulate possible situations to be considered.

K.2.8   Exercise 8   Exercise on protection concepts and inspection

Instructions: Based on the definitions of initial, periodic and visual inspection fill in the following:

'ALL' – To mean all apparatus
'SAMPLE' – To mean the size to be decided by authorized person of the plant.

A. Inspection schedule for apparatus relying on pressurizing/purging concept of protection

Check that	Types of Inspection		
	Initial	Periodic	Visual
Apparatus is appropriate to area classification			
Apparatus surface temperature class is correct			
Apparatus carries the correct circuit identification			
There are no unauthorized modifications			
Earthing connections, including any supplementary earthing connections, are clean and tight			
Earth loop impedance or resistance is satisfactory			
Lamp rating and type is correct			
Source of pressure/purge medium is free from contaminants			
Pressure/flow is as specified			
Pressure/flow indicators, alarms and interlocks function correctly			
Pre-energizing purge period is adequate			
Ducting, piping and enclosures are in good condition			
No undue external accumulation of dust and dirt			

B. Inspection schedule for apparatus relying on protection concept 'n', 's' or 'e'...

Check that	Types of Inspection		
	Initial	Periodic	Visual
Apparatus is appropriate to area classification			
Apparatus group (if any) is correct			
Apparatus surface temperature class is correct			
Apparatus carries the correct circuit identification			

(Continued)

Check that	Types of Inspection		
	Initial	Periodic	Visual
Enclosures, glasses and glass/metal parts are satisfactory			
There are no unauthorized modifications			
Earthing connections, including any supplementary earthing connections, are clean and tight			
Earth loop impedance or resistance is satisfactory			
Lamp rating and type is correct			
Bolts, glands and stoppers are of the correct type and are complete and tight			
Enclosed-break and hermetically sealed devices are undamaged			
Condition of enclosure gaskets is satisfactory			
Electrical connections are tight			
Apparatus is adequately protected against corrosion, the weather, vibrations and other adverse factors			
There is no obvious damage to cables			
Automatic electrical protection devices are set correctly			
Automatic electrical protection devices operate within permitted limits			
No undue external accumulation of dust and dirt			

C. Inspection schedule for intrinsically safe systems

Check that	Types of Inspection		
	Initial	Periodic	Visual
System and/or apparatus are appropriate to area classification			
System group or class is correct			
Apparatus surface temperature class is correct			
Installation is correctly labeled			
There are no unauthorized modifications (including readily accessible lamp and fuse ratings)			
Apparatus is adequately protected against corrosion, the weather, vibration and other adverse factors			

Check that	Types of Inspection		
	Initial	**Periodic**	**Visual**
Earthing connections are permanent and not made via plugs and sockets			
The intrinsically safe circuit is isolated from earth or earthed at one point only			
Cable screens are earthed in accordance with the approved drawing			
Barrier units are of the approved type, installed in accordance with the certification requirements and securely earthed			
Electrical connections are tight			
Point-to-point check of all connections			
Segregation is maintained between intrinsically safe and non-intrinsically safe circuits in common marshaling boxes or relay cubicles			
There is no obvious damage to apparatus and cables			

K.2.9   Exercise 9   Hazardous area classification for compressor house

Objectives

The compressor type could be either reciprocating or centrifugal. It is supposed to be handling gas, which is heavier than air. The sides of the building are open, and roof-ventilation is envisaged (Figure Ex. 9.1).

Indicate any area you wish to designate Zones 0, 1 or 2.

Supplementary instructions

It is appreciated that you may have insufficient information to complete the task. Where you require more information, note this need adjacent to a relevant part or area.

You may augment the drawing with additional information.

Any assumptions made should be recorded by appropriate annotations.

You should also try to estimate the extent of area falling under the influence of the zone classified.

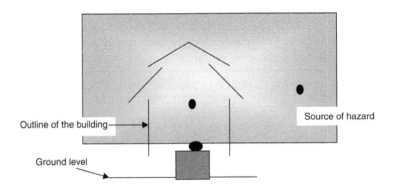

Figure Ex. 9.1
The elevation of the building housing the compressor

Method of approach

Use the principles and terminology outlined in Chapter 3 of this book. Assess the risk by taking account of the nature of the hazard. Consider likely scenarios and postulate possible situations to be considered.

K.2.10 Exercise 10 Exercise on protection concepts Ex 'd'

Introduction

The cabling is a vital part to maintain the integrity of the Ex 'd' protection. Figure Ex. 10.1 is a diagram of a cable gland in which a part is missing.

You are supposed to identify it and place it appropriately.

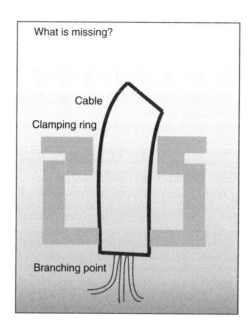

Figure Ex. 10.1
Cable gland for Ex 'd' protection

K.2.11   Exercise 11   Switch status input system calculations

Introduction

This exercise is based on the switch inputs to a high reliability shutdown system.

Description of system operation

In Figure Ex. 11.1, a single supply is used to power logic inputs to a PLC. The logic inputs provide details on the status of a plant and may be on or off independently. The supply is common to all inputs and is separately monitored for integrity (although this is not of concern here).

In this exercise, you are asked to commission the system which does not operate correctly. The resistors R1–5 have been specified as 1 KΩ. Check the calculations based on the design parameters given.

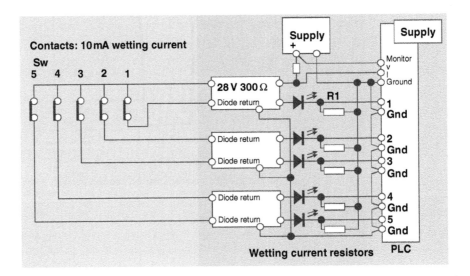

Figure Ex. 11.1
Switch input system

K.2.12   Exercise 12   Solenoid valve operational characteristics

Introduction

Solenoid valves require a minimum excitation current in order to operate. In this exercise, realistic values are provided. The calculations necessary for the correct operation of this type of loop requires knowledge of the ambient temperature of use which affects the operation (Figure Ex. 12.1).

Objective

From the characteristics given, determine whether the valve will operate from the solenoid details provided on the Euro block E 1230.

Characteristics

The Asco IM 12 details are given as typical figures:

Operational

Pull-in current (min)	25 mA
Resistance @ 25 °C max	496 Ω
Resistance @ 65 °C max	560 Ω
Cable	1.0 mm^2 @ 80 Ω/loop km

Safety

Apparatus group	IIC
Grade of safety	Ex 'ia'
T Class	T$_4$

Required duty

Distance	400 m
Operating maximum ambient	65 °C
Hazard	Ex 'ib' IIB T$_4$

Determine the longest length of cable that may be used, taking the 'Kerpen' cable characteristics.

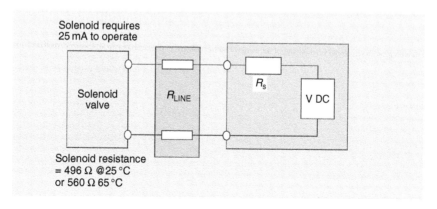

Figure Ex. 12.1
A sketch the aid the thinking process

Operating specification

- The minimum wetting current required for each field contact (for reliable operation) is 10.0 mA.
- The input threshold of the PLC input is 'Off' at 0–3.0 V and 'On' at >3.5–40 V.
- The supply is 24 V DC nominal but may be increased to 28 V DC maximum. Supply current capability is 500 mA maximum at 28 V DC.

How can this system be made to work?

Suggest an alternative design using isolating interfaces?

Are there any advantages or disadvantages to the use of isolators?

How many diode return channels are permitted with this system?

Calculations

K.3 Solutions to exercises

K.3.1 Exercise 1

For discussions.

K.3.2 Exercise 2

Answer to Exercise 2 is as follows:

2.1. Ex 'ia', Ex 'ib', Ex 'd', Ex 'e', Ex 'p', Ex 'm', Ex 's'.
2.2. (a) Pumps, Meters, Controllers, Lighting.
(b) Ex 'd' /'e', Ex 'i'/'d'/'e', Ex 'i'/'e', Ex 'd'/'e'.
2.3. Identify the individual chemicals. The worst compound determines the classification. Fuel and diesel fuel = IIA.
2.4. Inside the liquid (oxygen level too low for ignition) = safe area. Above the liquid there would be vapor that could be at LEL or above = Zone 0. The vapor concentration would depend on the ventilation of the tanks.
2.5. Simple apparatus like a capacitive sensor inside the liquid (same as motor vehicle), ultra sonic, laser, flow sensors. All these devices will have to be IS and suitable for use in this type of environment.

K.3.3 Exercise 3

Answers to Exercise 3 are:

3.5. Corrosive environments (chemicals, sea, etc.), high humidity, high lightning strikes, cable damage, accumulation of dust and dirt, mechanical damage, vibration, high temperatures, training and experience of people, likelihood of unauthorized modifications, inappropriate maintenance.
3.6. The light metal alloys are prone to sparking when struck with another metal object.
3.7. Inspect it after an explosion has occurred. Visual inspection of damage to the enclosure, glands, flanges, bolts, thread, seals, windows.
3.8. Repairs to IS devices must be done by people trained in IS requirements. The circuits must be repaired to the original type approval and all protection must be restored. The protection could include conformal coating, encapsulation and seals for IP ratings.

K.3.4 Exercise 4

Group discussion only.

K.3.5 Exercise 5

To be discussed in class.

K.3.6 Exercise 6

Answer to Exercise 6 are:

The appropriate methods of protection are –

- *Lighting*: Explosion-proof Ex 'd'
- *Enclosure*: Purged and pressurization technique by air from non-hazardous area

- *Valve position sensor*: Non-incentive position sensor
- *Motor*: Explosion-proof Ex 'd'
- *Motor starter*: Explosion-proof Ex 'd'
- *Speed monitor*: Intrinsically safe Ex 'i'
- *Push button with pilot light*: Intrinsically safe Ex 'i'
- *Transmitters*: Intrinsically safe Ex 'i'.

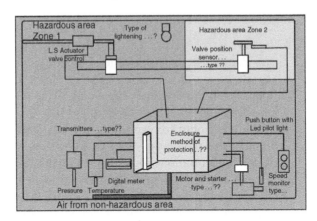

K.3.7 Exercise 7

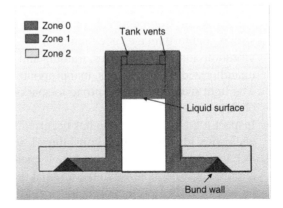

K.3.8 Exercise 8

A. Inspection schedule for apparatus relying on pressurizing/purging concept of protection

Check that	Types of Inspection		
	Initial	Periodic	Visual
Apparatus is appropriate to area classification	ALL	ALL	ALL
Apparatus surface temperature class is correct	ALL	SAMPLE	NONE
Apparatus carries the correct circuit identification	ALL	SAMPLE	NONE
There are no unauthorized modifications	ALL	SAMPLE	ALL

Check that	Types of Inspection		
	Initial	Periodic	Visual
Earthing connections, including any supplementary earthing connections, are clean and tight	ALL	SAMPLE	ALL
Earth loop impedance or resistance is satisfactory	ALL	SAMPLE	NONE
Lamp rating and type is correct	ALL	SAMPLE	NONE
Source of pressure/purge medium is free from contaminants	ALL	ALL	ALL
Pressure/flow is as specified	ALL	ALL	ALL
Pressure/flow indicators, alarms and interlocks function correctly	ALL	SAMPLE	NONE
Pre-energizing purge period is adequate	ALL	SAMPLE	NONE
Ducting, piping and enclosures are in good condition	ALL	ALL	ALL
No undue external accumulation of dust and dirt	ALL	ALL	ALL

B. Inspection schedule for apparatus relying on protection concept 'n', 's' or 'e'...

Check that	Types of Inspection		
	Initial	Periodic	Visual
Apparatus is appropriate to area classification	ALL	ALL	ALL
Apparatus group (if any) is correct	ALL	SAMPLE	NONE
Apparatus surface temperature class is correct	ALL	SAMPLE	NONE
Apparatus carries the correct circuit identification	ALL	SAMPLE	NONE
Enclosures, glasses and glass/metal parts are satisfactory	ALL	ALL	ALL
There are no unauthorized modifications	ALL	SAMPLE	ALL
Earthing connections, including any supplementary earthing connections, are clean and tight	ALL	SAMPLE	ALL

(Continued)

Check that	Types of Inspection		
	Initial	Periodic	Visual
Earth loop impedance or resistance is satisfactory	ALL	SAMPLE	NONE
Lamp rating and type is correct	ALL	SAMPLE	NONE
Bolts, glands and stoppers are of the correct type and are complete and tight	ALL	ALL	ALL
Enclosed-break and hermetically sealed devices are undamaged	ALL	SAMPLE	NONE
Condition of enclosure gaskets is satisfactory	ALL	SAMPLE	NONE
Electrical connections are tight	ALL	SAMPLE	NONE
Apparatus is adequately protected against corrosion, the weather, vibrations and other adverse factors	ALL	ALL	ALL
There is no obvious damage to cables	ALL	ALL	ALL
Automatic electrical protection devices are set correctly	ALL	SAMPLE	NONE
Automatic electrical protection devices operate within permitted limits	SAMPLE	SAMPLE	NONE
No undue external accumulation of dust and dirt	ALL	ALL	ALL

C. Inspection schedule for intrinsically safe systems

Check that	Types of Inspection		
	Initial	Periodic	Visual
System and/or apparatus is appropriate to area classification	ALL	ALL	ALL
System group or class is correct	ALL	SAMPLE	NONE
Apparatus surface temperature class is correct	ALL	SAMPLE	NONE
Installation is correctly labeled	ALL	SAMPLE	NONE
There are no unauthorized modifications (including readily accessible lamp and fuse ratings)	ALL	SAMPLE	ALL

Check that	Types of Inspection		
	Initial	Periodic	Visual
Apparatus is adequately protected against corrosion, the weather, vibration and other adverse factors	ALL	ALL	ALL
Earthing connections are permanent and not made via plugs and sockets	ALL	ALL	ALL
The intrinsically safe circuit is isolated from earth or earthed at one point only	ALL	ALL	NONE
Cable screens are earthed in accordance with the approved drawing	ALL	SAMPLE	NONE
Barrier units are of the approved type, installed in accordance with the certification requirements and securely earthed	ALL	ALL	ALL
Electrical connections are tight	ALL	SAMPLE	NONE
Point-to-point check of all connections	ALL	NONE	NONE
Segregation is maintained between intrinsically safe and non-intrinsically safe circuits in common marshaling boxes or relay cubicles	ALL	ALL	ALL
There is no obvious damage to apparatus and cables	ALL	ALL	ALL
No undue accumulation of dust and dirt	ALL	ALL	ALL

K.3.9   Exercise 9

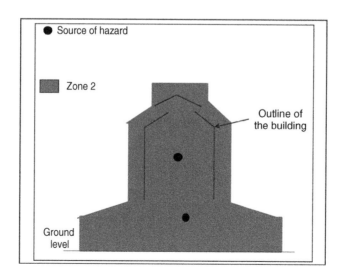

K.3.10   Exercise 10

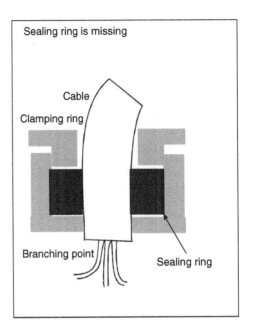

Index

THIS BOOK WAS DEVELOPED BY IDC TECHNOLOGIES

WHO ARE WE?

IDC Technologies is internationally acknowledged as the premier provider of practical, technical training for engineers and technicians.

We specialise in the fields of electrical systems, industrial data communications, telecommunications, automation & control, mechanical engineering, chemical and civil engineering, and are continually adding to our portfolio of over 60 different workshops. Our instructors are highly respected in their fields of expertise and in the last ten years have trained over 50,000 engineers, scientists and technicians.

With offices conveniently located worldwide, IDC Technologies has an enthusiastic team of professional engineers, technicians and support staff who are committed to providing the highest quality of training and consultancy.

TECHNICAL WORKSHOPS

TRAINING THAT WORKS

We deliver engineering and technology training that will maximise your business goals. In today's competitive environment, you require training that will help you and your organisation to achieve its goals and produce a large return on investment. With our "Training that Works" objective you and your organisation will:

- Get job-related skills that you need to achieve your business goals
- Improve the operation and design of your equipment and plant
- Improve your troubleshooting abilities
- Sharpen your competitive edge
- Boost morale and retain valuable staff
- Save time and money

EXPERT INSTRUCTORS

We search the world for good quality instructors who have three outstanding attributes:

1. Expert knowledge and experience – of the course topic
2. Superb training abilities – to ensure the know-how is transferred effectively and quickly to you in a practical hands-on way
3. Listening skills – they listen carefully to the needs of the participants and want to ensure that you benefit from the experience

Each and every instructor is evaluated by the delegates and we assess the presentation after each class to ensure that the instructor stays on track in presenting outstanding courses.

HANDS-ON APPROACH TO TRAINING

All IDC Technologies workshops include practical, hands-on sessions where the delegates are given the opportunity to apply in practice the theory they have learnt.

REFERENCE MATERIALS

A fully illustrated workshop book with hundreds of pages of tables, charts, figures and handy hints, plus considerable reference material is provided FREE of charge to each delegate.

ACCREDITATION AND CONTINUING EDUCATION

Satisfactory completion of all IDC workshops satisfies the requirements of the International Association for Continuing Education and Training for the award of 1.4 Continuing Education Units.

IDC workshops also satisfy criteria for Continuing Professional Development according to the requirements of the Institution of Electrical Engineers and Institution of Measurement and Control in the UK, Institution of Engineers in Australia, Institution of Engineers New Zealand, and others.

CERTIFICATE OF ATTENDANCE

Each delegate receives a Certificate of Attendance documenting their experience.

100% MONEY BACK GUARANTEE

IDC Technologies' engineers have put considerable time and experience into ensuring that you gain maximum value from each workshop. If by lunch time of the first day you decide that the workshop is not appropriate for your requirements, please let us know so that we can arrange a 100% refund of your fee.

ONSITE WORKSHOPS

All IDC Technologies Training Workshops are available on an on-site basis, presented at the venue of your choice, saving delegates travel time and expenses, thus providing your company with even greater savings.

OFFICE LOCATIONS

AUSTRALIA • CANADA • IRELAND • NEW ZEALAND • SINGAPORE • SOUTH AFRICA • UNITED KINGDOM • UNITED STATES

idc@idc-online.com • www.idc-online.com

Visit our Website for FREE Pocket Guides

IDC Technologies produce a set of 4 Pocket Guides used by thousands of engineers and technicians worldwide.

Vol. 1 - ELECTRONICS
Vol. 2 - ELECTRICAL
Vol. 3 - COMMUNICATIONS
Vol. 4 - INSTRUMENTATION

To download a **FREE copy** of these internationally best selling pocket guides go to:
www.idc-online.com/freedownload/

Printed and bound by CPI Group (UK) Ltd, Croydon, CR0 4YY

03/10/2024

01040336-0015